QE
570
N34
1991

INTRODUCTION TO THE PETROLOGY OF SOILS AND CHEMICAL WEATHERING

INTRODUCTION TO THE PETROLOGY OF SOILS AND CHEMICAL WEATHERING

DANIEL B. NAHON
University of Aix-Marseille 3
France

A Wiley-Interscience Publication
JOHN WILEY & SONS, Inc.
New York / Chichester / Brisbane / Toronto / Singapore

In recognition of the importance of preserving what has been written, it is a policy of John Wiley & Sons, Inc., to have books of enduring value published in the United States printed on acid-free paper, and we exert our best efforts to that end.

Copyright © 1991 by John Wiley & Sons, Inc.

All rights reserved. Published simultaneously in Canada.

Reproduction or translation of any part of this work beyond that permitted by Section 107 or 108 of the 1976 United States Copyright Act without the permission of the copyright owner is unlawful. Requests for permission or further information should be addressed to the Permissions Department, John Wiley & Sons, Inc.

***Library of Congress Cataloging in Publication Data*:**

Nahon, Daniel.
Introduction to the petrology of soils and chemical weathering/
Daniel B. Nahon.
p. cm.
"Translation into English of an original French manuscript" — Translator's foreword.
"A Wiley-Interscience publication."
Includes bibliographical references.
ISBN 0–471–50861–6
1. Chemical weathering. 2. Soils — Analysis. 3. Petrology. 4. Soil mineralogy.
5. Geochemistry. I. Title.
QE570.N34 1991 90–40173

551.3'05 — dc20 CIP

Printed in the United States of America

10 9 8 7 6 5 4 3 2 1

To Irène
To Delphine, on her eighteenth birthday.

TRANSLATOR'S FOREWORD

This book, quite unusually, represents a translation into English of an original French manuscript. Therefore, a comparison with the original is not available. Such would have been the case if Voltaire's *Candide* had been translated immediately into English and never published in French. The work of the translator has consequently increased in responsibility toward the reader in spite of the close and pleasant collaboration with the author, which spanned over several years. Together, author and translator discussed at length problems of organization, style, and meaning of the text, an approach that eventually reached a sentence by sentence analysis as ultimate symbiosis.

Although scientific terminology is essentially identical in English and French, a translation cannot be a simple automatic and technical conversion process. The translator felt at times frustrated when delicate nuances or stylistic forms of the French idiom had no equivalent in English — as is if he were faced with the loss of some untranslatable residue. However, in reality, no such loss really occurs, because to translate is to transpose a theme from one instrument of communication to another, which has its own and equally well-tuned keyboard. If this transmutation is accomplished with care, the essence of the message remains the same. This has been the most profound interest of this translator.

ALBERT V. CAROZZI

University of Illinois at Urbana-Champaign
October 1990

PREFACE

In 1964 R. Brewer published his benchmark book entitled *Fabric and Mineral Analysis of Soils*. A revised edition in 1976 provided additional examples together with an improved terminology.

Brewer's work represents the first synthesis of the macromorphological and, in particular, the micromorphological analysis of soils. It required at the time a complete new terminology to characterize the structures and textures observed in soils. This work was furthermore the first to introduce the concept of levels of organization and of pedologic features.

During the intervening 25 years, the petrographic, mineralogic, and geochemical analysis of soils and alterites has made great advances. Numerous investigations were undertaken, particularly on the weathering mantle in tropical areas, and physical and physicochemical analytical techniques steadily improved. Consequently, minerals and textures can be studied today in the most minute details, resulting in an increased sophistication of numerical modeling and of the monitoring of laboratory experiments. In other words, modern petrological analysis of soils and alterites at all levels of observation at last reveals genetic processes and allows their quantification.

Today, soils and alterites can no longer be studied without the interdisciplinary approach of petrography, mineralogy, and geochemistry. Soils and alterites represent the same biogeochemical system whose understanding is critical for the future of humanity. Therefore, the composition and "organization" of the pedologic cover (soils and alterites) is the basis for the study of this system.

The pedologic cover is a complex system that evolves through geological time by self-development or under the influence of the changes of external climatic and tectonic factors. This book does not pretend to cover everything in terms of subject matter and examples. For instance, it does not include the study of the various aspects of organic matter. Its aim is to stress the organization of the pedologic cover at all scales of observation and, whenever possible, to unravel its geochemical significance. From crystal to field sequence, this book shows the interweaving of organizations in time and space, the appearance of order among structures and textures, and how this order has changed to disorder and back again to order.

Another purpose of this book is to promote a closer collaboration between pedologists, geologists, and geochemists in a teamwork investigation of a unique geochemical system.

The terminology used in this book is essentially Brewer's, often simplified and illustrated for microscopic descriptions. At first glance it may appear esoteric, but it serves to characterize structures, textures, and interrelations between minerals in soils and alterites.

The reader might be surprised by the numerous case histories from Africa. Indeed, French geologists have worked in that continent for decades, at first within the colonial framework, and subsequently under a program of technical and scientific cooperation with the independent French-speaking countries of Africa. Nevertheless, a comparison between these examples from Africa and those from Brazil, the United States, Europe, and Australia insures a complete coverage that is representative of the modern approach to the study of soils and alterites.

DANIEL B. NAHON

Aix-en-Provence
October 1990

ACKNOWLEDGMENTS

The idea of writing this book was conceived several years ago following the numerous discussions I had with Gérard Bocquier. A portion of this work and of its scientific ideas were elaborated together as a consequence of our respective experiences in the field and in the laboratory. I am therefore deeply endebted to Gérard Bocquier for his generosity in letting me present our common early results.

My interest in the petrology of alterites and soils began 20 years ago, in 1969, when I met Georges Millot. He taught me in a masterly fashion the discipline and the serenity required by scientific endeavors. I would like to express to him, through this book, my deep affection and gratefulness.

I would like to thank all those who helped me through many years by their scientific collaboration, their fruitful discussions and constructive comments, and their friendship. I am thinking in particular of Hélène Paquet, Alain Ruellan, René Boulet, Yves Tardy, Armand Chauvel, and Roland Trompette.

I am very grateful to Yves Lucas, Jacques Schott, and Alain Meunier who read critically various chapters of this book at different stages of preparation.

Gratitude is also expressed to Bruno Boulangé, Jacques Muller, Yves Noack, Jean-Paul Ambrosi, Claude Parron, René Flicoteaux, Samuel Z. Altschuler, Enrique Merino, Robert A. Berner, the late Robert M. Garrels, Bertrand Fritz, Adolfo Melfi, Adilson Carvalho, Fabrice Colin, Anicet Beauvais, Milan Pavich, Jacques Boulègue, Alain Decarreau, Stephen P. Altaner, and Alain Manceau. It is in their laboratories that I was given the opportunity to work on some of the case histories presented in this book: They taught me a lot.

I met Albert V. Carozzi in the field, in the Ivory Coast, in 1977. Our joint paper on the weathering of glauconite grains was published ten years ago. When I began writing this book, I asked him if he would translate it into English. He accepted the challenge, although it turned out to be very time-consuming. During these past years, he was not only a translator, but also an unyielding critic who compelled me to express my ideas more clearly, to rewrite or reorganize entire sections, and, most important, to introduce a final synthesis that greatly enhanced the presentation. The friendship and constant support of Albert and Marguerite Carozzi were critical to me. I am deeply grateful to both of them.

I would like to thank Raymond Dassulle and Jean-Jacques Motte for their outstanding drafting skills in the preparation of all the illustrations. Nicole Pelegrin, Claude Augas, Jacqueline Lappartient and Anne-Marie Gandolfi have dedicated much time and care toward a perfect manuscript.

Thanks are expressed to Michael A. Velbel of Michigan State University for his very thorough critical reading of the final manuscript and for his constructive criticisms.

I deeply appreciate the keen interest of Charles S. Hutchinson, Consulting Editor, and the editorial staff, Scientific and Technical Division of John Wiley and Sons, Inc., who saw this book through to press.

DANIEL B. NAHON

CONTENTS

1 Introduction to Major Geochemical Processes of Weathering 1

Rates of Chemical Weathering / 2

Rates of Mineral Weathering / 3

Rates of Rock Weathering / 4

Processes of Hydrolysis / 7

Hydrolysis Reactions of Silicate Minerals / 8

Dissolution of Feldspars / 9

Dissolution of Magnesian and Calcic Silicates: Enstatite, Diopside, and Tremolite / 15

Dissolution of Ferrous Silicates: Bronzite and Fayalite / 24

Dissolution of Micas / 27

Processes of Hydrolysis Reactions of Crystals / 31

Exchange Reactions / 32

Dissolution Reaction and pH / 33

Stability of Exchange Sites / 34

Accessibility of Exchange Sites / 36

Role of Temperature on Dissolution Rates / 38

CO_2 Kinetics and Weathering / 38

Congruent and Incongruent Dissolutions / 40

Neoformation and Transformation / 43

Kinetic Inhibitors of Dissolutions / 45

Behavior of Ions in Solution / 46
Transport of Solutes / 48

2 Supergene Alteration of Minerals and Rocks: Preservation of Original Structures **51**

The Two Weathering Pathways of Parent Minerals / 52
Incongruent Dissolution: Pseudomorphosis of Weathering Products after Parent Minerals / 53
Optically Invisible Modifications / 53
Modifications Visible under the Petrographic Microscope and SEM / 54
Simple Dissolution of Parent Minerals: Absence of Pseudomorphosis / 57
Conclusions on Earliest Transformations due to Weathering / 61

Preservation of Parent Rock Structures: Generation of Weathering Plasmas / 63
Pseudomorphoses by Oxyhydroxide Septa: Crystalliplasmas / 64
Pseudomorphoses by Clay Minerals: Argilliplasmas and Complex Plasmas / 70
Parent Mineral Phases Undergoing Independent Weathering / 70
Parent Mineral Phases Undergoing Simultaneous Weathering and Mutual Reactions / 75
Parent Mineral Phases Undergoing Simultaneous Weathering and Whose Secondary Products Also Express an External Supply / 79
Weathering Argilliplasmas and Their Complexity / 81
Conclusions / 85

Relative Positions of Parent Relicts and of Weathering Plasmas: Mineral Parentages / 86
Morphology of Distributions / 86
Peripheral Distribution / 86
Patchy Distribution / 86
Nature of Contacts / 87

Relative Accumulation in Structure-Preserving Weathering / 90

3 Structural Transformations of Pedoturbation **97**

Microstructural Transformations of Isovolumetric Alterites / 98
Evolution toward Petrographic Structures of Vertisolic Horizons / 98
Alterite of Migmatite (Horizon C) / 99
Transition between Alterite (or Saprolite) and Pedoturbated Horizons / 99

Pedoturbated Horizon of Vertisolic Type / 101
Evolution toward Ferralitic Petrographic Structures / 102
Migmatite Alterite / 102
Ferralitic Pedoturbated Alterite / 102

Characteristics of Pedoturbation / 104
Increasing Weathering Plasmation and Pedoturbation / 104
Pedoturbation and Isovolume / 105
Pedoturbation of Nonargillaceous Horizons / 106

Plasmic Structures of Weathering and Pedoturbation / 106
First-Order Ultrastructures: Sheets, Layers, Crystallites, Particles and Association of Particles / 108
Introduction / 108
Layers and Crystallites / 108
Argillaceous Particles / 109
Association or Arrangement of Particles / 111
First-Order Argillaceous Structures and Their Relationship with Water / 114
Second-Order Microstructures: Juxtaposed Crystallites, Domains, and Networks of Tactoids / 116
Introduction / 116
Major Types of Microstructures Argillaceous Plasmas / 117
Relationship between First-Order and Second-Order Plasmic Structures / 118
Effects of Drying and Wetting on Microscopic Argillaceous Structures / 122
Third-Order Macroscopic Structures: S-Matrix and Peds / 126
Introduction / 126
Macroscopic Structures of Argillaceous Plasmas / 127
Relationship between Second-Order and Third-Order Plasmic Structures / 128
Effects of Drying and Wetting on Macroscopic Argillaceous Materials / 129

Conclusions / 132

4 Transfers and Accumulations **135**

Characteristics of Loss of Material / 135
Leached Horizons / 135
Differentiation of Microstructures of Leached Horizons / 136
Survival of Microsctructures of Leached Horizons / 139

Succession and Hierarchy of Microstructures Typical of Loss and Accumulation of Material / 147
Characteristics of Accumulation of Material / 148
Accumulations Developed in S-Matrix: Glaebulization / 149
Weakly Differentiated Glaebular Accumulations: Initial Concentrations / 149
Moderately Differentiated Glaebular Accumulations: Nodules / 150
Strongly Differentiated Glaebular Accumulations: Well-defined Nodules and Individualization of Cortex / 154
The Different Stages of Glaebular Microstructure / 158
Processes of Glaebulization / 160
Continuous Glaebular Structures: Duricrusts / 162
Evolutionary Mineralogical Sequences in Glaebular Accumulation / 167
Accumulations of Material in Micropore and Macropore Systems: Coatings / 170
The Different Types of Cutanic Accumulation / 170
Subcutanic Features: Neocutans and Quasicutans / 175
Transformations Affecting Cutans / 176
Historical and Geochemical Differentiations of Accumulation / 187
Structures Resulting from Successive Accumulations / 187
Structures Resulting from Simultaneous Accumulations / 188
Pseudobrecciated and Pseudoconglomeratic Structures of Accumulation / 191
Lateritic Profiles on Phosphatic Rocks of Lam Lam / 191
Other Examples of Pseudobrecciated or Pseudoconglomeratic Structures / 198
Degradation of Accumulation Structures / 199
Pisolitic Structures: Reorganization or Degradation? / 199
Banded Structures of Reorganization / 199
Banded Structures of Aggradation / 200
Degradation by Geochemical Fragmentation / 200
Banded Structures of Degradation / 204
Conclusions: Evolution of Microstructural Features of Accumulation and Transfer / 205
Inherited Microstructures / 205
Accumulation Microstructures / 206
Transformation Microstructures / 206
Leaching Microstructures / 206
Microstructures Are Progressive / 206

5 Differentiation and Evolution of Pedologic Mantles of Tropical and Subtropical Zones **207**

Pedologic Mantles of Tropical Humid Landscapes / 208
Distribution of Major Horizons / 208
Horizons with Polyhedric Structure and with Microaggregates of Tropical Humid Pedologic Mantles / 209
Data from the Pedologic Analysis / 210
Results of the Analytical and Experimental Approach / 212
Sequences with Lateral Differentiation Resulting from a Change in Drainage Conditions / 215
Derived Sequences Resulting from a Change in Climatic Conditions / 223

Pedologic Mantles of Tropical Landscapes with Alternating Dry and Rainy Seasons / 228

Pedologic Mantles of Tropical Landscapes with Predominant Extended Dry Season / 231
Garango I / 231
Garango II / 234
Upslope Domain / 234
Middle Domain / 236
Downslope Domain / 236
Comparison between Garango I and II Sequences / 238
General Conclusion / 240

Pedologic Mantles of Subarid Tropical Landscapes / 240
Pedologic Mantle in Temperate Climate / 245
Middle Domain / 246
Upslope Domain (Plateau) / 246
Downslope Domain / 246

Conclusions / 248
Original Horizons and Their Dynamic Evolution / 249
Horizons with Limited Transformation / 252
Horizons with Advanced Transformation / 253
Original Sequences with Vertical Differentiation / 254
Horizons with Lateral Transformation / 254
Leaching Transformations / 256
Accumulating Transformations / 257
Lateral Eluvial–Illuvial Transformations: Derived Sequences with Lateral Differentiations / 257

6 Conclusions 259

The Pedologic Cover: A Biogeochemical System / 259
The Pedologic Cover: Reflection of the Present and the Past / 260
Monophased Pedologic Covers of the Present / 260
Polyphased Pedologic Covers of the Past / 260
The Pedologic Cover: An Interference of Several Histories / 262
Weathering Rates and Young Pedologic Covers / 262
Weathering Rates and Old Pedologic Covers / 263

The Past Pedologic History Seen in the Light of Global Evolution / 264
Past Climates and Continental Areas / 264
Time Variation of the CO_2 Factor / 265
Sinking of Oxidation Front / 265
The Temperature Factor / 267

Interaction between Controlling Factors of Weathering / 267
Weathering and Structural Discontinuities / 267
Weathering and Isostasy / 268
Water Flushing and Reactivity of Rocks to Weathering / 268
Endless Feedbacks and Evolutive Weathering / 269

References 271

Index 299

INTRODUCTION TO THE PETROLOGY OF SOILS AND CHEMICAL WEATHERING

CHAPTER 1

INTRODUCTION TO MAJOR GEOCHEMICAL PROCESSES OF WEATHERING

Soils and weathering mantles that cover the surface of exposed lands are produced by the destruction of igneous, metamorphic, or sedimentary rocks under the action of external atmospheric and biological agents.

As compared to purely biological or mechanical processes, the process of chemical destruction of fresh, massive, and coherent rocks into unconsolidated materials is by far the most important weathering agent. Therefore, only geochemical processes of weathering are discussed here.

Most of the fresh igneous, metamorphic, and sedimentary rocks that form the epidermic portion of exposed lands are subjected to environmental conditions different from those that prevailed during their formation. Therefore, weathering consists of the thermodynamic readjustment of rocks to the environmental conditions of the surface where meteoric water and atmospheric gases prevail. The resulting chemical evolution leads to the formation of new weathering minerals that are in equilibrium with (or at any rate formed by) the new environmental conditions.

Weathering reactions are controlled by meteoric water and gases dissolved in it (mainly O_2 and CO_2) that permeate rocks and minerals. Rainwater falling on the surface of exposed lands is the essential vector of all interface reactions. Baumgarter and Reichel (1975) and later Tardy (1986) reported that annual average rainfall on exposed lands is 74.6 cm/year. These authors noticed that if the percentage of precipitation that reevaporates is compared to the percentage that percolates through the soil or flows at the surface, only 36% of the water is "effective," namely, the average annual precipitation that plays a role in chemical weathering is, in fact, only 26.6 cm/year. In reality, there are huge variations in these precipitations according to various climatic zones.

When rainwater penetrates into soil or rocks and reacts with their constituent minerals or organic matter, its chemical composition varies during its percolation.

The weathering geochemical system has, therefore, three main variables: rocks, rainwater, and dissolved gases. Organic matter enters into this system by means of its oxidation (consumption of O_2) and its mineralization (production of CO_2).

This geochemical system is in reality extremely complicated when each of its variables is considered in detail. For instance, a rock consists of several different mineral phases, each characterized by its accessibility to weathering according to its intimate crystallo-chemical structure and, hence, to the nature and organization of the elements and of the bonding forces that unite them. Furthermore, the alterability rate of each mineral phase varies according to, among other factors, its degree of mechanical fracturing. Similarly, in the process of becoming percolating water, rainwater, changes composition depending upon the climatic zone, the season, or simply its location at a given time in the weathering profile. Furthermore, its thermodynamic activity varies according to the size of the pores it enters. All these factors lead to an extreme complexity of the variables, and to an equally complex thermodynamic understanding of the system, which furthermore requires to take into consideration the kinetic factors of weathering.

RATES OF CHEMICAL WEATHERING

A general study of rocks with the naked eye and under the petrographic microscope frequently reveals that all constituent parent minerals do not weather at the same rate. This has been a well-known fact since Goldich's (1938) study first presented a *double series of differential weathering of parent minerals*, ranging from the most vulnerable to the least vulnerable (Fig. 1–1). Subsequent studies either refined this sequence or proposed other sequences for specific cases pertaining to given associations of minerals.

The order of differential weathering of feldspars was established by Graham (1949) and Hay (1959); that of minerals of acid rocks by Tardy (1969); of minerals of basic and ultrabasic rocks by Hotz (1961), Hendricks and Whittig (1968), and Trescases (1975); and of carbonates by Garrels and Mackenzie (1971). Numerous other studies demonstrate that minerals do not weather at the same rate. Consequently, with a few small exceptions, Goldich's series remains valid (Tardy, 1969; Lasaga, 1984). Secondary minerals produced by weathering should therefore appear in a definite order.

Evaluation of rates of chemical weathering of rocks and minerals was approached quantitatively by means of mass-balance studies pertaining par-

olivine Ca plagioclase
augite Ca-Na plagioclase
hornblende Na-Ca plagioclase
biotite Na plagioclase
potassic feldspars
muscovite
quartz

Figure 1.1. Double Sequence of Differential Weathering of Parent Minerals (from Goldich, 1938).

ticularly to amounts of silica released. Indeed, silica is the major element in waters that undergoes the least amount of random variations (Tardy, 1969). Evaluation of rates of chemical weathering was attempted by means of the following techniques (for a review, see Coleman and Dethier, 1986): experimental dissolution of minerals and rocks in the laboratory; thermodynamic simulation of weathering; comparison of geochemical composition of fresh parent rocks and of their isovolumetric weathering products; and chemical composition of natural waters produced by weathering reactions of minerals.

Rates of Mineral Weathering

Rates of mineral weathering show sometimes appreciable discrepancies when results from field data are compared with those from laboratory experiments completed under similar hydrogeochemical conditions.

For instance, Velbel (1986, p. 446), in his discussion of the results of geochemical mass balances obtained from small forested watersheds in the Southern Blue Ridge of North Carolina units, wrote that "the rates of garnet and plagioclase weathering are no more than approximately one order of magnitude lower than rates determined in laboratory experiments . . . , rates of biotite mica weathering are considerably different, due in part to a paucity of appropriate open-system laboratory experimental data for reasonably similar conditions of flow rate and solution pH."

Lasaga (1984) calculated the dissolution rate of silica of several minerals at 25° C and pH 5 on the basis of laboratory data provided by various authors. He was able to evaluate the required average time for complete dissolution of a crystal 10^{-3} m in size (Table 1.1). For instance, average weathering of anorthite would occur at a rate 300,000 times faster than that of quartz, whereas enstatite and diopside would respectively weather 68 and 88 times faster than forsterite, using for the latter mineral data from Grandstaff (1981).

The figures in Table 1.1 refer to minerals that are in continuous contact with undersaturated solutions, namely in weathering profiles of humid and leaching zones. Lasaga (1984) stated that in a situation where minerals would be in contact with such solutions only for a fraction x of time, the spans of time (Table 1.1) should be increased by a factor of $1/x$.

Table 1.1. Mean Lifetime of a 1 mm Crystal (from Lasaga, 1984, p. 4013, copyright by the American Geophysical Union).

Mineral	Lifetime
Quartz	34 m.y.
Muscovite	2.7 m.y.
Forsterite	600,000 years
K-feldspar	520,000 years
Albite	80,000 years
Enstatite	8,800 years
Diopside	6,800 years
Nepheline	211 years
Anorthite	112 years

Petrographic observation of natural samples of weathered rocks show that these figures do not always reflect reality. Although general trends of the weathering sequence of minerals, with the exception of forsterite, remain valid, one has to consider that quantification of weathering rates of minerals is difficult when minerals are studied by themselves outside their petrographic context. However, experimental dissolution of minerals is of great importance for the understanding of weathering processes.

Rates of Rock Weathering

Rates of rock weathering can be estimated by considering geochemical export via either percolating solutions or dissolved load of streams, or by studying the chemical composition of the isovolumetric weathering residue. Tardy (1969) studied the average concentration of waters transporting ions from watersheds and thus calculated the rate of chemical weathering of various rocks using the rate of silica removal each year per unit surface (Table 1.2).

The controling factor of the weathering rate is average rainfall. In temperate regions contents of silica are lower, but average rainfalls may be greater than in tropical areas. The time spans given in Table 1.2 are maximum times. It is not necessary to assume that all minerals in the rocks lead to the formation of kaolinite. In weathering profiles, gruss or saprolites contain an appreciable amount of residual minerals. This is the reason Tardy (1969, p. 116) estimated that in tropical or temperate countries, an order of magnitude of 30,000 years was necessary to transform 1 m of granite into a kaolinitic saprolite, that is, about 34 mm per 1000 years. Similarly, Velbel (1985)

Table 1.2. Time Required to Weather One Meter of Rock under Different Climates (modified from Tardy 1969, p. 116, reproduced with permission of Sciences Géologiques).

Location	Average rainfall	Average SiO_2 content of water	number of years required for the transformation of one meter of roch into kaolinite
Granite, Norway	1,250 mm	3 mg/l	85,000
Granite, Eastern France	850 mm	9.2 mg/l	52,000
Granite, Southern France	680 mm	11.5mg/l	41,000
Migmatite, Southern France	680 mm	5.9 mg/l	100,000
Migmatite, Northern Ivory Coast	540 mm	20 mg/l	65,000
Amphibolite, Southern France	640 mm	14 mg/l	68,000
Basalt, Madagascar	1,500 mm	16mg/l	40,000

estimated that in the Southern Blue Ridge of North Carolina, the rate of formation of saprolite is about 37 mm per 1000 years, which is of the same order of magnitude as the figure proposed by Tardy (1969).

A study undertaken by Dethier (1986) on 41 catchments in the Pacific Northwest involved a large variety of igneous, volcanic, and sedimentary rocks. His calculations indicated that, at present, saprolites form at the average rate of 30 mm per 1000 years, a value close to the averages of Tardy (1969) and Velbel (1985). However, Pavich (1986), in a study of a basin consisting of uniform igneous and metamorphic rocks drained by a tributary of the Accoquan River in the Northern Virginia Piedmont, calculated that weathering of granite into 1 m of kaolinitic saprolite required about 250,000 years, that is, a rate of 4 mm per 1000 years. In this particular case, formation of 1 m of saprolite occurred at a rate 8 to 9 times slower than the average proposed by Tardy (1969), Velbel (1985), and Dethier (1986).

Following a similar approach in the Ivory Coast, Leneuf (1959) estimated that the weathering rate of a granitic rock varied between 5 and 50 mm per 1000 years. In the same region and for similar rocks, Boulangé (1984) reached the figure of 14 mm per 1000 years, which is close to the 9.4 mm per 1000 years given by Gac (1979) for Chad. In New Caledonia, Trescases (1975) estimated the weathering rate of ultrabasic rocks to be between 29 and 47 mm per 1000 years, the variation being a function of the topographic position of the rocks in the landscape.

Of interest is the calculation of the mean lifetime of 1 mm of fresh rock into a kaolinitic saprolite (see Table 1.3 and Table 1.1 for comparison). These results show that in cold, temperate, or tropical humid climates, the average rainfall (that is, the flushing of water) probably controls the rate of weathering (Tardy, 1969). Furthermore, chemical weathering of basic and ultrabasic rocks is on the average 2.5 faster than that of acid rocks, showing the importance of reactivity of rock to weathering.

These data of rates of chemical weathering of rocks are given only for the record. They are difficult to apply to the entire thickness of a pedologic or weathering mantle because the latter was produced over a long span of time, sometimes lasting several million years. Indeed, the flow of reacting fluids and their chemical composition change over time (Aagaard and Helgeson, 1982); minerals generated by weathering also react while being formed with percolating fluids (Jackson, 1968); dissolution rates of parent minerals are

Table 1.3. Mean Lifetime of One Millimeter of Fresh Rock.

Rock	Climate	Lifetime
Acid Rocks	tropical semi-arid	65 to 200 years
	tropical humid	20 to 70 years
	temperate humid	41 to 250 years
	cold humid	35 years
Metamorphic rocks	temperate humid	33 years
Basic rocks	temperate humid	68 years
	tropical humid	40 years
Ultrabasic rocks	tropical humid	21 to 35 years

logarithmic functions of both time and depth beneath the surface of the soil (Locke, 1986; Hall and Martin, 1986); weathering of the surface of parent minerals does not occur uniformly (Wilson, 1975; Berner, 1978); and, finally, denudation and surficial mechanical erosion are not considered.

PROCESSES OF HYDROLYSIS

The aqueous phase during its penetration in soil and rocks reacts with minerals, and its composition is bound to change as a consequence of the reactions. Therefore, interfaces between solid phase (minerals) and aqueous phase (solvent) explain processes of supergene alteration.

Silicate minerals represent more than 75% of the rocks at the surface of exposed lands. With them, the aqueous phase induces the largest number of chemical reactions by hydrolitic decomposition, more commonly called hydrolysis.

Water that comes in contact with parent minerals presents an agressiveness related to the amount of CO_2 initially dissolved. Although weathering reactions depend upon this amount, they are also a function of the residence time of the aqueous phase, namely, of the nature of the site of weathering and of the temperature that increases thermal agitation and hence transfers elements toward the interface. Note that for reasons of simplification, weathering reactions are considered here at an ambient temperature of 25° C, although it should be clearly understood that such reactions are magnified in tropical areas compared to colder or temperate zones. Similarly, in desert areas, hot but not subject to the action of liquid water, soils or alterites of chemical origin do not develop at the surface. The aqueous phase is therefore the essential factor in weathering reactions.

Hydrolysis reactions, to be the most representative, have to take into account the important role played by dissolved CO_2. The most commonly quoted example is that of the hydrolysis of forsterite (Mg olivine) given by Krauskopf (1967). This reaction can be written in two ways:

$$Mg_2SiO_4 + 4CO_2 + 4H_2O \leftrightarrow 2Mg^{2+} + 4HCO_3^- + H_4SiO_4$$

or

$$Mg_2SiO_4 + 4H_2CO_3 \leftrightarrow 2Mg^{2+} + 4HCO_3^- + H_4SiO_4.$$

This reaction uses acid and therefore, the solution becomes increasingly alkaline during completion of hydrolysis reactions. This tendency toward increasing alkalinity of the solution depends on the nature of the hydrolized silicate.

In the case of a silicate containing alumina (aluminosilicate), which is considered to be a weakly mobile element, the hydrolysis reaction can lead to precipitation of a secondary mineral phase called *weathering phase*. The example given here is the hydrolysis of albite with formation of kaolinite, which can be expressed as follows:

$$2NaAlSi_3O_8 + 2CO_2 + 11H_2O \leftrightarrow Al_2Si_2O_5(OH)_4 + 2Na^+ + 2HCO_3^- + 4H_4SiO_4.$$

Here, too, acid is used and the remaining solution after the hydrolysis reaction is more alkaline.

In general, hydrolysis equilibria can be represented by the following equation:

$$\text{primary silicate} + H_2O + CO_2 \rightarrow \text{ions in solution } (Na^+;\ K^+;\ Ca^{2+};\ Mg^{2+};\ H_4SiO_4) + \text{mineral weathering phase.}$$

Under these conditions, silicates can be "considered as salts of weak acids and strong bases, hence the hydrolysis reaction is alkaline" (Millot, 1964, p. 83).

In the above-mentioned hydrolysis reactions, the primary mineral contains only a few kinds of cations. Therefore, the reaction is simpler than for a silicate containing numerous cations released at different rates.

This situation shows the complexity of certain hydrolysis reactions and raises the possibility of observing the formation of heterogeneous layers at the mineral-solution interface.

The following questions may be raised: How do exchanges occur between cations of the solid and the solution in contact? What happens to that solution? How are the Si–O and Al–O bonds broken? How are the ions of the crystal transported toward the solution, and then into the solution? How are the weathering phases precipitated?

Dissolution reactions at the interface of silicate minerals are discussed below in more detail.

Hydrolysis Reactions of Silicate Minerals

Recent advances have substantially increased our knowledge of dissolution processes of silicate minerals by means of: laboratory dissolution experiments; direct scanning electron microscope (SEM) observations; X-ray photoelectron spectroscopy (XPS), also called electron spectroscopy for chemical analysis (ESCA); analysis of surface conditions of minerals undergoing dissolution; auger spectroscopy; and secondary ion mass spectrometry (SIMS). The combined use of these techniques has provided a more precise understanding of these processes and, in particular, has eliminated a certain number of hypotheses based on purely chemical methods of analysis.

Several types of silicates, including aluminosilicates (feldspars), Ca-Mg silicates (pyroxenes, amphiboles), Fe-silicates (pyroxenes, olivines), are discussed below. Their dissolution modes are compared, using data from authors who studied them with similar techniques. Crystalline structures of parent minerals undergoing dissolution reactions obviously play a major role in these reactions, particularly in their rate, as given by Goldich (1938). This is the reason for reviewing the initial structure of each silicate.

Dissolution of Feldspars

Review of the Structure of Feldspars

Feldspars have a three-dimensional framework (infinite network) of SiO_4 and AlO_4 tetrahedra in which all oxygens of each tetrahedron are shared with the adjacent tetrahedron. This structure derives from a network of silica tetrahedra by substitution of a portion of silicon by aluminum. Compensation of charges resulting from these substitutions are accounted for by incorporation into the available spaces of monovalent cations (Na^+, K^+) or bivalent cations (Ca^{2+}).

In the potassic feldspars $KA1Si_3O_8$, distribution of Si and Al in the two distinct tetrahedral sites, T_1 and T_2 depends on the temperature of formation of these minerals. In the high-temperature polymorph (sanidine), distribution of Al and Si is random (disordered). The K^+ ion is bonded to the ten closest oxygens and occupies specific positions (Fig. 1.2) on the mirror planes perpendicular to the *b* axis (Papike and Cameron, 1976).

The arrangement is comparable to four rings of tetrahedra linked in chains parallel to the *a* axis. In the low-temperature polymorph (microcline), distribution of Al and Si is ordered, K^+ ions are no longer housed in particular position, and the symmetry is triclinic ($C\bar{1}$). Orthoclase crystallizes at a temperature intermediate between those at which the two polymorphs microcline and sanidine form. Hence its structural characteristics are intermediate (Smith, 1974).

In the plagioclase series, sodic feldspars $NaAlSi_3O_8$ (albite) are triclinic ($C\bar{1}$ group space) with a low-temperature form in which distribution of Al and Si is highly ordered, and a high-temperature form in which the distribution is disordered and may lead to a monoclinic variety.

Si−O bonds are 50% ionic and 50% covalent, whereas Al−O bonds are 63% ionic. Covalent bonding is known to be the strongest chemical bond. Therefore, bonding energies between Si and O are stronger than those between Al and O.

Consequently, feldspars containing the greatest amount of Al−O bonds are the most reactive to chemical weathering. Indeed, calcic plagioclases $CaAl_2Si_2O_8$ (anorthite), which are the most aluminous, are theoretically the most alterable. This is confirmed by the alterability sequence of minerals given by Goldich (1938), in which Ca plagioclases are the first to be dissolved.

Data on the Experimental Dissolution of Feldspars

Kinetics and mechanisms of dissolution of feldspars have been the subjects of investigations that sometimes produced contradictory results. This is

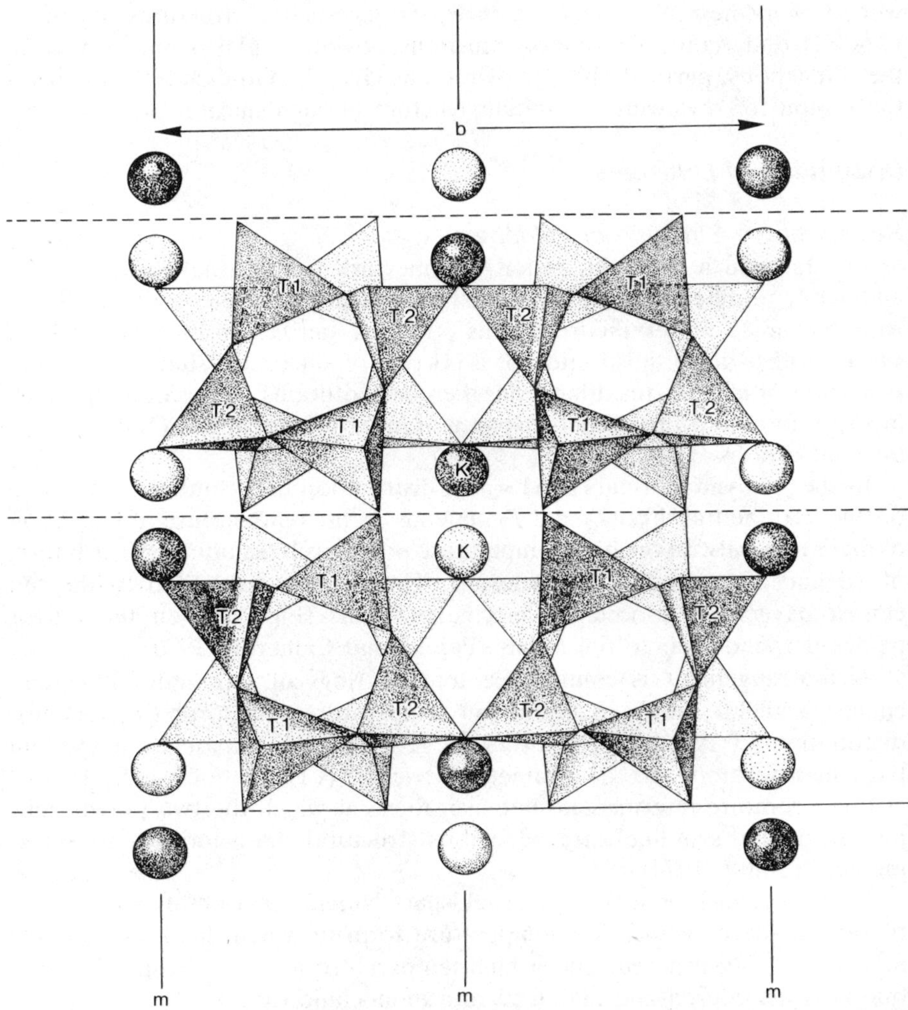

Figure 1.2. Structure of Sanidine (K-Feldspar) Projected on (201) (modified from Papike and Cameron 1976). m = mirror planes, T_1 and T_2 are distinct tetrahedral sites, (from Hulburt and Klein, 1977, Fig. 10-75, p. 422, reprinted by permission of John Wiley & Sons, Inc.).

particularly the case of nonstoichiometry or stoichiometry of dissolution reactions, namely, of existence or absence of a protective layer built at the surface of the feldspar.

ABSENCE OF PROTECTIVE LAYER

The analytical procedure used by Petrovic et al. (1976), Berner and Holdren (1977, 1979), and Holdren and Berner (1979) for the study of feldspars have also been applied to the study of other silicates (pyroxenes, amphiboles, and olivine). It consists of an experimental aqueous dissolution combined with both XPS analysis of the surfaces of minerals and SEM observation. Experiments were performed under conditions close to those most frequently encountered in weathering environments, namely, pH 6 at room temperature and normally oxidizing conditions. When experiments were completed at a lower pH, this was specified in each case. Minerals studied in the laboratory were compared with those collected in soil horizons. Some experiments dealt with raw minerals and others with previously treated minerals (ultrasound cleaning, treatment with HF + H_2SO_4) called etched samples.

The most important results obtained from orthoclases and albites can be summarized as follows: Experiments of progressive dissolution on raw materials indicate that it does not proceed in a uniform manner and that a preferential attack occurs at the places of emergence of dislocations (points of excess energy). Shallow corrosion cupules with lenticular shapes elongated along (010) appear first. They are followed by prismatic corrosion pits with an almost rectangular section developing from the cupules. These pits become deeper and eventually coalesce, forming a beehive type of structure cutting across the crystal. Dissolution structures obtained in the laboratory occur in feldspars collected in situ in weathering horizons. This preferential dissolution at the level of dislocations might not be in agreement with the theory of the so-called protective layer, according to which surficial and continuous protective coatings assumed to surround feldspar grains would regulate dissolution kinetics (Correns and Engelhardt, 1938; Wollast, 1967; Helgeson, 1971; Busenberg and Clemency, 1976). In fact, the rate of hydrolysis reactions would depend upon the density of dislocations.

Dissolution experiments by aqueous means reject the existence of this adherent protective coating. Indeed, feldspar samples previously etched with HF$-H_2SO_4$ show a linear kinetics of the hydrolysis reaction (Fig. 1.3). Nontreated samples show only an initial parabolic segment. Actually, this situation does not express dissolution of the bulk feldspar, but of ultrafine particles generated during the grinding of samples that remain attached to the surface of the feldspar. These particles are very reactive and the first to dissolve, as shown under SEM. Another demonstration is afforded by XPS analysis of samples collected directly from the soil, of those treated in the laboratory and then submitted to dissolution, and of fresh feldspars: Analysis of surfaces shows little or no loss of cations and no change of value of the ratio Al/Si.

The rate of the hydrolysis reaction is described by linear kinetics. Under normal conditions of weathering (pH 6, room temperature), feldspar dissolves in a congruent manner without the development of a protective coating that would control this dissolution by diffusion. Furthermore, any observed parabolic kinetics would not indicate a hypothetical protecting surface layer, but would be an artifact resulting from initial dissolution of ultrafine particles that remained attached to the sample.

All SEM observations made on natural samples (Wilson, 1975; Berner

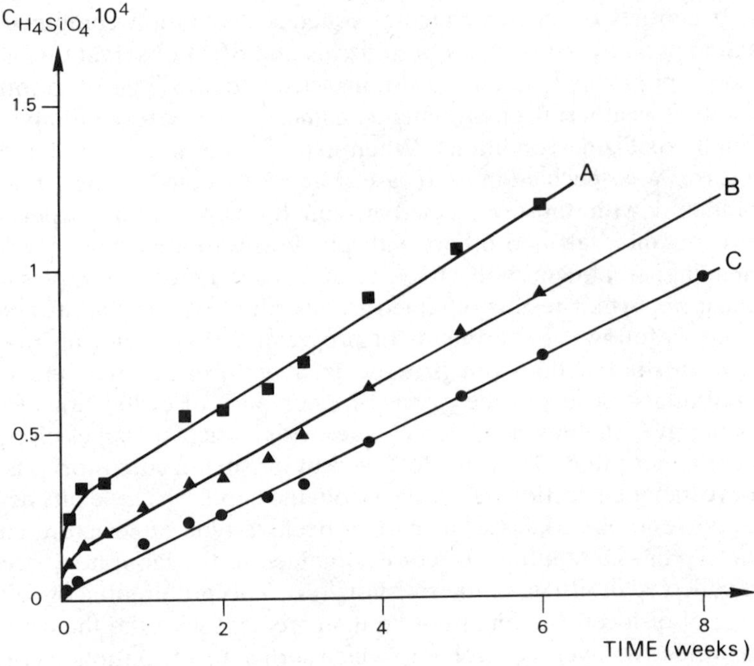

Figure 1.3. Dissolution of Albite in a Buffered Solution at pH 6 and under Normal Conditions: Curve of Release of Silica as a Function of Time (modified from Holdren and Berner, 1979, Fig. 4, p. 1166; reprinted by permission of Pergamon Press plc). A is crushed and sieved sample; B is a crushed, sieved, and water-washed sample; C is a crushed, sieved, and weak HF-treated sample.

and Holdren, 1979; Velbel, 1984a and b) point in the same direction: Dissolution cavities are devoid of any weathering material, implying congruent dissolution of the feldspar and removal of the elements by solutions.

PRESENCE OF A RESIDUAL LAYER

Recent studies by Chou and Wollast (1984; 1985) and Holdren and Speyer (1985a; 1985b; 1986) suggest that dissolution processes of feldspars might be more complex than previously assumed (Schott and Petit, 1987).

Earlier experiments on the rates and mechanisms of feldspar weathering (particularly those mentioned above) have been done by using a batch-type reactor where concentrations of dissolved species are measured as a function of time. Under such conditions the composition of the reacting solution changes continuously with time, making difficult the identification and quantification of individual steps of complex reactions. This pertains, in particular, to the influence of individual dissolved compounds promoting precipitation of Al- or Fe-bearing secondary phases within the system.

To avoid these difficulties, Chou and Wollast (1984, p. 2206) studied the weathering of albite (from Amelia Courthouse, Virginia) using a continuous-flow reactor based on the fluidized bed technique in which the solid is "continuously and vigorously mixed and there are thus no strong gradients of concentration in the aqueous and solid phases as exhibited in the packed column" used in previous experiments conducted by Correns and Von Engelhardt (1938). In summary, reactions were conducted in an open system.

Holdren and Speyer (1985b), using similar techniques (flow-cell reaction chamber) studied a sodium-rich alkaline feldspar from Hybla, Ontario, and confirmed most of Chou and Wollast's results. Ultrafine particles electrostatically bound to the surfaces of the Amelia albite and the Hybla feldspar samples were removed by appropriate treatments in order to eliminate the artifact pointed out by Holdren and Berner (1979), namely, high initial rate of the first step of dissolution.

Weathering experiments completed on the Amelia albite and the Hybla feldspar show that dissolution is not stoichiometric and is strongly dependent on the pH of the solution. A residual layer several tens of Å thick formed at the surface of the feldspar. Under mildly acidic conditions the weathering reaction preferentially releases aluminum from the crystalline framework and thus a residual layer enriched in silica (Fig. 1.4 A) is formed, as schematically represented by Chou and Wollast (1984, p. 2212).

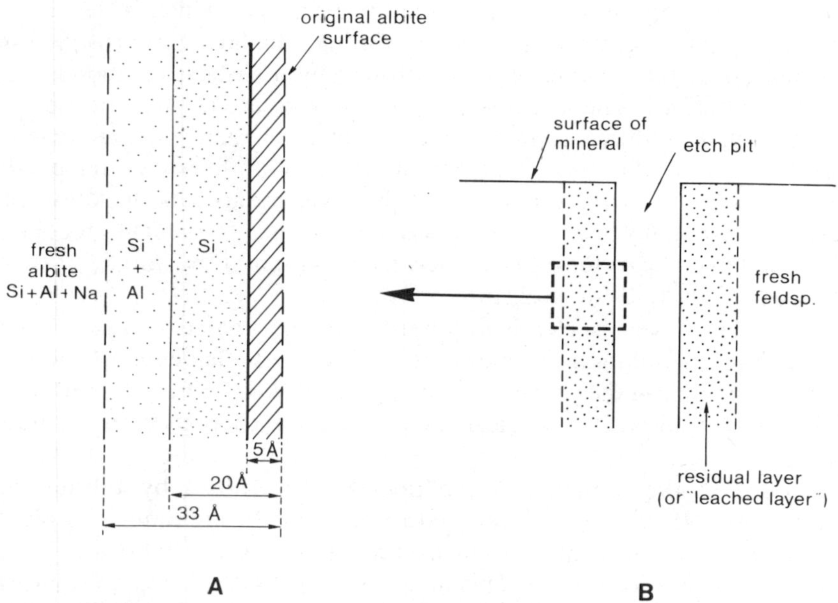

Figure 1.4. Schematic Representation of an Albite Grain After 500 Hours of Dissolution (modified from Chou and Wollast, 1984). **A**: The shaded area of 5 Å is where the three main components of albite have been dissolved; the heavy line is the surface of the remaining solid after dissolution; the dashed line represents the boundary between the residual layer and the fresh albite. **B**: Localization of the residual layer on the walls of an etch pit (modified from Schott and Petit, 1987).

Holdren and Speyer (1985b) observed that at high pH values the reaction is always nonstoichiometric and consists of removal of the siliceous residual layer previously formed at low pH. At pH values close to neutral, leaching of aluminum is reduced and precipitation reactions decrease the amount of aluminum in solution, making the detection of stoichiometry difficult. Holdren and Speyer (1985b) pointed out that the experiments of Holdren and Berner (1979), which led to the assumption that the dissolution reactions of feldspars are stoichiometric, were conducted at pH of the order of 6–8, namely, in a range where weathering reactions of the Hybla feldspar appeared congruent. Consequently, the nonstoichiometry of weathering reactions of feldspars is greater at low pHs. Aluminum and alkaline cations are removed preferentially to silica. This relative leaching decreases when the pH increases, leading to an almost stoichiometric dissolution reaction when the pH reaches approximately 5.5–6.0. Holdren and Speyer (1985b) considered pH the predominant factor in the stoichiometric weathering reaction of feldspars, whereas for Chou and Wollast (1984, 1985) the residual layer plays the fundamental role. During differentiation of the residual layer, "the rate of dissolution decreases rapidly until it reaches a quasi-steady state .... The residual layer formed at the surface of the feldspar tends to remain constant in thickness. The quasi-steady state may then be due to a balance between: (i) the rate of alteration of the fresh feldspar which is dependent on the diffusion of reactants and products through the layer, and (ii) the rate of dissolution of the residual layer controlled by a surface reaction at the solid-liquid interface. The first reaction tends to increase the thickness of the layer and thus to reduce the rate of mass transfer of the elements depleted in the layer from the fresh feldspar interface. On the other hand, the external dissolution of the outer part of the layer reduces its thickness and releases preferentially the elements enriched in that layer. This feedback mechanism finally leads to a stoichiometric dissolution of the initial feldspar" (Chou and Wollast, 1984, p. 2216).

The residual layer is clearly different from a protective layer, which would control dissolution reactions by diffusion of elements. Results of recent studies confirm the conclusions of Lagache (1965, 1976) by emphasizing that the dissolution process of feldspars is limited by reactions at the mineral-solution interface.

Reactions at the mineral-solution interface, as shown by Berner and Holdren (1977, 1979) and Holdren and Berner (1979), are limited to places of excess energy (edges, corners, cracks, scratches, holes, point defects, twin boundaries, dislocations, etc.). This indicates that dissolution is not uniform over the entire surface and that the sketch (Fig. 1.4 A) proposed by Chou and Wollast (1984) is limited to the surfaces of the mineral bordering the preferential zones where reactions occur. Hence, the distribution of residual layers is discontinuous (walls of etch pits of Fig. 1.4 B). The places of excess energy where dissolutions concentrate intersect only a small portion of the surface of the mineral (Berner et al., 1985). This situation has a direct consequence on the estimation of the thickness of residual or leached layers whenever it is derived from the chemical analysis of the solution from the experiment. Schott and Petit (1987, p. 303) stated, for instance, that "the preferential Al loss during albite dissolution which Chou and Wollast (1984) interpret in terms of a 20 Å-thick uniform Al–free surface layer, can just as well be interpreted as representing Al removal along a number of holes, microcracks, and so on, extending 200 Å into the crystal and intersecting only 10% of the outermost crystal surface."

Dissolution of Magnesian and Calcic Silicates: Enstatite, Diopside, and Tremolite

Using the same techniques as for feldspars, Berner et al. (1980), Schott et al. (1981), Berner and Schott (1982), and Schott and Berner (1985) investigated the dissolution process of several magnesian and/or calcic and magnesian pyroxenes and amphiboles. Here, too, the structural framework of the silicate and, in particular, the sites occupied by cations play a fundamental role in their reactivity with regard to aqueous solutions. This is the reason for reviewing these crystalline structures first.

Review of the Structure of Pyroxenes and Amphiboles

In orthopyroxenes, SiO_4 tetrahedra are linked to one another by a corner and associated in simple chains elongated along the *c* axis (inosilicates). Therefore, each tetrahedron shares two of its oxygens with the two adjacent tetrahedra. Laterally, these chains are linked to one another by cations that occupy sites called M_1 and M_2 (Fig. 1.5). Site M_1 is a relatively regular octahedron. Site M_2 is irregular and varies according to the ion it houses. In enstatite ($MgSiO_3$), M_1 and M_2 are completely occupied by Mg; in bronzite $(Fe^{2+}_{0.15}Mg_{0.85})SiO_3$, Mg^{2+}, and Fe^{2+} cations are distributed in sites M_1 and M_2 with the larger Fe^{2+} ion showing a clear preference for the M_2 which are larger and more distorted (Ohashi and Burnham, 1972). In the case of diopside $CaMgSi_2O_6$, a clinopyroxene, the large Ca^{2+} cation occupies site M_2 in such a manner that the coordination number of oxygens around it is close to 8.

In the most distorted sites, M_2 large cations are clearly the least bonded to the silicate structure. Therefore, these sites are the most fragile and hence the most accessible to hydrolysis reactions.

In amphiboles, SiO_4 tetrahedra are linked in double chains $(Si_4O_{11})_n$ developed along the "c" axis. These double chains are linked one to another by cations located in several sites called *A*, M_4, M_3, M_2 M_1 (Fig. 1.6). The nature of the cation varies according to the types considered. Site *A* presents with O and the OH anions a coordination number that varies from 10 to 12, the site actually contains the Na^+ cation whenever present. In tremolite $(Ca_{1.91}Mg_{5.09}Si_8O_{22}(OH)_2)$, a monoclinic amphibole, site M_4 has a coordination number of 6–8 and houses all Ca^{2+} cations. Sites M_1, M_2 and M_3 are octahedra that house Mg^{2+} cations. These sites share their margins in order

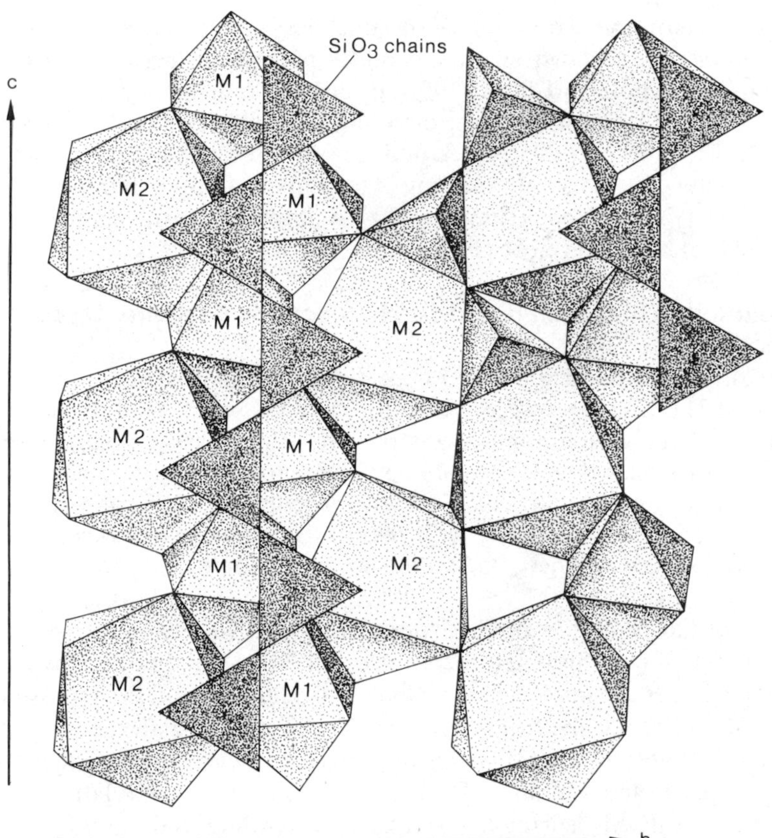

Figure 1.5. Structure of Monoclinic Pyroxene (Jadeite) Projected on (100) (modified from Prewitt and Burnham, 1966). M_2 sites are occupied by Na^+; M_1 sites are occupied by Al^{3+} (from Hurlbut and Klein, 1977, Fig. 10-37, p. 373, reproduced by permission of John Wiley & Sons, Inc.).

to form octahedral bonds parallel to the "c" axis. Sites M_1 and M_3 are coordinated by four oxygens and two OH groupings; site M_2 is coordinated by six oxygens. It is, therefore, easy to predict that cations will be more or less strongly bonded to the structure. Obviously, sites M_2 are the most stable, followed by sites M_1 and M_3, and finally by site M_4. It is evident that Ca^{2+} cations housed in M_4 sites are the least bonded, and hence the most mobile during dissolution reactions.

Site M_4 of amphibole can therefore be considered as equivalent to site M_2 of pyroxene (Berner and Schott, 1982). Indeed, they house the largest cations that are the least bonded with the structure. Conversely, sites M_1, M_2, and M_3 are comparable to site M_1 of pyroxene; they house Mg.

Data on the Experimental Dissolution of Enstatite, Diopside, and Tremolite
Samples studied by Schott et al. (1981), Berner and Schott (1982), and

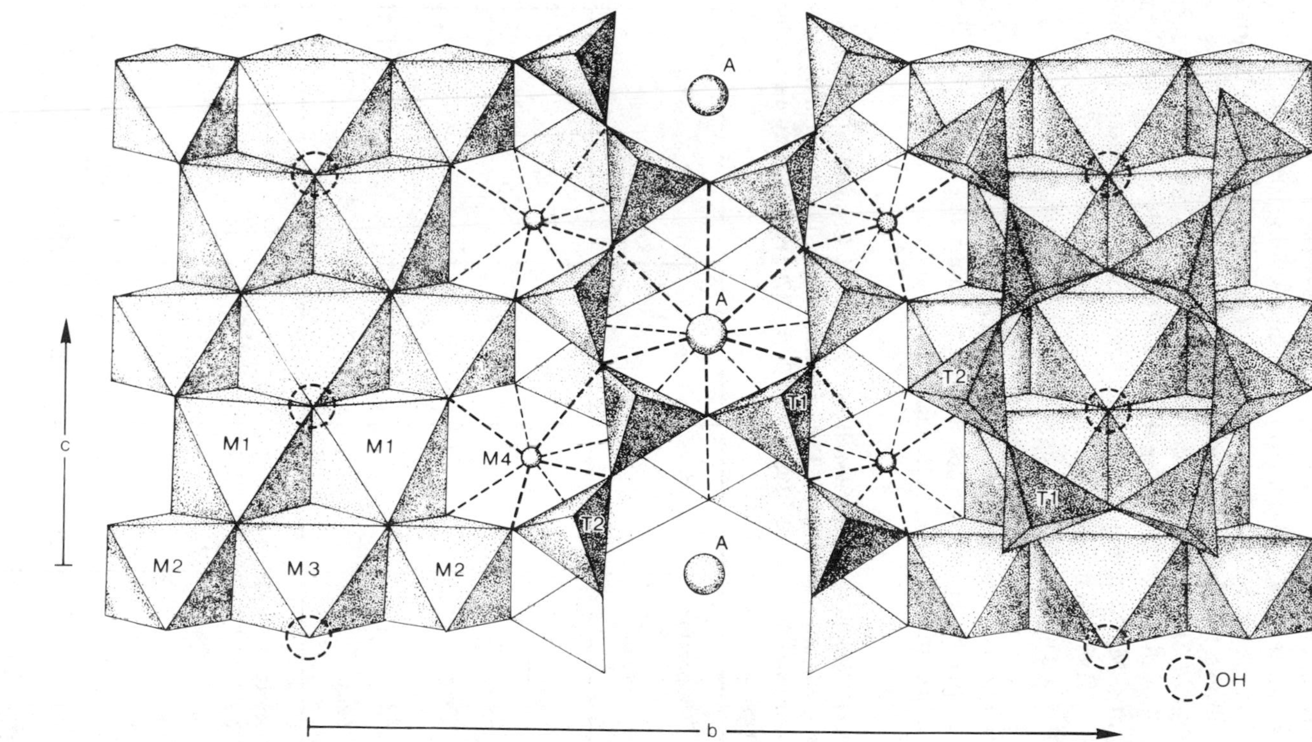

Figure 1.6. Structure of a Monoclinic Amphibole Projected Down the A Axis (modified from Papike et al., 1969). OH groups are located in large holes of the rings in the chains. M_1, M_2 and M_3 sites are occupied by Y cations, which are 6 coordinated; M· sites are occupied by larger X cations, which are 6–8 coordinated. A sites are occupied by ions, which are 10–12 coordinated (from Hurlbut and Klein, 1977, Fig. 10-48, p. 386, reproduced by permission of John Wiley & Sons, Inc.).

Schott and Berner (1985) are as follows: enstatites of Bamble (Norway); diopsides of Pitcairn (New York) and of Gouverneur (Quebec); and tremolites from Canaan (Connecticut). Analytical procedures were the same as those previously described for feldspars.

When these three types of minerals are undergoing initial stages of dissolution (pH 6, room temperature), the behavior of Ca and Mg cations relative to Si shows the following results: (1) For enstatite, loss of Mg is greater than that of Si but the release of Mg is nonstoichiometric (Fig. 1.7, Table 1.4). (2) For diopside, Ca is eliminated preferentially to Mg and Si; in this case the ratio Mg/Si remains constant, suggesting a stoichiometric dissolution of these two elements (Fig. 1.8). (3) For tremolite, preferential elimination of Ca and Mg versus Si occurs with a faster release of Ca (Fig. 1.9).

At pH 1, and at room temperature, these three silicates display the same processes as at pH 6. However, dissolution reactions are faster and the relative elimination of cations is better developed.

Table 1.4. Surface Compositions Established by XPS for the Dissolution of Etched Enstatite and Diopside at pH 6. Ratios of Elements Are Referred to Ratios of Surfaces of Peaks (modified from Schott et al., 1981, table 2, p. 2129, reprinted by permission of Pergamon Press plc).

	Temperature °C	Time (days)	Mg_{25}/Si_{2p}	$Ca_{2p3/2}/Si_{2p}$
Enstatite	initial material		0.61	-
	20	22	0.56	-
	20	28	0.54	-
	60	28	0.52	-
Diopside	initial material		0.32	1.90
	20	2	0.33	1.62
	20	22	0.32	1.62
	60	22	0.32	1.62

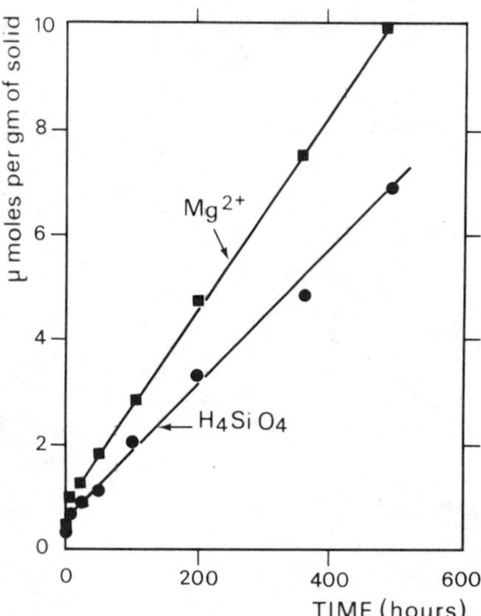

Figure 1.7. Plot of Released Mg^{2+} and H_4SiO_4 versus Time for the Dissolution of Etched Enstatite at pH 6, T20° C (modified from Schott et al., 1981, Fig. 8, p. 2129, reprinted by permission of Pergamon Press plc).

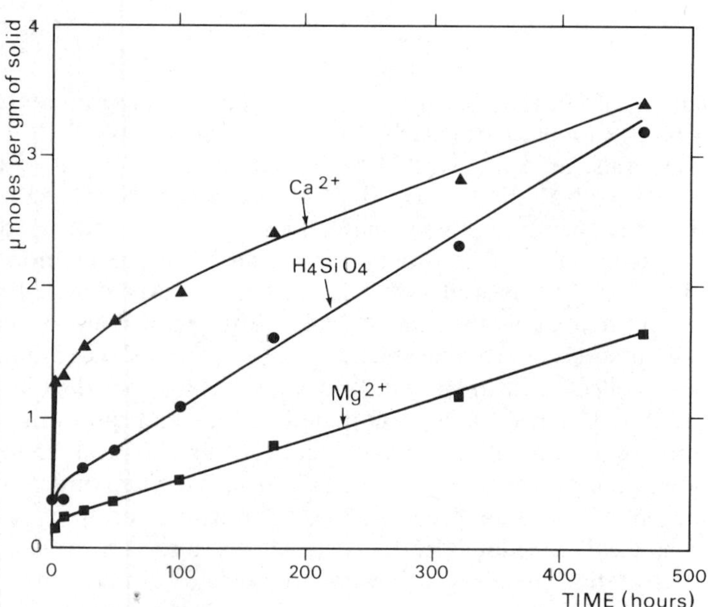

Figure 1.8. Plot of Released Ca^{2+}, Mg^{2+} and H_4SiO_4 versus Time for the Dissolution of Etched Diopside at pH 6, T20° C (modified from Schott et al., 1981, Fig. 10, p. 2130, reprinted by permission of Pergamon Press plc).

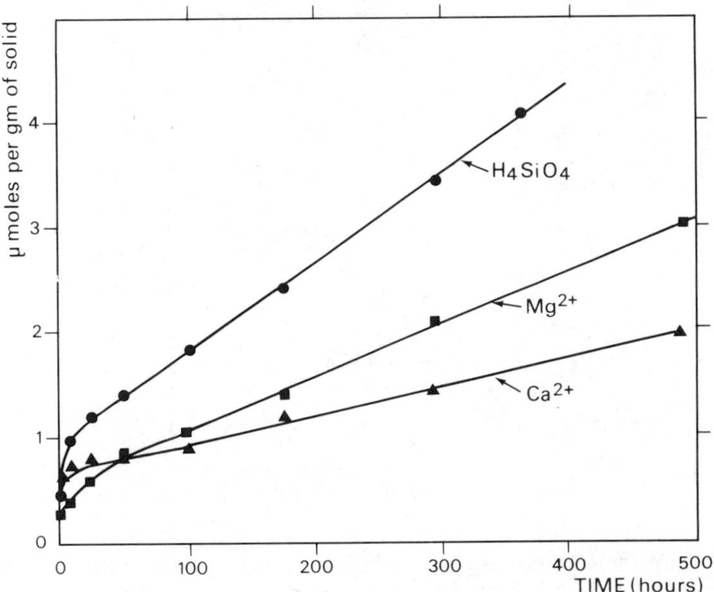

Figure 1.9. Plot of Released Ca^{2+}, Mg^{2+}, and H_4SiO_4 versus Time for the Dissolution of Etched Tremolite at pH 6, T20°C (modified from Schott et al., 1981, Fig. 13, p. 2132, reprinted by permission of Pergamon Press plc).

Continuation of these experiments clearly shows that preferential dissolution of basic cations, particularly Ca^{2+} versus Si, is limited to the very surface of minerals, as demonstrated by XPS analysis. This loss of cations results from their substitution by H^+ or H_3O^+ over a thickness of the order of 10 Å without formation of a new phase. This protonated surface adherent to the silicate is unstable. This situation is shown by the dissolution curves (Figs 1.7, 1.8, and 1.9) as well as by the fact that the protonated surface always keeps approximately the same thickness after a few hours of reaction, regardless of the analyzed sample (etched samples or those not treated and fresh or those collected in soils). This indicates a congruent dissolution.

However, the existence of the protonated surface and the evaluation of its thickness are based on indirect evidence from chemical analyses of elements other than hydrogen. Petit et al. (1987) gave the first direct evidence that dissolution of diopside proceeds via a surficial hydration, by using a resonant nuclear reaction (RNR) that allows direct hydrogen depth profiling. Dissolution experiments were conducted on centimeter-sized polished sections of two monocrystalline diopsides from Rothenkopf, Germany (Sample 1 = Fig. 1.10 A) and from Val d'Alla, Italy (Sample 2 = Fig. 1.10 B), the chemical compositions of which are, respectively:

$$Ca_{0.98}Mg_{0.93}Fe_{0.07}Si_2O_6$$

and

$$Ca_{0.98}Mg_{0.97}Fe_{0.045}Si_2O_6.$$

The profiles obtained are almost identical for the two diopsides and the two pHs (Fig. 1.10 A). They present a maximum hydrogen concentration at ~ 250 Å (~ 8.5 atom %) for samples leached for 75 days, while the hydrogen profile extends to ~ 800 Å from the surface of the diopside. In order to discriminate between the chemical forms of hydrogen (H^+, H_3O^+, H_2O) the diopside 1 was dried at 150° C for 1 hour and then examined by means of RNR (Fig. 1.10 B). The resulting profile shows a partial loss of hydrogen of ~ 25%. This is in favor of H penetrating into the mineral partly in the form of molecular water. Moreover there is no marked effect of the pH of the reacting solution on the amplitude of H penetration in the mineral.

Corrosion figures observed on natural grains were reproduced in the laboratory by attack with a concentrated solution of HF−HCl. Observed figures were controlled by the crystallographic structure of minerals and aligned parallel to specific axes leading to deeply striated surfaces when they joined one another end to end, or to tooth-combed shapes when they joined each other side by side. Obviously, no isotropic dissolution took place leading to formation of a uniform protective coating surrounding the silicate grain and controlling dissolution reactions. Any interpretation of parabolic kinetics was an artifact due to the dissolution of ultrafine particles that remained attached to the surface of the mineral during preparation of the sample (particularly during grinding). The best proof was afforded by comparing etched samples with nontreated ones and subjecting both to aqueous dissolution (Fig. 1.11). Furthermore, since the samples of enstatite studied by Luce et al. (1972) originated from the same place as those studied by Schott et al. (1981), the latter could plot on their diagrams the curves provided by the authors who had not previously treated their samples and had reached the conclusion of a nonlinear kinetics of dissolution.

In conclusion, as for feldspars, the dissolution of enstatites, diopsides, and tremolites (Mg silicates and Mg−Ca silicates) follows a linear kinetics. These minerals show the differentiation of a protonated stage consisting of a very thin surface layer whose thickness does not increase with the degree of advancement of hydrolysis reactions. Moreover, Petit et al. (1987, p. 706) showed that, for diopside, "penetration of hydrogen into the mineral could exist in the form of molecular water even in acidic solutions and not only as H^+ ions replacing cations within the crystal as suggested by the protonated surface layer theories." The density of defects should play a major role in water diffusion up to a depth of a few hundred Å. This penetration reflects an increase of porosity over this thickness. Thus, water diffusion permits

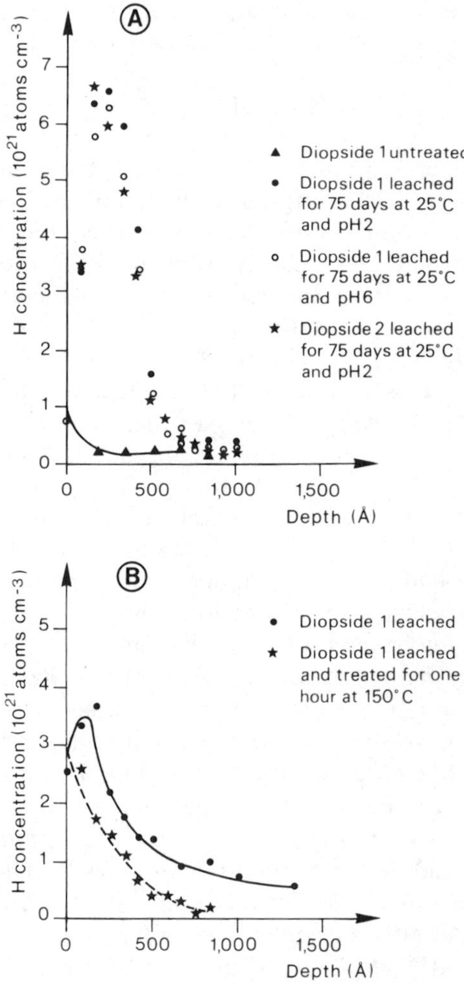

Figure 1.10. Evidence for Hydrogen Penetration in Leached Diopside (modified from Petit et al. 1987. **A**: RNR H-depth profiles before and after leaching in various conditions. **B**: The effect of a one-hour postleaching thermal treatment at 150° C on the RNR H-profile of sample 1. (Reprinted by permission from Nature, Vol. 325, Fig. 1a and b, p. 705, copyright 1987 Macmillan Magazine Ltd.)

water molecules to jump from one interstice to another (Petit et al., 1987; Schott and Petit, 1987).

Mogk and Locke (1988), using auger electron spectroscopy (AES), reached results close to those of Petit et al. (1987). They analysed naturally weathered hornblendes from the B horizon of a sandy soil of Rocky Mountain National Park, Colorado. The profiles obtained, in considering Al as an invariant element, showed a decrease of all the mobile cations up to a thickness of about 1200 Å inside the weathered mineral (Fig. 1.12). This loss of cations

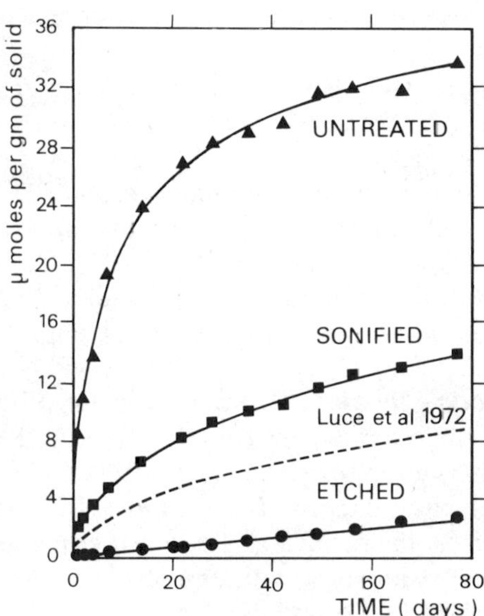

Figure 1.11. Silica Release during the Dissolution of Untreated, Sonified, and Etched Enstatite in Buffer Solution at pH 6, T20° C. The curve labelled Luce et al. (1972) refers to washed but untreated material from the same location (modified from Schott et al. 1981, Fig. 2, p. 2125, reprinted by permission of Pergamon Press plc).

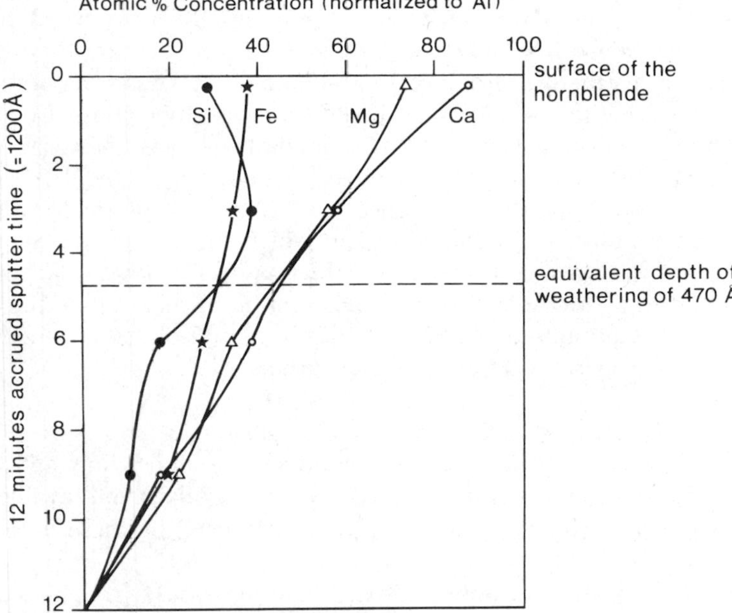

Figure 1.12. Cation Depletion, Assuming Al is Immobile Relative to Hornblende Exposed after 12 Minutes Accrued Sputter Time (= 1200 Å) Interpreted as Depth to the Unweathered Mineral (modified from Mogk and Locke, 1988, Fig. 4, p. 2539, reprinted by permission of Pergamon Press plc).

was greatest near the surface of the mineral and decreased in a nonlinear fashion toward the inside. Ca and Mg were leached preferentially to Si and Fe up to depths of the order of 500 Å beneath the surface of the mineral. This example shows that the weathering mechanism of hornblende does not result only from surface-controlled reactions but also from diffusions of cations through a thickness of the order of 1000–1200 Å. According to Mogk and Locke (1988) there are no facts that lead us to assume that such a layer remains constant while weathering is proceeding.

Dissolution of Ferrous Silicates: Bronzite and Fayalite

These minerals were investigated by Berner and Schott (1982) and Schott and Berner (1983, 1985) using the same analytical techniques as for feldspars and for silicates with basic cations. But in the above studies, in order to determine more precisely the behavior of iron, experiments were undertaken not only under normal oxidizing conditions of pH 6, 20° C, but also under anoxic conditions at pHs 1–1.5 and pH 6.

Because an understanding of crystal structures is critical for the interpretation of hydrolysis processes, a few essential data are viewed below.

Review of the Structure of Bronzites and Olivines

The structure of orthopyroxenes has been discussed previously and it is only necessary to recall that Mg^{2+} and Fe^{2+} cations of bronzite $(Fe^{2+}_{0.15}Mg_{0.85})SiO_3$ are located at M_1 and M_2 sites with Fe^{2+} preferentially at M_2, which is larger and distorted; hence iron is less well bonded to the structure than Mg^{2+}.

Olivine has a structure with layers parallel to (100), consisting of independent SiO_4 tetrahedra (neosilicates) linked crosswise by chains of octahedra bonded by divalent ions (Fig. 1.13). Octahedral sites are known as M_1 and M_2, with M_1 distorted and M_2 relatively regular. Mg^{2+} and Fe^{2+} occupy these sites without specific preference for M_1 or M_2, particularly for the investigated fayalite $(Fe^{2+}_{1.9}Mn^{2+}_{0.1})SiO_4$, in which these sites are mainly occupied by Fe^{2+}. However, in olivines of the monticellite type, Ca^{2+} occupies M_2 sites and Mg^{2+} occupies M_1 sites. Dissolution of minerals of the olivine type does not require breaking Si–O bonds because isolated tetrahedra lead directly by hydrolyisis to H_4SiO_4 silanol groups.

Data on the Dissolution of Bronzites and Fayalites

The samples studied by Berner and Schott (1982) and Schott and Berner (1983, 1985) are bronzites from Webster (North Carolina) and fayalites from Rockport (Massachussets). The major results reached by these authors are summarized below.

In the case of bronzite under anoxic conditions (pO_2 = O atmosphere) and regardless of the pH after a slight initial reaction, the ratio of Si release is constant during the entire duration of the experiment (Fig. 1.14), whereas under oxygenated conditions (pO_2 = 0.2 atmosphere) at pH 6, dissolution follows a law of parabolic types with a slower speed of dissolution

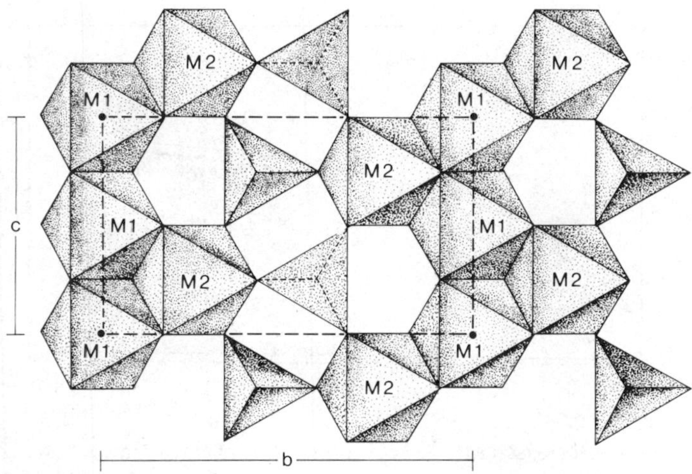

Figure 1.13. Portion of the Idealized Structure of Olivine Projected on (100). M_1, M_2 are octahedral sites (modified from Hurlbut and Klein, 1977, Fig. 10-3, p. 341, reprinted by permission of John Wiley & Sons, Inc.).

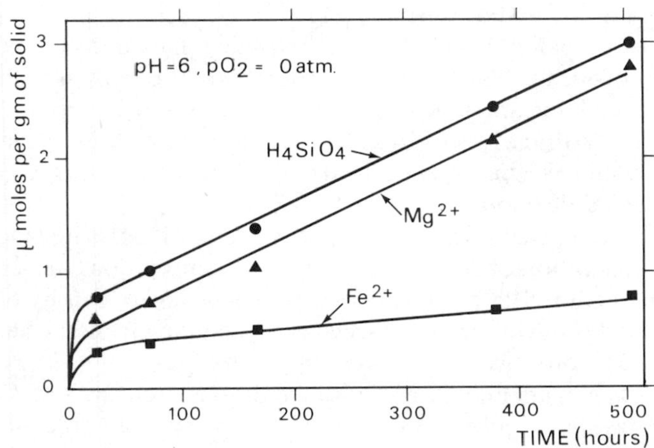

Figure 1.14. Plot of Released Mg^{2+}, Fe^{2+}, and H_4SiO_4 as a Function of Time for the Dissolution of Etched Webster Bronzite at pH 6, T20° C, pO_2 = O atm. (Anoxic Conditions) (modified from Schott and Berner, 1983, Fig. 2, p. 2235, reprinted by permission of Pergamon Press plc).

(Fig. 1.15). At pH 6, Mg shows a slight depression with respect to silica regardless of the conditions of oxidoreduction of the solution. However, Fe has, for a given pH, a behavior very dependent upon oxygen concentration so that, under well-established oxidizing conditions, iron appears appreciably less mobile than Si and Mg to the extent of coating bronzite with an

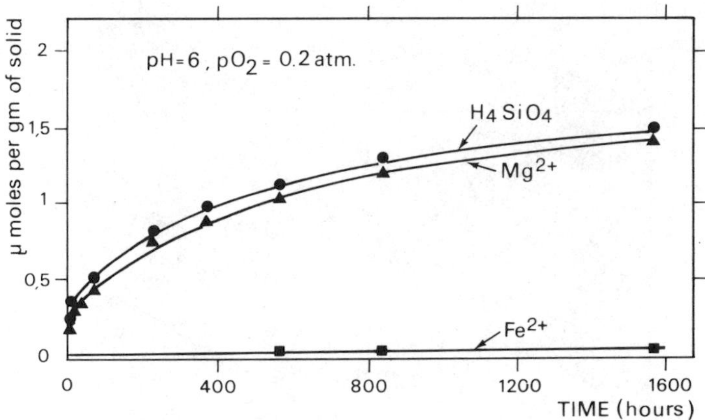

Figure 1.15. Plot of Released Mg^{2+}, Fe^{2+}, and H_4SiO_4 versus Time for the Dissolution of Bronzite at pH 6, pO_2 = 0.2 atm. (Oxidizing Conditions) (modified from Schott and Berner, 1983, Fig. 32, p. 2236, reprinted by permission of Pergamon Press plc).

iron-rich layer. Conversely, under anoxic conditions, iron undergoes a differential release compared to Si and Mg.

In the case of fayalite, iron behaves in the same manner as in experiments dealing with bronzite. Furthermore, under anoxic conditions and pH 1, losses of Mg and Fe are much more pronounced than at pH 6 with, in particular, an important loss of iron. Thus, while the reaction proceeds, the surface of fayalite is almost entirely devoid of iron whereas a coating of precipitated silica develops.

Differential removal of cations at the surface of bronzite and fayalite leads to the generation of surface layers whose compositions are established by measurements of XPS binding energies. These layers consist either of a new precipitated product, or of a leached zone resulting from the substitution of cations by H^+ protons.

Thus, at pH 6 and under oxidizing conditions, removal of Fe^{2+} from bronzite and fayalite induces, at the surface of the silicate, precipitation of a thin hydrated layer of ferric oxide, easily removed with ultrasound. However, beneath this ferruginous layer, another one is differentiated, whose thickness does not reach 10 Å, and that properly belongs (is adherent) to the silicate mineral. In this particular layer (called ferric silicate surface), iron is oxidized in Fe^{3+}, Mg is impoverished, and a portion of Mg is substituted by H^+ (or hydroxilized). This shows that oxidation of Fe^{2+} does occur not only during its release to solution, but also prior to release inside the silicate structure. This ferric silicate surface is certainly less reactive than the surface of the fresh ferrous silicate and therefore protective since it is adherent. However, this surface does not continue to thicken with time. Results of laboratory experiments are confirmed by observation of pyroxenes collected in soils.

These grains, which have undergone an alteration of long duration measurable in thousands and tens of thousands of years, do not display adherent layers thicker than those observed during laboratory experiments, with a weathering time measured only in hundreds of hours (600 at the most). This adherent layer is therefore unstable and rapidly destabilized in favor of a hydrated layer of ferric oxides. Experimental dissolutions of forsterite from Hanauma Bay, Hawaii (Grandstaff, 1978; 1983) confirm the congruent dissolution of olivine at an increasing rate when pH decreases.

At pH 1 and under anoxic conditions, none of the above-mentioned layers are formed because they are not stable under these conditions. However, an adherent layer is formed, slightly thicker than 10 Å, impoverished in Mg^{2+} and Fe^{2+}, and protonated. The thickness of this layer does not increase while the weathering reaction proceeds. It destabilizes and leads to a precipitate of amorphous silica.

Dissolution of Micas

Review of the Structure of Micas

In micas, SiO_4 tetrahedra are arranged between a sheet structure of infinite extent. Within this sheet, each tetrahedron shares three oxygens with its neighbour. The structure of muscovite is a typical example (Fig. 1.16).

Data on the Dissolution of Micas

The study under the microscope of the various stages of meteoric weathering of mica crystals (Rausell-Colom et al., 1965; Bisdom, 1967; Rimsaite, 1967; Sawhney and Voigt, 1969; Fayolle, 1979) shows several changes that can occur simultaneously, such as deformation, opening of cleavages, and modification of polarization colors.

These earliest transformations visible under the petrographic microscope and considered to be an expression of the gradual loss of interlayer potassium. However, according to the crystallo-chemical nature of mica, large K^+ cations are freed in various ways. Indeed, it is known since Lapparent (1909), that dioctahedral micas (muscovite, margarite) have a greater resistance to weathering than trioctahedral micas (phlogopite, biotite, annite). Millot (1964), after Basset (1960), explained this different sensitivity to weathering as resulting from the difference in orientation of hydroxyls with respect to the basal cleavage of micas. Fripiat (1960), Norrish (1973), and Fanning and Keramidas (1977) also used the above interpretation to explain the higher sensitivity to weathering of trioctahedral micas.

Serratosa and Bradley's (1958) work on the position of hydroxyls with respect to sheets of mica were confirmed numerous times, in particular by Vedder and MacDonald (1963); Newman (1969); Rothbauer (1971); and Giese (1979).

In trioctahedral micas, OH radicals are oriented at right angles to the sheets because the proton of the OH is repelled by divalent cations that

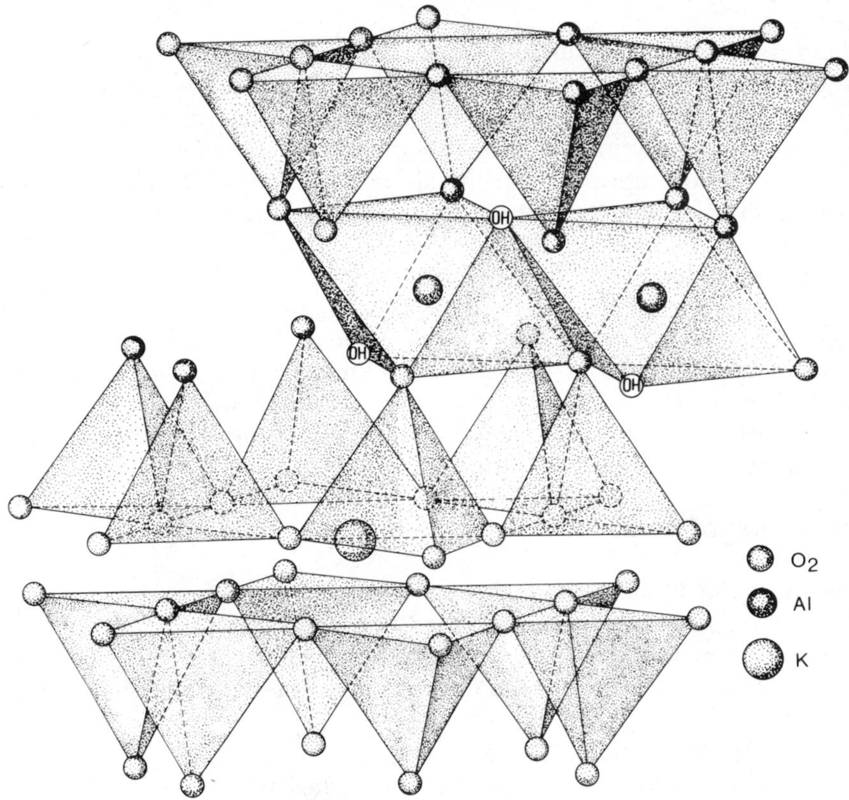

Figure 1.16. Sketch of the Structure of Muscovite (modified from Grim, 1968). (Reproduced from Hurlbut and Klein, 1977, Fig. 10-56, p. 394, by permission of John Wiley & Sons, Inc.)

occupy all the octahedral sites. Therefore, the proton is located very close to K^+ and a process of repulsion takes place between H^+ and K^+. In fact, in dioctahedral micas, only two octahedral sites out of three are occupied by trivalent cations. The proton is repelled by these cations and the hydroxyl has, therefore, a tendency to 'plunge' toward the open octahedral gap and to become oriented parallel to the plane of the sheets. Because the proton is located far away, the process of repulsion with respect to K^+ cations no longer exists. Thus, the large K^+ ions appear more strongly connected within the structures of dioctahedral micas than in those of trioctahedral micas, thus accounting for the greater sensitivity to weathering of the latter.

Kinetics of dissolution of dioctahedral micas (muscovite) and of trioctahedral micas (phlogopite) were studied by Lin and Clemency (1981a, 1981b) at room temperature, in the presence of distilled water, and in a closed system with a partial pressure of CO_2 of 1 atmosphere. These laboratory experiments confirmed the incongruent dissolution of the micas. The slower rate of release of silica with respect to that of other cations showed that tetrahedra were affected last. The phlogopite used in these experiments was from Madagascar, and the weight percentages of K, Mg, and Si released by the phlogopite were plotted as a function of time (Fig. 1.17). Durations of reactions lasted from a few minutes to a thousand hours. Under the conditions of the experiment, octahedral sheets containing Mg were twice as soluble as tetrahedral sheets containing Si. This fact was confirmed in similar dissolution experiments for other layer-type Mg minerals, such as brucite, antigorite, and talc (Lin and Clemency, 1981c).

A comparison of results obtained for brucite, antigorite, phlogopite, and talc indicates that in each case the weight percentage of Mg released as a function of time (Fig. 1.18) is proportional to the number of octahedral sheets compared to tetrahedral sheets of the mineral. Brucite contains no silica, it is completely dissolved after three days of reaction and the released magnesium reprecipitates as hydromagnesite before reaching equilibrium concentration with respect to brucite (see the flat curve of brucite in Fig. 1.18). In the dissolution of phlogopite, the concentration of H_4SiO_4 released in solution never reaches saturation in amorphous silica, whereas for talc

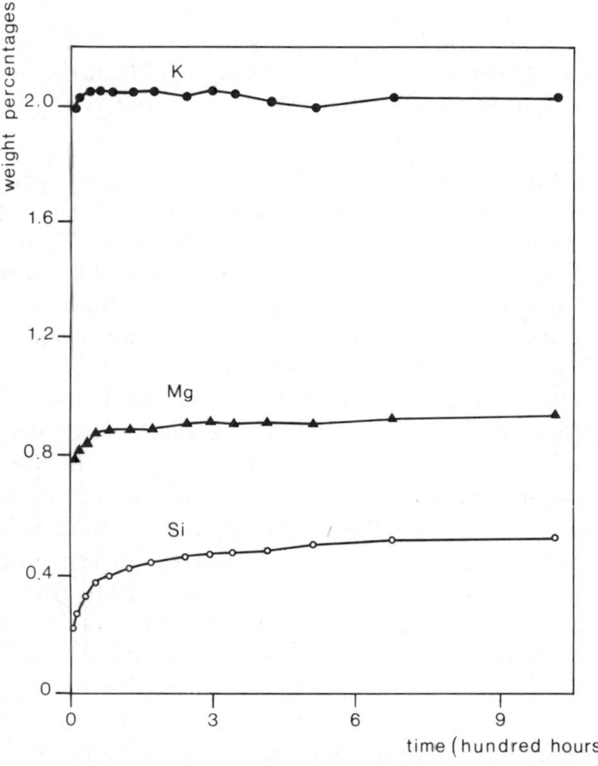

Figure 1.17. The Weight Percentages of K, Mg, and Si Released Phlogopite versus Time (Hundred Hours) (modified from Lin and Clemency, 1981c, Fig. 1, p. 803, copyright by the Mineralogical Society of America).

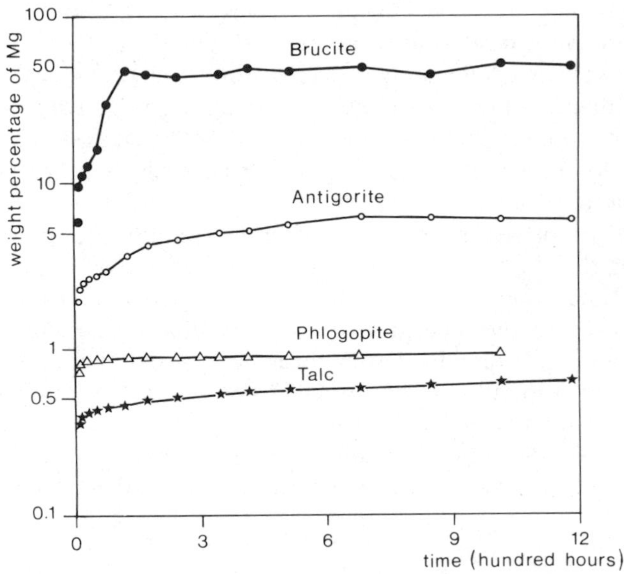

Figure 1.18. The Weight Percentages of Mg Released from Brucite, Antigorite, Phlogopite, and Talc Plotted versus Time (modified from Lin and Clemency, 1981c, Fig. 4, p. 805, copyright by the Mineralogical Society of America).

and antigorite, saturation with respect to amorphous silica is reached after 740 hours and 30 minutes, respectively. For antigorite, released magnesium reaches saturation with respect to magnesite (although the latter was never detected by X-ray diffraction analysis of the solvant), and this would explain the flattening of the antigorite curve in Fig. 1.18. In the case of talc, the solution remains undersaturated with respect to all magnesian minerals and no secondary magnesian solid phase precipitates.

Lin and Clemency (1981c) have also shown that the amount of dissolved mineral is a function of the proportion of the surface area of that mineral in contact with water. By comparing the weight percentage of Mg released in solution with an equivalent specific surface area in contact with water, these authors noticed (Fig. 1.19) that the solubility order remained the same for brucite, antigorite, phlogopite, and talc. However, in the initial stage of dissolution, phlogopite showed an amount of released Mg greater than that of antigorite. This might be explained by the fact that interlayer cations of phlogopite are more rapidly and more easily replaced by hydrogen ions.

In conclusion, dissolution of micas depends first on the orientation of hydroxyls, which allows a stronger bonding to the structure of interlayer potassium cations in dioctahedral micas than in trioctahedral micas; second, in trioctahedral phyllosilicates, it depends on the number of octahedral sheets with respect to tetrahedral sheets.

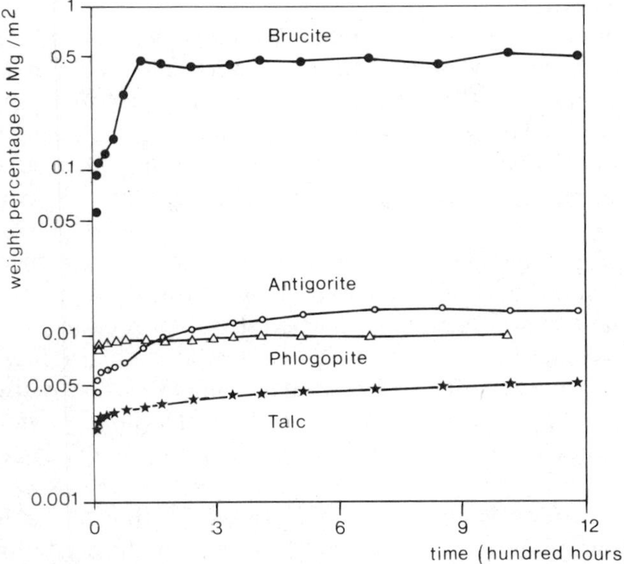

Figure 1.19. The Weight Percentages of Mg Released from Brucite, Antigorite, Phlogopite, and Talc Based on the Same Surface Area versus Time (modified from Lin and Clemency, 1981c, Fig. 5, p. 805, copyright by the Mineralogical Society of America).

Processes of Hydrolysis Reactions of Crystals

The examples just studied tell us the following:

1. For identical conditions of the weathering environment, kinetics of dissolution reactions vary from one mineral to another. This situation, generally called differential weathering of minerals, did not escape the earliest petrographers, in particular Lacroix (1914).
2. For a given mineral, certain areas undergo preferentially dissolution reactions.
3. In these areas, sites containing cations (coordination polyhedra) do not react in the same manner to hydrolysis.

Therefore, hydrolysis reactions that affect minerals should be visualized at several structural levels, from the atomic scale to that of the mineral, and eventually to that of the rock.

This chapter attempts to analyze the process of weathering at all scales of observation, namely at all structural levels. This approach begins with the smallest structures, namely, the hydrolysis of crystalline frameworks. It was investigated by Millot (1964) and Loughnan (1969) after the work by Frederickson (1951).

Exchange Reactions

Although frameworks of silicate minerals follow the rule of electron neutrality, the surface of these minerals displays atoms and ions with incompletely saturated valences. Therefore, in contact with water, charged surfaces allow polarization of water molecules and hence separation, at the mineral-solution interface, of the H_2O dipoles into H^+ and OH^- through respective attraction of the nonsaturated negative and positive charges of the surface of the crystal.

The H^+ proton attaches itself to one or several water molecules and forms a hydronium ion H_3O^+. The H^+ proton with its very small diameter and monovalent charge generates a very strong attraction on the oxygen ions of the surface of the crystal. Thus, development of these chemical reactions between substrate and adsorbate generates a chemisorption for which corresponding adsorption heats are about ten times greater than for a simple physical adsorption generating only Van der Walls forces (Touray, 1980).

This chemisorption is accompanied by a substitution of Ca^{2+}, Na^+, K^+ cations whose attraction with respect to the oxygens of the crystalline structure is weaker because of their large diameter. Following Lasaga (1981) and taking as an example the weathering of pyroxenes, Schott and Berner (1985) proposed (chemisorption and substitution) the following expression for this surface reaction:

$$M_1M_2Si_2O_6 + 2H^+ \leftrightarrow M_2 + M_1H_2Si_2O_6.$$

The progression rate of the reaction can be set as:

$$r^+ = \frac{dn_{M_2}}{dt} = k_+\chi_{M_2}\theta_H^2 \quad r^+ = k_+\chi_{M_2}\theta_H^2$$

with θ_H expressed in Langmuir isotherm.

$$K = \frac{\theta_H}{m_H - \theta_E}$$

θ_H and χ_E being, respectively, the fraction of the adsorption sites occupied by H^+ and vacant; $\chi_{M_2} = 1 - \chi_H$ = fraction of surface of M_2 sites occupied by M_2 cations.

K = equilibrium constant for adsorption reaction.

m_H = concentration of H^+ in the solution.

χ_{M_2} and χ_H can be expressed as a function of the number of moles of M_2 cations (n_{M_2px}) and of hydrogens (n_{Hpx}) contained in the M_2 sites of the surface of the pyroxene with $n_{px} = n_{M_2px} + n_{Hpx}$. Assuming that the number of moles of pyroxenes protonated at the surface is given by:

$$n_{Hpx} \doteq n_{M_{2s}} - n_{M_{1_s}}$$

where $n_{M_{2_s}}$ and $n_{M_{I_s}}$ are the number of moles of cations released in solution from sites M_I and M_2, Schott and Berner (1985) reached the following expression:

$$r^+ = k_+ \cdot K^2\theta_E^2 m_H^2 - \frac{k_+ \cdot K^2\theta_E^2 m_H^2}{n_{px}}(n_{M_{2_s}} - n_{M_{1_s}}).$$

For a constant pH, this equation implies a linear relationship between the rate of the exchange reaction and the number of moles of pyroxenes protonated at the surface. This is in agreement with experiments completed by these authors on pyroxenes.

Investigations by Petit et al. (1987), Schott and Petit (1987), and Mogk and Locke (1988) showed that the penetration of protons and even of water molecules can reach depths of the order of a thousand Å. The density of defects reaching the surface of the mineral must play an important role in this penetration and in the generation of the weathered superficial layer.

Dissolution Reaction and pH

The rate of reaction, as shown in the previous equation, depends on the pH, namely on the concentration of hydrogen ions, also expressed as the activity of H^+ ions or of H_3O^+, designated respectively $a_H{}^+$ or $a_{H_3O^+}$.

In the case of the congruent dissolution of mineral θ, the rate of weathering (or of dissolution) can be written in another form (Lasaga, 1981):

$$\frac{dC_i}{dt}\,(\text{dissolution}) = \frac{A_\theta}{V}\,\nu_i k_\theta.$$

in which: ν_i is the stoichiometric content of i in mineral θ and k_θ the overall dissolution rate of mineral θ; A_θ is the surface area of mineral θ; V is the volume of solution in contact with mineral θ; the ratio dCi/dt expresses the change of concentration of element i due only to mineral θ. Rates of dissolution of silicates increase with the activity of H^+ ions if the pH is lower than 7. In fact, the pH dependence of the rate constants in this equation can be written as (Lasaga, 1984):

$$k_\theta \alpha\ (a_{H^+})^{n_\theta} \quad 0 \le n_\theta \le 1.$$

This author gave values of n_θ for several minerals (Table 1.5).

Holdren and Speyer (1985b) showed that, at the scale of initial stages of

Table 1.5. The pH Dependence of Mineral Dissolution Rates (modified from Lasaga, 1981; and from Schott and Petit, 1987).

minerals	k_θ	pH Range	References
K-feldspar	$a_{H+}^{1.0}$	pH < 7	Helgeson et al. 1984
	$a_{H+}^{1.0}$	pH < 5	Lasaga 1984
Nephelite	$a_{H+}^{-0.20}$	pH > 7	Lasaga 1984
Augite	$a_{H+}^{0.7}$	1 ≤ pH ≤ 6	Schott and Berner 1985
Diopside	$a_{H+}^{0.5}$	2 ≤ pH < 6	Schott et al. 1981
Enstatite	$a_{H+}^{0.8}$	2 < pH < 6	Schott et al., 1981
Forsterite	$a_{H+}^{1.0}$	3 ≤ pH ≤ 5	Grandstaff, 1981
Quartz	$a_{H+}^{0.0}$	pH < 7	Rimstidt and Barnes 1980
Anorthite	$a_{H+}^{0.54}$	2 < pH < 5.6	Fleer 1982 in Lasaga 1984

weathering reactions of alkaline feldspar, dissolution is stoichiometric only at pH of the order of 5.5–6. Furthermore, when reactions are not stoichiometric, release of a given element is variable and depends on the more acidic or more basic environment in which dissolution occurs.

Stability of Exchange Sites

In the above-discussed pyroxene, cations of M_2 sites preferentially undergo hydrolysis reactions. In general, among the coordination polyhedra forming silicates, some display a greater stability than others with respect to hydrolysis processes. These polyhedra are increasingly stable with the decreasing number of their apexes, that is the cations occupying their center have a strong ionization potential (high valence and small diameter). Conversely, large cations with weak ionization potential, which are located in the center of polyhedra with high coordination numbers, display the weakest bonding forces and hence are the most vulnerable.

This situation was verified in all the summaries of studies presented in this chapter. Thus, for pyroxenes, sites M_2, in which coordination can reach eight large cations, are the least bonded to the structure with respect to sites M_1, which form regular octahedra occupied by the smallest cations. For amphiboles, sites M_4 with coordination numbers that can reach 8, and that carry the large cations, are the most unstable.

These stabilities of coordination polyhedra may be expressed in terms of site energies. (Madelung site energies), which represent the electrostatic work required to pull away an ion from its site to an infinite distance. Schott et al. (1981) quoted in their study values obtained by Ohashi and Burnham (1972) for sites M_1 and M_2 of pyroxenes, and values obtained by Whittaker (1971) for sites M_1, M_2, M_3, and M_4 of tremolites (Table 1.6).

In this table, Ca cations, which show the lowest values of site energies, that is, the least negative, are the most mobile (Dowty, 1980).

During weathering reactions, cations of coordination polyhedra are replaced by H^+ or H_3O^+ and change from the state of reactant to that of product. As seen above this transition stage is defined by a free energy maximum (ΔG) schematically illustrated for the weathering of K-feldspar (Fig. 1.20) by Aagaard and Helgeson (1982).

Table 1.6. Occupation Site of Cations and "Madelung." Energy Sites for Several Pyroxenes (from Schott and Berner, 1983, tables 3 and 4, p. 2131 and 2132, reprinted by permission of Pergamon Press plc).

Mineral	Site	Cation	Energy (KJ mol^{-1})
diopside	M1	Mg	-4346
	M2	Ca	-3157
orthoenstatite	M1	Mg	-4318
	M2	Mg	-3940
tremolite	M1	Mg	-4560
	M2	Mg	-4686
	M3	Mg	-4519
	M4	Ca	-3975

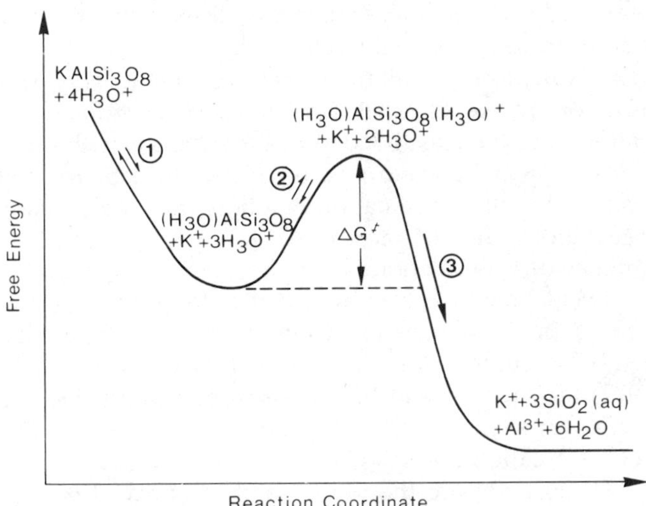

Figure 1.20. Sketch Illustrating the Free Energy Maximum Which Reactants Must Pass to Become Products: Example of the Hydrolysis of K-Feldspar in Acidic Aqueous Solution (modified from Aagaard and Helgeson, 1982, Fig. 2, p. 256, reproduced by permission of American Journal of Science).

Accessibility of Exchange Sites

In crystalline structures, coordination polyhedra are arranged according to well-known laws of atomic organization. However, the most vulnerable polyhedra are not all accessible simultaneously. As shown above, hydrolysis reactions depend upon the density of sites of high surface energy: edges, corners, cracks, holes, and, at smaller scale, point defects, twin boundaries, and dislocations. Thus, whereas the dissolution front does not affect the entire surface of the mineral, it has a preferred orientation.

An excellent example was given by Velbel (1983, 1984a and b) and by Velbel and Dowd (1983). They showed in a weathering profile of the southern Appalachians that andesine-oligoclase grains are twinned (albite twinning) and that only a particular series of lamellae shows a greater sensitivity to dissolution reactions. This difference of reactivity between series of twinned lamellae (Fig. 1.21 A, B and C) does not result from a difference in chemical composition but from a difference in density of dislocations. This density of dislocations may be related to the history of the deformation and to the tensions undergone, but it depends also upon the crystallographic orientation of the crystal in the pressure field.

Lasaga (1984) emphasizes the importance of the surface area of the mineral in contact with the solution during dissolution reactions: The greater surface area of the mineral, the greater its dissolution. Lin and Clemency (1981c) reached a similar conclusion for the dissolution of magnesian phyllosilicates. As mentioned above, the surface of dissolution of the mineral is

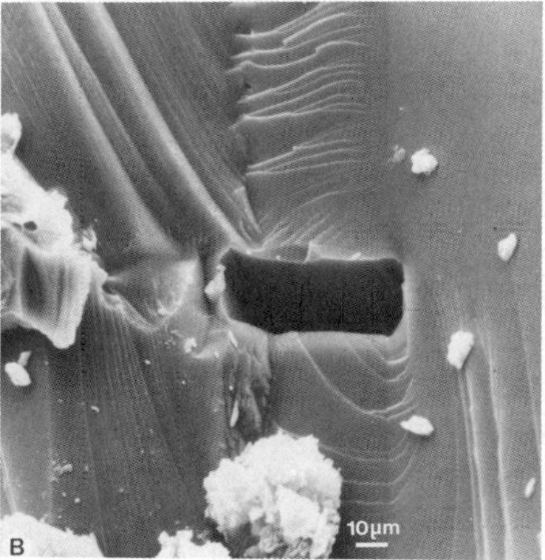

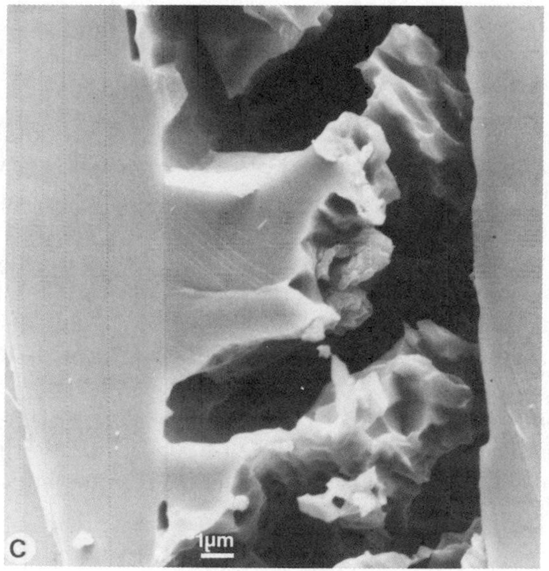

Figure 1.21. Feldspar Weathering and Accessibility of Exchange Sites (photographs courtesy M. A. Velbel). **A**: Vacuoles and regularly-spaced etch pits along one twin set of the feldspar. SEM photograph (from Velbel 1983). **B**: Rectangular etch pit in twin set, detail of **A**; SEM photograph (from Velbel, 1983). **C**: Vacuoles in twin set, detail of A. SEM photograph (from Velbel, 1983). (Reprinted by permission of Sciences Géologiques.)

not uniform and dissolution concentrates on points of excess surface enegy. Holdren and Speyer (1985a) defined the relationships between the specific surface of the mineral, defects reaching its surface, and the dissolution rate as follows: (1) If minerals have a large size compared to the distance between adjacent exposed defects, dissolution rates vary in a linear fashion with respect to the specific surface of the mineral; (2) if the size of the mineral is of the same order of magnitude as the distance between adjacent exposed defects, the above relationship tends to disappear. Therefore, Holdren and Speyer (1985a, p. 675) suggested a distinction between the "surface reaction controlled mechanism and a surface area controlled model which is frequently and incorrectly, assumed to be equivalent."

Furthermore, the surface of the mineral in contact with the weathering solution changes while the mineral is being weathered, decreases in size, becomes subdivided, and controls reaction rates.

Role of Temperature on Dissolution Rates

As shown by Lasaga (1984), temperature is a parameter that influences the dissolution rate of a mineral, it is represented in equations by the symbols r or k_θ. For most silicate minerals, the temperature and dissolution rate follow the law of Arrhenius:

$$k_\theta = Ae^{-\Delta E/RT}$$

where ΔE is the activation energy. Several values of ΔE obtained at steady state for dissolution reactions of minerals (Table 1.7) show a variation from 35 to 89 KJ/mol^{-1}. These values are higher than those of activation energy measured for transfer in solution [~ 21 KJ/mol^{-1}, Tsao and Pask (1982)], and on the average lower than those required to break bonds in crystals [80–400 KJ/mol^{-1}, Schott and Petit (1987)]. It looks as if catalytic effects of adsorption on surfaces (Stumm et al., 1985) and the role of surface defects (Lasaga, 1984) reduce activation energy to these intermediate values.

CO_2 Kinetics and Weathering

According to Lasaga (1981, 1984) the role of CO_2 in weathering reactions is as follows. According to Henry's law, CO_2 present in the atmosphere and in the soil dissolves in water as follows:

$$m_{CO_2} = k_H P_{CO_2} = 10^{-1.5} P_{CO_2}$$

in which m_{CO_2} is the molality of CO_2 in the solution and P_{CO_2} its partial pressure. CO_2 in the soil originates mainly from decomposition of organic matter, its concentration ($P_{CO_2} = 10^{-2}$ bars ~ 1000 Pa) is higher than that in the atmosphere ($P_{CO_2} = 10^{-3.5}$ bars ~ $10^{1.5}$ Pa). Hence, according to the above equation, water percolating through the soil reaches a maximum equilibrium concentration of $m_{CO_2} = 10^{-3.5}$ molal = 3.2×10^{-1} mol/m^3. In order to play the role of an acid, aqueous CO_2 at 25°C should first be converted into carbonic acid:

$$H_2O + CO_{2aq} \underset{k-1}{\overset{k1}{\rightleftharpoons}} H_2CO_3 \overset{fast}{\rightleftharpoons} H^+ + HCO_3^-.$$

Table 1.7. Activation Energy for Mineral Dissolution Reactions (modified from Lasaga 1984, p. 4012; and from Schott and Petit, 1987, table 11-1, p. 297, reprinted by permission of John Wiley & Sons, Inc.).

Minerals	ΔE (kJ mol^{-1})	References
Diopside	38 ($4 \leq pH \leq 6$) 81 (pH ~ 2)	Schott *et al.*(1981)
Enstatite	50	Berner *et al.* (1980)
Bronzite	45	Grandstaff (1977)
Olivine (FO_{83})	38	Grandstaff (1981)
Augite	78	Schott and Berner (1985)
Nephelite	54-71	Lasaga (1984)
Anorthite	35	Fleer (1982) in Lasaga (1984)
K-feldspar	38 ($3 \leq pH \leq 8$) 82 ($pH \leq 3$)	Helgeson *et al.* (1984)
Albite	38 ($3 \leq pH \leq 8$) 88.5 ($pH \leq 3$)	Helgeson *et al.* (1984)
Wollastonite	74	Murphy (1985)
Quartz	77	Rimstidt and Barnes (1980)

On the basis of the studies of Van Eldik and Palmer (1982), Lasaga (1984) showed that at 25° C and for an ionic strength of 0.5:

$$k_1 \quad = 4.37 \times 10^{-2} S^{-1}$$
$$k-1 = 19.2 S^{-1}$$

and hence obtained $m_{H_2CO_3} = 0.0023\ m_{CO_2(aq)}$, namely less than 1% of dissolved CO_2 in the state of H_2CO_3. The maximum rate of generated CO_2 is given by:

$$\frac{dm_{H_2CO_3}}{dt} = k_1\ m_{CO_2} = (4.37 \times 10^{-2} S^{-1})\ (3{,}2 \times 10^{-1}\ \text{mol/m}^3)$$
$$= 1.4 \times 10^{-2}\ \text{mol/m}^3\text{/S}.$$

Thus, 1 m^3 of solution generates acid at the rate of 1.4×10^{-2} mols/s. The proton rate to dissolve a given mineral can be expressed by the following equation:

$$\text{Rate} = \nu_{H^+} \frac{A}{V} \bar{\text{k}}$$

in which ν_{H^+} is the stoichiometric number of H^+ required for the dissolution of the mineral; A/V is the surface area (it is understood that this surface is not uniformly etched) by unit of volume of solution; and $\bar{k}$ is the intrinsic rate of dissolution of the mineral (see Table 1.1 for the order of magnitude).

The conversion rate of CO_{2aq} is an important factor in the dissolution of minerals, particularly for those that are the most subject to weathering. However, in nature most of the dissolved CO_2 remains in the state of aqueous CO_2.

In conclusion, it seems that organic matter acts in the weathering of minerals mainly by its role of reducing agent (consumption of oxygen for its mineralization in CO_2) and by the organic solutes it produces, which in turn act as complex-forming ligands on surfaces of minerals and on metallic ions in solution (Furrer and Stumm, 1986).

Congruent and Incongruent Dissolutions

Substitution of cations by protons (H^+ or H_3O^+) occurs without disruption of the crystalline framework. But this substitution introduces a weakening of Si−O and Al−O bonds because H^+ protons are smaller than the replaced cations, and thus have a greater attraction for oxygen and lead certainly to some deformation of the network near the surface. Cations released within the thin aqueous layer of the surface generate a weak field on the hydroxyls [ions of Na−OH, $Ca(OH)_2$ type], which give an alkaline character to the solution in contact with the surface of the solid. The association of weakened Al−O and Si−O bonds and increased alkaline character of the aqueous phase in contact with the solid is eventually responsible for the breaking of bonds. Although occupation sites of cations differ from one case to another, a parent mineral weathers more rapidly when it is richer in cations.

Alkalinity increase of the solution in contact with the parent solid has been well-known since the works by Stevens and Carron (1948), Keller et al. (1963), and Loughnan (1969). It can be expressed in terms of abrasion pH. Thus olivines, pyroxenes, and amphiboles rich in cations have an abrasion pH of 10–11 and plagioclase feldspar, 8–9; for micas, on the other hand, it is only 7–9.

Siloxane bonds Si–O–Si are the least vulnerable. Their breaking, which gives silanol bonds Si–OH, could be facilitated by the higher alkalinity of the solution in contact with the solid. According to Iler (1979), the reaction could be as follows:

$$SiO_2 + H_2O + HO^- \rightarrow Si(OH)_3O^-.$$

These surface reactions between solid and solution are probably limited to thin layers and the residence time of the solution at the contact varies with the degree of renewal of solutions at weathering sites. Away from the mineral-solution contact or as soon as circulation of solutions increases, renewal of protons is assured, and cations as well as silica are eliminated in the form of $Si(OH)_4$ because:

$$Si(OH)_3O^- + H^+ \rightarrow Si(OH)_4.$$

However, Rimstidt and Dove (1986), in their study of the dissolution of wollastonite, stressed the strong interactions between hydrogen ions (in the state of H_3O^+) and siloxane bonds. In particular, the interaction of H_3O^+ + $\equiv$ Si–O–Si $\equiv$ would be stronger than that of H_3O^+ + $\equiv$ Si–O–Al $\equiv$, which in turn would be stronger than H_2O + $\equiv$ Si–O–Al $\equiv$, which in turn would be still stronger than H_2O + $\equiv$ Si–O–Si $\equiv$. The corollary is that minerals with the largest number of siloxane bonds have the greatest affinity with respect to the hydrogen ion, and hence a large adsorption coefficient, whereas minerals with the largest number of $\equiv$ Si–O–Al $\equiv$ bonds have a lower adsorption coefficient for hydrogen ions. This statement contradicts what was said above on the basis of numerous observations.

It is generally accepted that the rate of dissolution of silicate minerals is controlled by reactions at the interface between solid and aqueous solution (Berner, 1978). These reactions occur in several stages, at specific sites, and not uniformly over the entirely exposed area. Two stages can be distinguished:

1. The first stage of weathering of a silicate mineral appears to be an exchange reaction between the basic cation, whenever present, and hydrogen ions as H^+ or H_3O^+. This protonated surface can reach 1000 Å thickness. This reaction is not rate-limiting and does not affect the rates of subsequent reactions.

2. The second stage corresponds to the breaking of Al–O–Si bonds (Holdren and Speyer, 1985b). According to the transition state theory (Aagaard and Helgeson, 1982), an activated complex would form corresponding to a free energy maximum (see Fig. 1.20) in which Al and Si participate. For feldspar (Holdren and Speyer, 1985b) it appears that the dissolution reaction breaks at first in Al–O–Si bonds to generate

O–Si–OH bonds which remain integrant parts of the feldspar. This situation allows, in turn, a subsequent hydrolysis of silicon at these sites. The assumed configuration of the activated complex would be "a highly energetic pentahedral complex in which ion (H_3O^+) becomes coordinated with the (tetrahedrally coordinated) aluminum in the feldspar structure" (Holdren and Speyer, 1985b, p. 1016). This transitional stage would force aluminum to become complexed in an octahedral shape, which is precisely the one in which aluminum occurs in secondary weathering products. From a chemical viewpoint, dissolution and formation of a secondary solid phase would be characterized by "a change in the coordinative environments of the reactants" (Furrer and Stumm, 1986, p. 1847). Complexes of each species have different affinities for the hydrogen ion (as H^+ or H_3O^+), a situation that allows aluminum or silicon to be removed differentially according to the activity of the protons involved. Thus, at a neutral or close to neutral pH, aluminum and silica are released at approximately the same rates because the formation of aluminum complexes is not favored over that of silicon (Holdren and Speyer, 1985b).

In conclusion, the weathering of minerals can be visualized as solid-solution interface reactions with more or less well-defined stages, depending upon the structural composition of the parent mineral and the activity of H^+ of the solution. Four cases can occur:

1. Parent minerals consist of only one type of bond (Si–O–Si, Na–Cl) and its breaking leads to simple dissolution of the mineral. This is the case of quartz.

2. The substitution of cations by protons and the breaking of Al–O–Al and Si–O–Si bonds follow each other very rapidly; hence they cannot be distinguished by known investigative techniques and both processes appear to be simultaneous. This is the case of feldspars dissolved in near neutral solutions.

3. The substitution of cations by protons is sufficiently distinct from the breaking of Al–O–Al and Si–O–Si bonds so that a thin protonated layer can be individualized with a very constant thickness, regardless of the stage of progress of weathering. Disruption of the framework of the parent mineral occurs before formation of a secondary weathering product. This is the case of some amphiboles, pyroxenes, and of feldspars.

4. The substitution of cations at certain crystalline sites by protons is distinct, in time, from the breaking of the other Al–O–Al and Si–O–Si bonds. A weathering mineral using the essential portion of the parent framework can develop. This is the case of the weathering of micas, and sometimes of pyroxenes or amphiboles.

In the first three cases, there is a rapid succession of dissociation reactions of the bonds between constituent ions of the parent crystalline framework before the appearance of a secondary mineral. This is the *congruent dissolution* of parent minerals. In the last case, the succession of these reactions takes place during a span of time sufficient to allow the differentiation of weathering minerals, which takes into account a portion of the parental crystalline framework. This is *incongruent dissolution*.

Neoformation and Transformation

According to the types of dissolution of parent minerals, two distinct possibilities of formation of secondary products occur: (1) These products are formed under equilibrium conditions from ions released in the weathering solution and transported for a variably short distance; this is the process of *neoformation*. (2) These products are formed in situ from the nondissolved parent crystalline structure; this is the process of *transformation*.

Neoformation is most probably the predominant process in environments where leaching by solutions is important, particularly in humid tropical environments. Noack et al. (1986) studied the weathering of pyroxenes into talc in the ultramafic bodies of Moyango and Sipilou in the Ivory Coast. Talc can precipitate as pseudomorph of parent pyroxenes or as coating of fissural voids. Talc from coatings was microsampled and analyzed by the oxygen isotopes method following the technique suggested by Escande (1983). Samples were first dried, then subject to a treatment of isotopic exchange with marked waters (Labeyrie and Juillet, 1982) in order to eliminate the isotopic effect of nonstructural oxygens. Oxygen was extracted by reaction with BrF_5, converted into CO_2, and analyzed with a Micromass 602D mass spectrometer. Results were given as δO^{18} with respect to standard sea water (SMOW). Values obtained from the talc of Moyango and Sipilou have a $\delta O^{18} = +20‰$.

These values do not result from reequilibrations following the formation of talc because such exchanges with meteoric waters and phyllosilicates are weak and restricted to interlayer water (O'Neil and Kharaka, 1976).

Therefore these isotopic values express the generation of talc. Such values of $\delta O^{18} = +20‰$ require a previous breaking of all the bonds with the oxygens of the parent mineral and isotopic reequilibration with the meteoric solutions of weathering. Unquestionably, this indicates congruent dissolution of the parent mineral and precipitation of talc from the solution. This is neoformation of talc. These isotopic values of talc due to weathering are in agreement with values obtained from weathering clays by Lawrence and Taylor (1971, 1972). Furthermore, this particular talc, analyzed by Mössbauer spectrometry, infrared spectrometry, and by optical spectrometry with diffused reflectance, indicates that it is ferriferous. Iron is located in the octahedral M_2 sites and is both ferric and ferrous. The ration Fe^{2+}/total iron is low and does not exceed 28%. There is no evidence of tetrahedral iron. The presence of ferric iron in the octahedral site clearly separates meteoric talc from hydrothermal talc, in which iron is only in the form of Fe^{2+} and the $\delta O^{18} = +10‰$.

Results obtained by Noack et al. (1986) not only provide arguments to distinguish hydrothermal talc from due to weathering, but they also demonstrate the neoformed character of the latter.

The transformation of phyllosilicate minerals into other argillaceous phyllosilicates by degradation of the former was demonstrated by Jackson et al. (1948), Walker (1949), Millot (1949), and Jackson (1959).

The most commonly mentioned case by these authors is transformation of biotite into vermiculite or smectite. Other examples are quoted, particularly the transformation of chlorite into vermiculite or smectite (Ross and Kodama, 1974; Mejsner, 1978) by modification or disappearance of the brucitic layer. Millot stated (1964, p. 367) in his analysis of transformation by degradation: "All these degradations take place sheet by sheet in a progressive manner which allows us to understand all their stages by means of the interstratified ones". Thus, transformation of phyllosilicate minerals is common; it can also appear during weathering of inosilicates. Basham (1974), in a study of the pseudomorphosis of an orthopyroxene by a dioctahedral vermiculite during weathering of a gabbro (Aberdeenshire), raised the possibility of a close structural control by the parent mineral. Eggleton (1975) and Eggleton and Boland (1982) demonstrated that weathering of pyroxenes into phyllosilicate products (nontronite or talc) takes place gradually, with a minimum disruption of the parent crystalline structure and by means of intermediate phases. This transformation by degradation is called here *topotactism*. It was studied in particular in the weathering of an orthoenstatite into talc in the region of Waratah, Northwest Tasmania, Australia (Eggleton and Boland, 1982). Observations were made by a transmission electron microscope (TEM) in sections cut normal to (100) of a pyroxene partially weathered and ionbeam thinned. Several stages were distinguished in the weathering of the orthopyroxene. The first stage corresponds to a 9 Å layer silicate displaying a structural coherence with the parent pyroxene (Eggleton and Boland, 1982, p. 75) as follows:

> (001) of the pyroxene becomes (100) of the layer silicate, (100) becomes (100) of the layer silicate, and (010) retains those indices. Essentially, the I-beam chains of the pyroxene (parallel to Z) recombine following a relative shift of b/y parallel to (010) to form an infinite sheet parallel to the original (100) plane. The enlargement reveals columns of vacancies, or channels, parallel to (001) of the orthopyroxene where strips of various chain widths terminate .... Growth of "talc" from the strip of 3-chain structure can be achieved by exchanging the vacancy for the adjacent 3-chain strip of I-beams. This growth mechanism requires no coordination change for the tetrahedral or octahedral cations, no noncrystalline intermediate phase, and minimum "bond-breaking."

The second stage of weathering consists of a mixture of 7–9 Å and 14 Å layers originating from recrystallization of the 9 Å phase. In this second stage, all structural continuity with the parent pyroxene disappears. Finally, a last stage appears as a mixture of talc and iron oxides.

Thus, only the first stage of weathering of orthopyroxene into talc appears really topotactic with the biopyriboles described by Veblen and Buseck (1980) as possible intermediate. In the case studied by Eggleton and Boland (1982), the chemical composition of the biopyriboles can be expressed as follows:

$$(Si_{3.89}\ Al_{0.11})\ (Fe^{3+}_{0.65}\ Cr_{0.02}\ Mg_2)\ Ca_{0.05}\ O_{10}\ (OH)_2.$$

This composition would actually be very close to that of the parent pyroxene. It would result only from a loss of MgO without volume change. Weathering reactions responsible for this transformation take place at the scale of the crystalline structure. Eggleton and Boland (1982) suggested that these reactions took place under the effect of water vapor diffusing through the crystal. The 'protonation' of the parent pyroxene in this initial stage of weathering could be expressed as follows:

$$\underset{\text{(orthopyroxene)}}{Mg_8\ Si_8\ O_{24}} + 2H^+ \rightarrow \underset{\text{(amphibole)}}{Mg_7\ Si_8\ O_{22}\ (OH)_2} + Mg^{2+}$$

$$\underset{\text{(amphibole)}}{Mg_7\ Si_8\ O_{22}} + 2H^+ \rightarrow \underset{\text{(talc)}}{Mg_6\ Si_8\ O_{20}\ (OH)_4} + Mg^{2+}.$$

This evolution of weathering processes can be compared to the development of biopyriboles described in metamorphic processes by Veblen and Buseck (1980) and Nakajima and Ribbe (1980). These processes generated an entire sequence of intermediate biopyriboles, with variable chain width, and all in topotactic relationship with each other. Studies by Eggleton (1975) and by Eggleton and Boland (1982) suggest that these topotactic transformations also develop in weathering zones.

These transformations, which take place with a minimum disruption of parent crystalline structure and generate a weathering product, pertain to phyllosilicates and inosilicates whose structural configuration can easily accommodate such rearrangements. These topotactic transformations are in contrast to frequent congruent dissolution, in particular in inosilicates as mentioned earlier. Gradual hydrolysis is probably responsible for transformations and degradations that affect parent minerals. Access to crystalline structures is made difficult by the lack of sufficient opening of alteration sites and thus occurs in successive stages. Initial weathering reactions leading to differentiated products amount therefore to substitutions of cations by H^+ or by water layers, or at the most to very reduced substitutions in the octahedral or sometimes tetrahedral parental sites.

Kinetic Inhibitors of Dissolutions

Because the negative valences of atoms and ions at the surface of minerals are not entirely compensated, they allow positive ions taken from the weathering solution to saturate the surface of the solid in contact with the solution. Protons play this role for weathering solutions close to pure water. However, weathering solutions may be charged with other cations, particularly at the base of soils and weathering profiles. Under such conditions, occupation of superficial sites of the solid can occur by means of other ions present in the solution, and hence in competition with protons. These other ions can be metallic ions and particularly transition metals. Such metallic ions behave as inhibitors of dissolution reactions of minerals (Touray, 1980; Garcia Hernandez, 1981) mainly at points of high surface energy, such as dislocations, which are preferred places for absorption of cations of the solution (Drever, 1982). Therefore, the weathering solution displays variations of composition at the point of contact with the surface of the solid. Upon completion of the absorption, the liquid layer of contact, called "superficial layer with diffused boundary," shows an equilibrium concentration. From this layer ions diffuse (Touray, 1980, p. 16).

Behavior of Ions in Solution

With progress of incongruent dissolution, or of congruent dissolution of the parent mineral, elements that leave crystalline networks are transferred in solution by means of the superficial layer with a diffused boundary. Hence, there occurs in the solution a competition between Me^{n+} cations released from the solid and H^+ protons of the water molecules. The behavior of $O-H^+$ bonds with respect to $Me^{2n+}-O$ bonds directs the behavior of ions in solution. In the case of $M^{2+}-O$ bonds stronger than the $O-H^+$ bonds, cations take on oxygen-liberating protons, such as:

$$S^{6+} + 4H_2O \rightarrow SO_4^{--} + 8\ H^+.$$

In the case of equivalent bonds, hydroxides are formed, such as:

$$Zn^{2+} + H_2O \rightarrow Zn\ (OH)_2.$$

Finally, in the case of $O-H^+$ bonds stronger than $M^{2+}-O$ bonds, the cation remains dry.

The behavior of ions in weathering reactions was called *ionic potential* by Goldschmidt (1934, 1937), after Cartledge (1928). The ionic potential corresponds to the relation $Q = Z/r$ in which Z is the charge in valence units, and r the radius of the ion. Goldschmidt (1934) drafted a diagram (Fig. 1.22) that divides the elements into three main groups of behavior. This diagram, used by several authors, particularly Millot (1964), expresses the following.

1. Elements with weak ionic potential (Z/r not exceeding 3), which correspond to ions of large diameter and whose charge of 1 or 2 is distributed

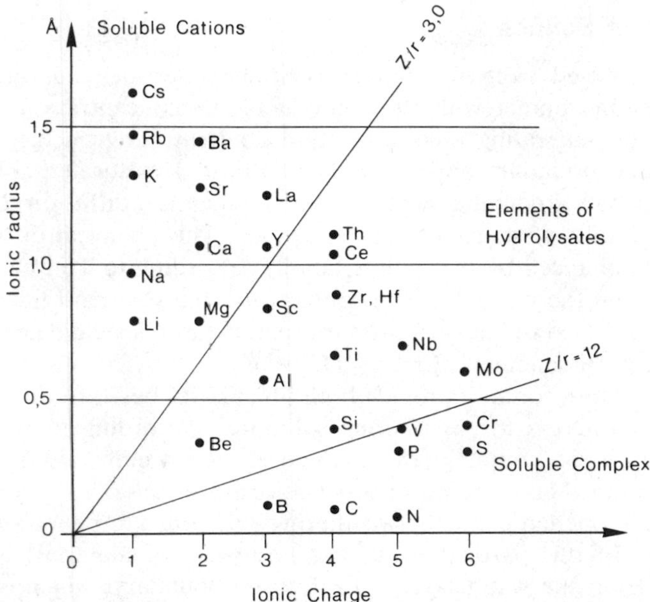

Figure 1.22. Distribution of Elements According to Ionic Potential (modified from Goldschmidt, 1934).

over a wide surface, hence with weak polarizing power on the H_2O dipole. Among them, K, Rb, and Cs ions have a very large diameter and develop over their extensive surface a field which is too weak. Their polarizing power on water is insufficient so that these ions remain dry.

2. Elements with ionic potential between 3 and 12, which correspond to ions of average diameter with a charge of 2 or 3. Their polarizing power is stronger and hence attracts hydroxyls in order to build amphoteric hydroxides, which can precipitate easily and can function as acids or bases.

3. Elements with ionic potential higher than 12, whose ions display a small diameter for a high charge of 3, 5, or 6. Their polarizing power is very strong; it allows them to dissociate hydroxyl ions, and to associate themselves in turn to oxygens in complex anions leaving a solution rich in H^+.

In summary, the ionic potential is "a simple expression of the polarization power" (Millot, 1964, p. 71). All cases exist between ions that have an insufficient polarization power and that, in solution, remain dry ions and those whose polarization power is such that it allows dissociation of hydroxyls (break of $O-H^+$ bonds) to the benefit of complex anions (generation of $Me^{n+}-O$ bonds).

The ionic potential, therefore, allows us to predict the behavior of ions in water after their release by weathering from crystalline structures.

Transport of Solutes

Any ion released from the parent crystalline network migrates into the water layer in contact with the solid; here, its concentration strongly increases, thus generating a concentration gradient between the water layer with diffused boundary and the bulk of the undersaturated solution. The existence of this gradient generates a flux of chemical diffusion that tends to eliminate the dissymmetry of concentrations. This chemical diffusion of an element is enhanced by thermal agitation. According to Touray (1980), the flux J between the water layer in contact with the solid and the bulk of the solution can be considered as proportional to the respective concentrations C' and C of this element: $J = K_{+} (C' - C)$.

In rocks, grain contacts provide both boundaries between different parent minerals and access for weathering solutions. According to the degree of opening of these contacts, the amount of the weathering solution that remains within them is more or less stationary.

In weakly opened contacts, weathering solutions are penetrating but not circulating. In this particular situation, transfers of materials by chemical diffusion, from the water layer with diffused boundary, which is in contact with the different parent minerals, into the bulk of the stationary solution, allow a fast homogenization of chemical concentrations. Furthermore, if parent minerals separated by grain contacts are not of the same composition, concentration gradients may reverse for certain elements. Thus, chemical diffusions may exist in the opposite direction, from the bulk of the solution toward the water layer with diffused boundary in the case of any element which, in comparison with the bulk of the solution, occurs in small amount or is missing in that layer. In other words, for a given amount of stationary solution, chemical diffusion may take place from the surface of a parent mineral toward the surface of another parent mineral, adjacent but of different composition. Thus, a tendency exists toward equalization of the chemical potentials of each constituent within the system (Hénin and Pédro, 1979; Pédro and Delmas, 1980). These geochemical interactions between minerals undergoing weathering are very common. They tend to homogenize secondary products of weathering (called *weathering plasmas*) of different parent minerals, beginning with the early stages of supergene alteration (Colin et al., 1985).

In open contacts or open weathering sites, solutions circulate. The rate of displacement of the bulk of the solution becomes greater than the rate of chemical diffusion. The environment becomes sufficiently leaching so that ions generated by diffusion are gradually removed together with the bulk of the percolating solution. Alkaline and calcoalkaline earths are eliminated first; this happens not only because they are the first to be released from parent crystalline structures (they have a weak ionization potential and are bonded to large and vulnerable polyhedra), but also because they diffuse easier. Conversely, the least mobile ions, called "inert" (Korzhinski, 1965), precipitate rapidly as weakly soluble hydrolysates such as hydroxides or oxyhydroxides.

CHAPTER 2

SUPERGENE ALTERATION OF MINERALS AND ROCKS: PRESERVATION OF ORIGINAL STRUCTURES

Chemical kinetics of interactions between minerals and water, as discussed in Chapter 1, were established mainly through numerous laboratory experiments, sometimes contradictory or complementary, completed for the most part well below equilibrium conditions. Lasaga (1981, 1984) and Aagaard and Helgeson (1982) used thermodynamics and the principle of detailed balancing to fill the gap between kinetic data obtained from laboratory experiments and a general law that would also account for precipitation rates close to equilibrium or to steady state.

In nature, dissolution of parent minerals and growth of secondary weathering products occur simultaneously. In the initial stages of weathering, newly formed secondary minerals can replace primary minerals from which they are derived. Numerous studies have been carried out addressing weathering minerals as a function of parent rocks and climatic zones. This chapter does not discuss processes, but, rather, how weathering products can be investigated by petrology. This approach relies mainly on megascopic and microscopic observations. However, it also uses, particularly in the case of minerals precipitated at room temperature and hence of small size and often poorly crystallized, high resolution microscopic techniques and new diffractometric and spectrometric techniques for the observation and identification of phases.

Certain minerals weather without generating secondary products; therefore, this study is limited to the aspects of their dissolution at different scales of observation. However, most silicates and aluminum silicates produce weathering products. For such a case, Eggleton (1986) distinguished two possibilities: Either the weathering mineral has an exact lattice coherence with the parent mineral, or no coherence exists.

Perfect lattice coherence between secondary product and parent mineral is possible whenever the latter displays either nucleation sites from which epitactic growth of the secondary mineral can occur, or structural building blocks acting as nucleation for the weathering product. This is the case of tetrahedral and octahedral chains of primary inosilicates that can rearrange themselves to allow the formation of weathering phyllosilicates. This was designated in Chapter 1 as a transformation process. Structural continuity between primary phase and secondary product is visible under the petrographic microscope through similarity of crystallographic orientations of both phases. The growth of such a weathering phase should indeed be fast and not require a supersaturation of nucleation.

Absence of lattice coherence between primary phase and weathering product leads to the assumption that the latter should be entirely reconstructed from the solution without an inherited parent structural element. The growth of such a weathering mineral should be much slower and belong to a neoformation process.

Congruent and incongruent dissolutions were considered in Chapter 1 at the scale of weathering mechanisms involving the surfaces of minerals. At the scale of observation of the petrographic microscope, weathering is designated as congruent if no secondary mineral is observed within the original boundaries of the parent mineral. Weathering is designated as incongruent when the secondary product replaces the parent mineral within its original boundaries. In such a case, contradiction results from a different scale of observation. Observations under the petrographic microscope are of essential importance in this study; hence, incongruent weathering includes mechanisms both of transformation and of neoformation of secondary products — in other words, processes of congruent and incongruent dissolution affecting the surface of parent minerals.

THE TWO WEATHERING PATHWAYS OF PARENT MINERALS

At each weathering site, solutions circulate along contacts between minerals. Therefore, the characteristics of alteration are regulated by the nature of parent constituents, the chemical composition of solutions, and also by the structure of the site where reactions and equilibria take place. At each weathering site, the first changes of parent minerals appear at the mineral-solution interface. Since these voids allow the circulation of solutions, weathering should proceed from a site toward a fresh mineral or at least from a given site toward a less accessible one.

On account of their size, their degree of interconnection, and hierarchy, sites become a critical factor in the transformation processes of parent minerals. Three types of voids — transcrystalline, intercrystalline, and intracrystalline — control the following aspects:

rates of circulation of solutions;
hydrodynamic characteristics of each microenvironment of weathering;
variable accessibility of minerals to chemical weathering;
coexistence of different secondary products.

Two major types of reactions may take place at the expense of parent minerals:

1. Generation of transformation products chemically related to the parent mineral. This evolution is called incongruent or selective (Touray, 1980). In this case there is generation of *alteration products that are pseudomorphic* after the parent mineral.

2. Total direct dissolution of the parent mineral *without generation of alteration products*. This evolution is called congruent and occurs *without pseudomorphism*.

Incongruent Dissolution: Pseudomorphosis of Weathering Products after Parent Minerals

In most cases, the first changes of parent minerals are visible under the petrographic microscope by *modification of their optical properties*. The most spectacular are changes of birefringence colors, of relief, and of pleochroism. Other more delicate modifications are enlargement of microfissures and sinuous aspect of certain cleavage planes. Finally, still other changes are *optically invisible* and can be detected only by detailed chemical analysis (see also Chapter 1).

Among the numerous examples illustrating this type of weathering, a type of mineral was chosen with a crystallo-chemical organization very sensitive to processes of superficial alteration (Robert, 1972), namely a mineral that displays clear optical or chemical changes: *mica*.

Optically Invisible Modifications

Analysis by thermogravimetry and by electron probe microanalysis of fresh biotite collected along the same vertical profile and in successive zones from the parent rock to the most weathered zones by Meunier (1980) showed the following:

1. *Hydration* of biotite can vary by a ratio of 1 to 6 without optically visible change. This process can be compared to the stage of "parabiotite" defined by Seddoh (1973), and to the stage of hydrobiotite (Wilson, 1970).

Packets of hydrobiotite develop within biotite, and these weather into other secondary phases (Gilkes and Suddhiprakarn, 1979; Eggleton, 1986).

2. There is a *progressive loss of interlayered potassium* that, if important, can lead to separation of sheets.

In conclusion, parent minerals can undergo crystallo-chemical transformations without changes of their optical properties visible under the petrographic microscope. Therefore, initial modifications have to be detected by using adequate analytical techniques.

Modifications Visible under the Petrographic Microscope and SEM

Microscopic examination of the various stages of meteoric alteration of mica crystals (Rausell-Colom et al., 1965; Bisdom, 1967; Rimsaite, 1967; Sawhney and Voigt, 1969; Curmi, 1979; Fayolle, 1979; Churchman, 1980) shows that several modifications may appear simultaneously.

Deformation and Opening of Cleavages

The regular and tight aspect of the cleavages of fresh mica is first replaced by a sinuous deformation and then by their opening (faces 010 and 100). These earliest transformations visible under the petrographic microscope are considered to be an expression of the *gradual loss of interlayer potassium*.

This separation of the sheets extends from their margins toward the inside of micas. Weathering may involve simultaneously several margins of sheets, or develop preferentially between two sheets only, leaving the other essentially unaffected (Jackson et al., 1952; Mortland, 1958; Scott and Smith, 1967; Norrish, 1973; Fanning and Keramidas, 1977). This process leads to gross exfoliation of micas and the opening up of "diffusion avenues" (Eggleton, 1986).

Modifications of Polarization Colors and Relief

The highest polarization colors of micas (second and third order) are replaced by grays and yellows of first order. Relief decreases, cleavages become faint, and sometimes a new cleavage with a different orientation appears. These transformations result from a change of the original micaceous structures into a new mineral phase (clay minerals).

The clay minerals studied here represent transformation products of micas. They are phyllosillicates of the 2/1 type such as vermiculite and smectites (Grüner, 1934; Barshad, 1948; Caillère and Hénin, 1951). The essential feature of this type of transformation is a replacement of the large interlayer K^+ cations by *hydrated cations*.

The example studied by Curmi (1979) showed (Fig. 2.1) that the modification of the interlayer region (loss of potassium), in the case of biotite, is immediately followed by changes of the octahedral sheet with loss of iron. Iron is not, as potassium, necessarily lost from the environment of weathering. It can precipitate in place as amorphous or crystallized oxyhydroxide. It

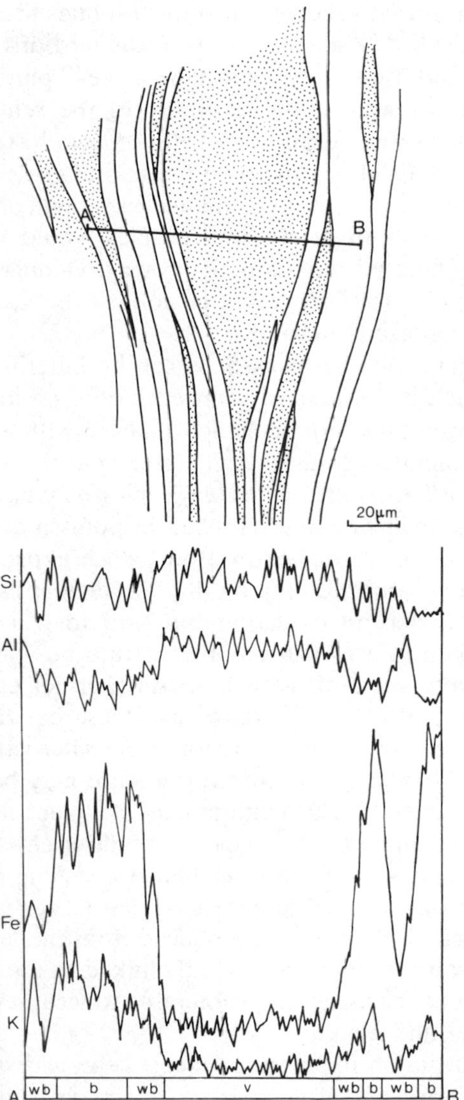

Figure 2.1. Semiquantitative Analysis of the Distribution of Constituents of a Zone of Exfoliated Biotite (modified from Curmi, 1979). Opening of the layers is accompanied by transformation of biotite into vermiculite (v). A transition zone (wb) occurs between fresh biotite (b) and vermiculite (v). In this zone potassium is eliminated by loss of ferrous iron. The width of the transition zone (wb) is about 6 μm. (From Curmi and Fayolle, 1981, reproduced by permission of Centre for Agricultural Publishing and Documentation, Wageningen, The Netherlands.)

corresponds to the diffused yellowish or brownish hues that either appear on the surface of vermiculite or smectite, or coat the portions not yet altered of parent mica. In the latter case, this "ferruginous veil" plays a protective role that diminishes considerably the accessibility of the remaining portion of black mica to chemical weathering. Fe oxidation also has an effect on mica weatherability through its effect on OH orientation (written communication, Velbel, 1989). Thus, from a thermodynamic viewpoint, the persistence of mica in the upper zones of weathering profiles would not result from a greater stability but simply from a kinetic process. Dennison et al. (1929) in Fanning and Keramidas (1977) previously noticed the greater resistance to weathering of "ferruginized" mica.

However, ferruginization of black mica can be inherited from hypogene transformations. Such is the case of bronzed biotite ("biotites mordorées") observed by Lapparent (1909) in Millot (1964), with surfaces of layers "covered by very minute crystals of goethite generated by exudation of iron." This *hypogene ferruginization* of black mica may have been caused by hydrothermal fluids, oxidizing and enriched in potassium. In this case, the large interlayer K^+ ions are not eliminated, which explains why hypogene ferruginizations of black mica do not modify the regular aspect of the sheets of mica. Therefore, such an oxidation can lead to a modification of the octahedral sheet when transformation of Fe^{2+} into Fe^{3+} is expressed by an expulsion of iron from some of the octahedral (Farmer et al., 1971; Gilkes et al., 1972; *Gilkes*, 1973). In certain instances, exudations have been erroneously attributed to the presence of iron. Microanalysis shows that crystallized titanium oxides in the form of anatase may be present.

In summary, hypogene transformations can take place directly at the level of octahedral sheets and in that respect are different from weathering, which, at first, eliminates interlayer ions before reaching octahedral sheets. Expulsion of iron from octahedral sheets of biotite generates gaps toward which the proton can move, thus getting away from the large interlayer K^+ ions. The latter cations are thus more solidly linked to the structure of black mica, which, therefore, becomes less vulnerable to weathering (Barshad and Fawzy, 1968; Gilkes and Young, 1974).

Meteoric ferruginization of biotite decreases the accessibility of the still fresh mica by generating a protecting screen: It is a matter of kinetics. However, "hydrothermal" ferruginization of biotite not only plays the role of protecting film, but also affects directly the structure of mica; that is, it increases its stability with respect to weathering. This case shows the important role that hypogene heritage can play in the mineral evolution of the surface.

Microfissuration, Followed by Modification of Polarization Colors of External Faces (001)

There is no immediate deformation of the stacks of sheets in this case. In fact, the various stages of alteration of mica described above belong to a simple *transformation* ("degradation" in the sense of Millot, 1964), with preservation of the type of the parent phyllosilicate structure: The layers of vermiculite and of smectite have kept the same crystallographic orientation as that of the parent micaceous layers from which they originated. Transformations involve only interlayer and octahedral sheets. Tetrahedral sheets are not affected, or if they are, modifications are limited to a few substitutions. These mineral evolutions can be observed in environments of moderate weathering because chemical conditions of weathering were not aggressive, because mica weathered in a not entirely open environment, or because the newly formed phases occupied the early formed "diffusion avenues," decreasing the weathering rate (Eggleton, 1986). However, environmental conditions can become more aggressive and quickly lead to phyllosilicate secondary products of type 1/1 (kaolinite), or even to oxyhydroxide of iron or aluminum (goethite and gibbsite). Simple transformation is hence replaced by true destruction of the parent crystalline lattice, and by *neoformation* of secondary products. This neoformation may display obvious relationships with the associated parent mineral. Such is the case of kaolinite developed on the altered margin of muscovite shown by Fayolle (1979) (Fig. 2.2). In order to explain the removal of interlayer potassium of muscovite, an influx in solution of extracrystalline origin is required to account for the amounts of silica and aluminium of kaolinite.

To kaolinite, which displays parental relationships with muscovite, one can contrast kaolinite located between the stacks of layers of biotite showing no parentage with the latter. This neoformed kaolinite is oriented perpendicular to the mica layers. When submitted to microanalysis, biotite appears unaltered; furthermore, no chemical transition exists between it and kaolinite (Curmi, 1979; Sarazin et al., 1982). In this case, all the elements necessary to the neogenesis of kaolinite can be assumed to be of extracrystalline origin. Biotite acts only as a receptive structure for *secondary accumulation* of kaolinite.

A schematic attempt to abstract the major modifications of biotite as seen under the petrographic microscope is shown in Fig. 2.3.

Simple Dissolution of Parent Minerals: Absence of Pseudomorphosis

Weathering of some chemically simple minerals, such as quartz, gypsum, calcite, etc., leads to a stoichiometric dissolution. Simple dissolution means dissolution of oxides among which quartz is the first; congruent dissolution, that of minerals consisting of several oxides, the most important being calcite and carbonates, gypsum and sulfates, and, in particular, silicates. Calcite or gypsum have often been dealt with by geologists at the scale of a landscape, namely sinkholes scattered in gypsiferous areas or karstic reliefs in limestone terranes. Dissolution of quartz appears more spectacular at the scale of the thin section. This example is discussed below.

Quartz is abundant in many rocks. Its weathering under surficial conditions always occurs in the same manner, namely dissolution. Silica of quartz is

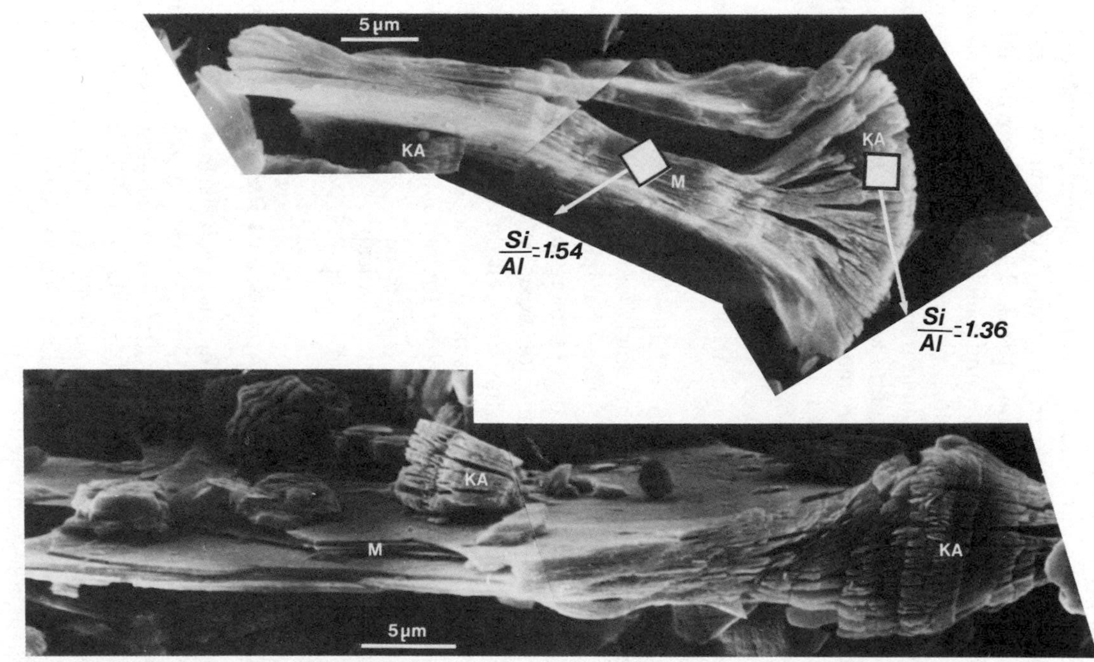

Figure 2.2. Weathering of Muscovite into Kaolinite after Fayolle (1979). Muscovite displays a fan-shaped expanded margin that is weathered into kaolinite. Microanalyses were completed in the nonweathered muscovite, in the fan-shaped margin weathered into kaolinite, and in an adjacent kaolinite vermicule. The ratios of measured X-ray intensities increase from the centre toward the margin of muscovite: 13.6% for the ratio Al/K and 11% for the ratio Si/K. The ratio Si/Al is 1.54 for muscovite (M) and 1.36 for kaolinite (KA). (Reprinted by permission of Department of Geosciences of the University of Paris 7.)

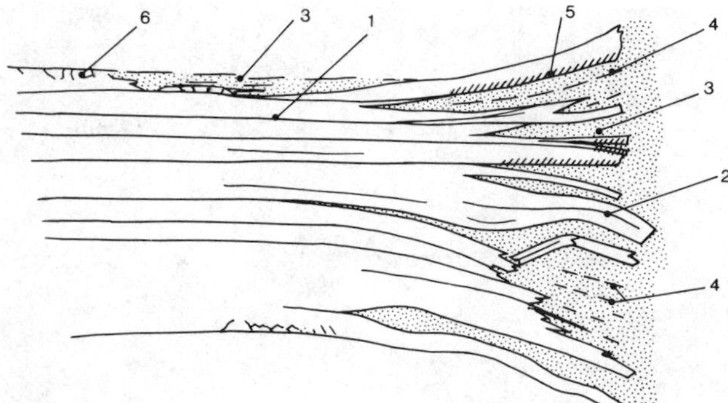

Figure 2.3. Schematic Representation of the Weathering of Biotite as Observed under the Petrographic Microscope. (1) Regular cleavages of fresh biotite; (2) deformation of biotite layers; (3) secondary products with optical characteristics different from those of biotite; (4) "ghost" of original cleavage; (5) diffused segregation of iron oxides along margin of biotite layers; (6) microfissures on faces (001) of biotite.

removed in solution, leaving a void where a portion of quartz was dissolved.

Observation under the petrographic microscope shows that instead of a sharp contact between fresh quartz and surrounding minerals, or along the margins of fine fissures that may crisscross it, weathered quartz displays a very thin transition zone in which birefringence colors are very low, or almost absent. This transition zone, when present, forms the edge of non-modified quartz and shows the same positions of extinction. This situation can be explained by a thinning of quartz along the margin of the dissolution void ("steplike" structure under SEM). The arrangement of these dissolution voids gives to quartz a corroded aspect with random shapes.

Observation under SEM of the surface aspects of quartz having undergone dissolution (Krinsley and Doornkamp, 1973; Le Ribault, 1977), including the morphology of embayments and protrusions, reveals the following (Fig. 2.4): a steplike striated aspect of the surface; striae which may be superimposed by negative micropunctuations of triangular or pyramidal shape; cavities of pyramidal shape, uniformly developed and revealing a prismatic relief of the surfaces; and cavities of pyramidal shape irregularly developed and deepening or increasing in size while keeping the same shape (Nahon et al., 1979).

Moreover, studies by Moss et al. (1973) and by Moss and Green (1975) show that quartz in igneous rocks displays *crisscrossing microfissures* breaking crystals into minute fragments, often in the shape of sheets and a few microns in thickness. These microfissures in parent minerals are inherited characteristics from the long history of igneous rocks. They may have resulted either from tension forces due to contractions undergone by quartz during

Figure 2.4. Quartz Weathered by Simple Dissolution. SEM picture of the surface of a quartz grain from a weathered Maestrichtian argillaceous sandstone (Ndias Massif, Western Senegal). (From Nahon et al. 1979, reprinted by permission of Sciences Géologiques.)

its cooling, that is, during its enantriotropic transformation from high to low temperature (Smalley, 1974), or simply from tectonic stresses. The occurrence of these microfissures in parent quartz and crystallographic imperfections indicate that this mineral can be included among those very accessible to weathering solutions. All these structural defects provide pathways to weathering and can explain some steplike surface aspects resulting from this process.

Experimental dissolution of quartz grains and the formation of etch pits were reproduced under experimental hydrothermal conditions by Joshi and Paul (1977) and Brantley et al. (1986). Simulated etch pits were similar to those observed in quartz grains weathered in soils. Pyramidal pits would result from dissolution along line defects, whereas shallow triangular features would originate from dissolution of surface defects. In order to test if etch pits of quartz grains were controlled by the chemistry of the solutions percolating through the soil, Brantley et al. (1986) studied quartz grains collected along a weathering profile of about 1 m thickness developed over a Proterozoic granite of Venezuela. These authors showed that the morphology of quartz grains and of their dissolution features was function of depth in the weathering profile as follows: The density of etch pits at the surface of quartz grains, in particular of triangular or pyramidal etch pits, decreases with depth; around 60 cm depth dissolution affected only edges and margins of quartz grains leading to a thin rounded aspect. Upon combining their observations of the above-mentioned weathering profile with their experimental dissolution of quartz grains, the authors suggested that concentration of silica could control pitting. The formation of etch pits indicates that concentration in silica of the solutions filling the porosity of the soil is lower than the critical concentration at 25°C. At a depth greater than 60 cm, concentration of silica in the pores is greater than the initial concentration and therefore quartz surfaces are not pitted.

Lasaga (1983) showed that the formation of etch pits at dislocation sites of the surface of silicates can be a very good indicator of the degree of undersaturation of the solution that percolated through the weathering profile. Lasaga and Blum (1986) and Blum and Lasaga (1987) proposed a method of calculation (Monte Carlo method) for predicting "when etch pits around dislocations should form, what the etch pit morphology should be, and the relative role of dissolution at dislocation sites versus dissolution in dislocation-free regions of surfaces on the overall dissolution of crystals in the laboratory and in nature" Blum and Lasaga, 1987 p. 255). This method can be used to monitor surface reactions when the thermodynamic state of a given system changes from a strong undersaturation, through equilibrium, to a strong supersaturation.

Highly corroded quartz is fragile and cannot withstand any mechanical transportation subsequent to its weathering. It provides an excellent demonstration of the in situ evolution by weathering of fresh original quartz.

Simple dissolution of minerals such as quartz leads to an increase of large-scale porosity in rocks. A similar situation would arise from the congruent dissolution of some silicates in a very aggressive environment.

Conclusions on Earliest Transformations due to Weathering

The two main pathways of earliest transformations at the sites of alteration have been described above. Their main characteristics and a definition of their limits are now presented.

Generation of secondary products is either direct, that is, the incongruent pathway of alteration; or there is absence of weathering products, that is, simple dissolution or congruent dissolution.

In the first case, and choosing mica as parent mineral, three main stages can be observed. In A_1 (Fig. 2.5), slight *intracrystalline transformations* may appear without visible modifications of optical properties. These transformations are restricted to weak exchanges at the level of interlayer ions. In A_2 (Fig. 2.5), the process becomes optically visible, interlayer ions are replaced by hydroxyls and octahedral sheets, and tetrahedral sheets sometimes undergo substitutions. It is *degradation* in the sense of Millot (1964). In A_3 (Fig. 2.5), the process is more advanced and disorganization of the parent mineral and formation of a new silicate of the type 1/1 occurs with orientation of its layers related to that in the original mica. This *neogenesis* requires total elimination of alkalis and alkaline earths, and a moderate influx of silicon and perhaps also of aluminum by weathering solutions. For these three stages, A_1, A_2, and A_3, chemical parentages between fresh mica and secondary product are gradual.

When influxes in solution increase significantly, they lead in A_4 by neo-

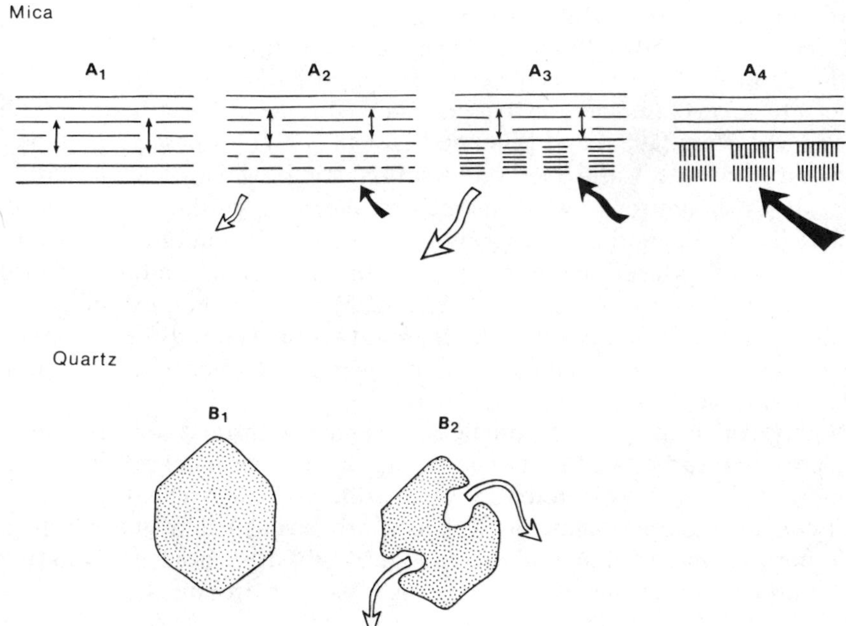

Figure 2.5. Schematic Diagram of Initial Transformations Due to Weathering. In mica, pseudomorphosis by degradation or neoformation should be distinguished from secondary accumulation by neoformation. *Types of transfers*: (A_1) nanotransfer (intracrystalline modification); (A_2 and A_3) predominant exporting microtransfers (intra- and intercrystalline modifications); (A_4) importing micro- and macrotransfers (inter- and extracrystalline modifications). *Origin of secondary products*: (A_2) pseudomorphosis by degradation; (A_3) pseudomorphosis by neoformation; (A_4) secondary accumulation by neoformation. In quartz, weathering by simple dissolution corresponds to exporting micro- or macrotransfers.

formation to an accumulation of secondary products that grow on the fresh parent mica. The chemical transition between the two types of minerals is abrupt and the orientations of the sheets are different. This is no longer a simple transformation of a parent mineral, but neoformation of a new mineral after *microtransfer* of elements. It is no longer a pseudomorphosis.

In the second case, and choosing quartz (B_1) as parent mineral, weathering is expressed by a microtransfer of elements in solution (B_2, Fig. 2.5).

In conclusion, earliest transformations due to weathering are of two types:

1. In incongruent dissolution, structures are preserved by weathering. Secondary minerals that are pseudomorphs after parent minerals or are added to them are generated. This weathering results from *nanotransfers* of elements inside parent minerals or from *microtransfers* between one mineral and another. Exporting microtransfers must be predominant over importing microtransfers.

2. In simple dissolution without insoluble residues, structures are rapidly destroyed by weathering. No pseudomorphosis occurs. If there is any neoformation of secondary minerals, it is by *absolute accumulation* after microtransfer of elements.

PRESERVATION OF PARENT ROCK STRUCTURES: GENERATION OF WEATHERING PLASMAS

In the field of structure-preserving weathering, two main processes of transformation of parent minerals have been described: the incongruent type, which generates secondary phases in parentage with relicts of parent minerals; and the simple dissolution type, which does not generate *secondary phases*, taking the place of the parent mineral, but leaves instead a residual portion of variable importance of the parent mineral.

These two processes of weathering have been examined in minerals that are isolated from their petrologic environment, whereas, in reality, minerals are generally associated to form rocks. In these heterogeneous natural environments, differential weathering is the main process. It is determined in a complex manner by the crystallo-chemical nature of the constituents, the characteristics of the weathering site, and the nature of the solutions. Thus, even in the earliest stages of transformation, a supergene alteration material becomes differentiated from the rock. It consists of a juxtaposition of two kinds of constituents: secondary products called *weathering plasmas* and relicts of primary minerals called *skeleton grains*. To investigate the major compositional and structural relationships, that may exist between these two types of constituents, one must use *distribution* and *orientation* analysis of weathering products with respect to those of parent mineral relicts.

Within the heterogeneous environments that correspond to rocks, distribution and orientation may be divided according to the order of alterability of primary minerals (that is, their susceptibility to reaction) and according to the nature of weathering plasmas, which may be further subdivided into *argilliplasmas* (consisting of clay minerals) and *crystalliplasmas* (consisting of other crystallized products).

The evolution of primary minerals into weathering minerals may produce either variably complex *polyphase products* (for instance, a population of different smectites), or directly *monomineralic products* (for instance, goethite or gibbsite). The former may consist either of an assemblage of minerals of different nature and intimately mixed (different clay minerals), or of a group of minerals with a composition ranging between two or several poles, such as solid solutions (alumina substituted for iron in hydroxides). The latter appear as relatively simple secondary minerals. It seems appropriate to begin the analysis of the distribution of weathering products with respect to their parent minerals by describing the simplest parentages first, namely those in which the secondary product is monomineralic.

Pseudomorphoses by Oxyhydroxide Septa: Crystalliplasmas

In some environments where weathering of minerals is very active (in particular in tropical ferralitic environments), fully evolved secondary products, generally *oxides and hydroxides of aluminum, iron, or manganese*, may be directly generated. These are, in fact, the less soluble products in the "aggressive" conditions of those particular environments. The primary mineral, by means of selective dissolution, generates only one single type of residual product, namely *weathering crystalliplasma*.

Following the studies of Lacroix (1914) and the works of Millot and Bonifas (1955), Leneuf (1959), Delvigne (1965), Novikoff (1974), Trescases (1975), Delvigne and Boulangé (1973), Eswaran and Heng (1976), Eswaran and Bin (1978), and Velbel (1989), it is possible to reconstruct petrographically (Fig. 2.6) the individualization and development of secondary products of oxyhydroxide type, which are organized in *septa* and thus account for the pseudomorphosis of the mineral.

In general, and regardless of the kind of oxyhydroxides involved, transformation of the primary mineral also begins here in different types of original voids, such as intracrystalline, intercrystalline, or transcrystalline (A_1 in Fig. 2.6) that isolate or break up the mineral. From these voids, and as the mineral is being dissolved, a septum of oxyhydroxides is generated. It consists of residual products from the destruction of the parent mineral located along its periphery in (A_2 in Fig. 2.6).

A *contact void* develops between the oxyhydroxide septum and the parent mineral (relict), whose width increases as the dissolution proceeds. The study of these septa indicates that they consist of crystallized oxyhydroxides of iron, aluminum, or manganese. Therefore, we are dealing with a *weathering crystalliplasma*. The crystals forming the septa are often oriented perpendicular to the wall of the void. Their size and crystallinity increase in a centripetal way toward the contact void with the parent relict. Therefore, wherever a septum consists of several successive generations of oxyhydroxides (A_3 in Fig. 2.6), the last is generally the best developed. A relationship can be assumed between this greater crystallinity and the increase in size of the contact voids that allows an increasing rate of circulation and hence a dilution of the solutions.

As these simultaneous processes of dissolution of the parent mineral and of crystallization of oxides operate, an anastomosed-network of septa consisting of oxyhydroxides develops. This network isolates alveoles in the center of which very irregularly shaped relicts of the fresh mineral remain. These relicts, when observed under the petrographic microscope, appear to float inside their alveoles. The relicts of fresh mineral may seem slightly displaced with respect to their original crystallographic orientation when the contact void that separates them from oxide septa is continuous. In reality, SEM observations show that these floating relicts are almost always connected to the septa by very delicate bridges of oxyhydroxides that cut across them and keep them in their original crystallographic position. In fact, during thin section preparation, these very delicate and fragile bridges break easily, allowing a slight rotation of the fresh relicts.

In more advanced stages, the fresh mineral is completely dissolved and only a framework of oxyhydroxides remains, outlining numerous alveolar voids. The latter represent the extreme stage of development of contact voids. The geometry of this framework and of the alveolar voids has often been directed during the entire transformation by the crystallographic

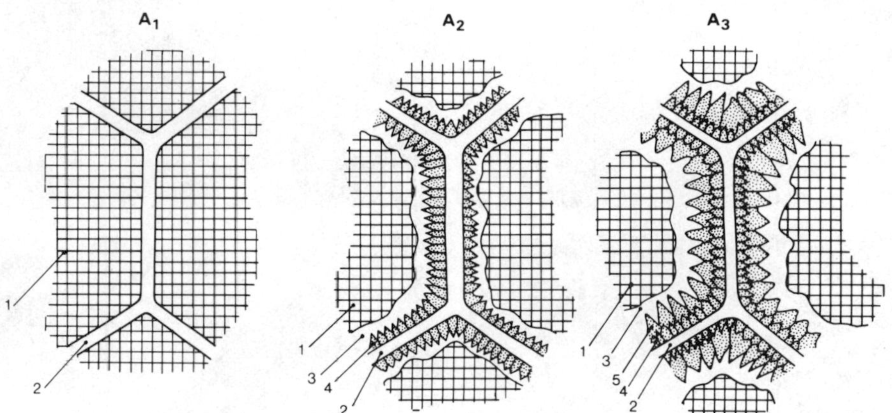

Figure 2.6. Sketch of Pseudomorphoses by Oxyhydroxides Septa. A_1, A_2, A_3 = Different Stages of Septa Formation. (1) Parent mineral or parent relict; (2) original fissural void; (3) contact void; (4) oxyhydroxide septa (first generation of crystals); (5) oxyhydroxide septa (second generation of crystals). (From Nahon and Bocquier, 1983, reprinted by permission of Sciences Géologiques.)

characteristics of the primary mineral. This is the reason the oxyhydroxide framework represents the "ghost" of the transformed mineral, and the oxide septa, the older voids in which are concentrated the insoluble secondary products. A pseudomorphosis has taken place starting along the discontinuities of the mineral. When these pseudomorphoses occur close to each other, they insure the preservation of the original mineral structures, and often that of the rock itself. This is the basis of the isovolumetric reasoning (Millot and Bonifas, 1955) that explains the preservation of lithologic struc-

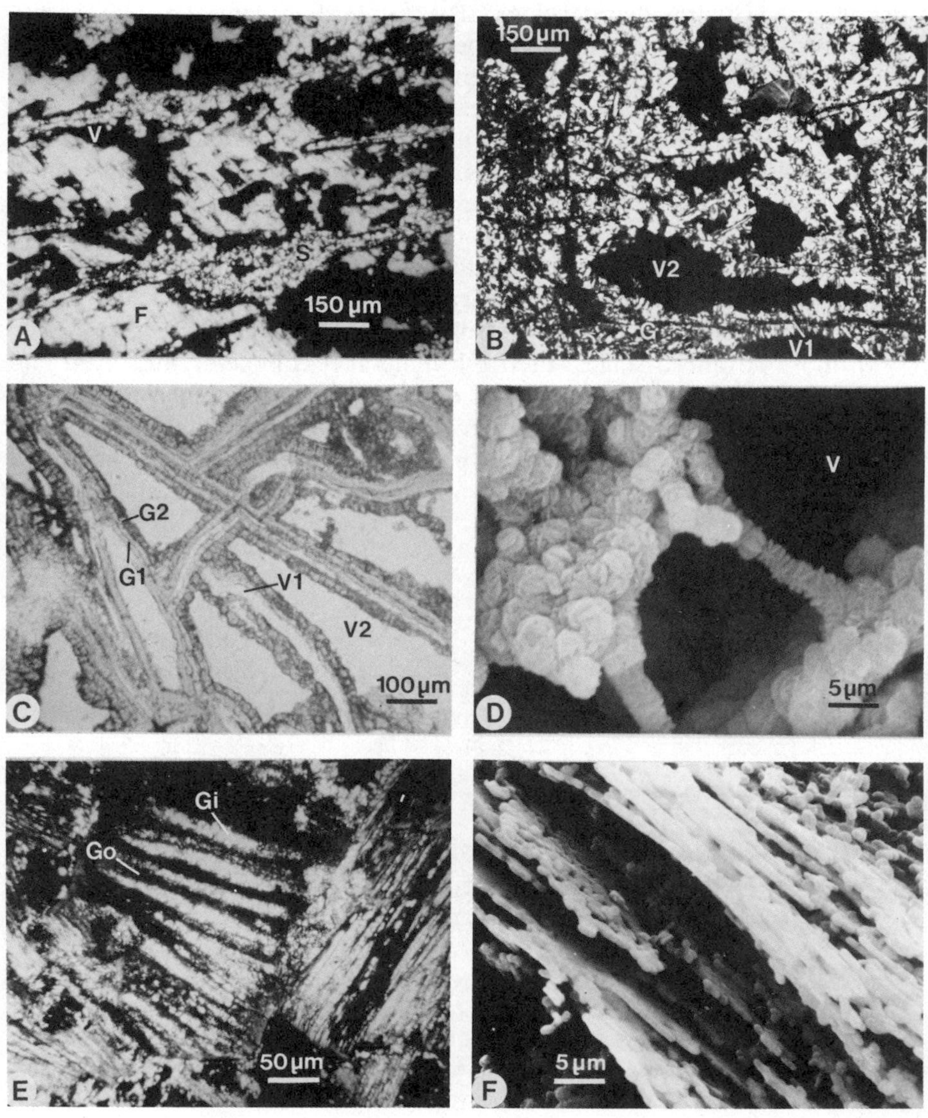

tures very high in the weathering profiles, namely in the oldest zones. When this preserved structure of the rock is eventually destroyed, it may survive for a long period of time in lithorelicts.

Individualization of these different types of oxides and hydroxides is related to the geochemical dominance of the parent mineral (Paquet, 1969). Indeed, the evolution of tectosilicates high in aluminum, such as feldspars and feldspathoids, leads to the generation of gibbsite septa (Fig. 2.7 A and B). On the other hand, the transformation of ferromagnesian minerals, such as olivine, pyroxenes, garnets, amphiboles, etc. (Fig. 2.7 C and D), generates a goethite framework. Whenever these ferromagnesian minerals contain alumina, mixed septa may result: The septa consist either of aluminous-goethite (Nahon and Colin, 1982) or of alternating goethite and gibbsite (Boulangé, 1984, Fig. 2.7 E). The evolution of the manganese-bearing minerals spessartite garnet, mangano-calcite similarly generates septa of lithiophorite or manganite, respectively (Nziengui Mapangou, 1981). According to geochemical conditions of the environment of transformation, weathering can, nevertheless, be selective among the dominant elements that constitute the parent mineral. As seen above, when biotite is submitted to leaching oxidizing conditions of a ferralitic environment, silicon is eliminated and the septa consist of aluminum and iron oxides (Fig. 2.7 E). When weathering of the pyritic shales of Sain-Bel occurs (Sornein, 1980), the evolution of muscovite is just the reverse. Indeed, these are very acidic environments where alumina, much more soluble than silica (Gardner, 1970), generates septa of cryptocrystalline lamellar silica (about 0.2 μm thick) oriented parallel to the 001 planes of the original muscovite (Fig. 7–2, F).

Figure 2.7. Examples of Pseudomorphoses by Oxyhydroxide and Oxide Septa. **A**: Partial pseudomorphosis of plagioclase by gibbsite septas (nepheline syenite, Island of Loos, Guinea). Petrographic microscope, crossed nicols. F: feldspar relict; V: contact void; S: gibbsite septa. **B**: Complete pseudomorphosis of microcline by gibbsite septa (Lakota granite, Ivory Coast). Petrographic microscope, crossed nicols. VI: original fissural void; G: gibbsite septa; V2: alveolar void of dissolution. Notice the perpendicular orientation of gibbsite crystals and their increasing size in the direction of the alveolar void. (Photograph courtesy B. Boulangé) **C**: Complete pseudomorphosis of almandine garnet by goethite septa (Yaoundé gneiss, Cameroon). Petrographic microscope, plane polarized light. VI: original fissural void; G1: first goethite generation of the septa; G2: second generation; V2: alveolar void. (Photograph courtesy G. Bocquier) **D**: Complete pseudomorphosis of hornblende by goethite septa (Urumbo Boka amphibolite, Ivory Coast). SEM. V: alveolar void. Notice spherulitic habit of goethite. (Photograph courtesy G. Bocquier) **E**: Pseudomorphosis of biotite by mixed goethite and gibbsite septa (Lakota granite, Ivory Coast). Petrographic microscope, crossed nicols. Go: goethite; Gi: gibbsite. (Photograph courtesy B. Boulangé) **F**: Pseudomorphosis of muscovite by cryptocrystalline silica septa (pyritic shales of Sain-Bel, France). SEM. (Photograph courtesy J. F. Sornein). (Fig. 7-D is from Boulangé, 1984, reprinted by permission of ORSTOM Editions; Fig. 7-F is from Sornein, 1980, reprinted by permission of Ecole Nationale des Mines de Paris.)

This selectivity of oxyhydroxides with respect to the chemical composition of the original mineral demonstrates that individualization of oxides is essentially a concentration in situ, a *relative accumulation* in the sense of D'Hoore (1954). However, the formation of these septa can only be explained, at the scale of the mineral, by *nanotransfers* of these elements across the contact void, from parent mineral to septa, which thus can grow in a centripetal direction owing to accumulation of different generations of oxides.

This relative accumulation of oxides as septa greatly increases the porosity of the rock (Fig. 2.8A). All these voids (dissolution voids) are interpreted as pathways for weathering solutions. Elements dissolved out of the parent mineral are eliminated through these voids, which also receive elements necessary for *late absolute accumulations*. Indeed, all these voids can act, during the history of the weathered rock, as receptive structures for subsequent accumulations, generally neoformed and of the same nature as the minerals forming septa, by originating from elsewhere in the profiles (Fig. 2.8B). These *macrotransfers* of elements, whenever important, may erase a major portion of the porosity formed by contact voids or even by original fissural voids (Fig. 2.9A, B, C, and D). At this stage, it is no longer possible to distinguish crystals generated by relative accumulation, namely crystals that build septa in the strict sense of the term, from late ones, which are precipitated by absolute accumulation. In order to make a distinction, late absolute accumulations, either particulated (translocated) or neoformed, must be mineralogically different from those forming septa or structurally discordant (Fig. 2.8 C).

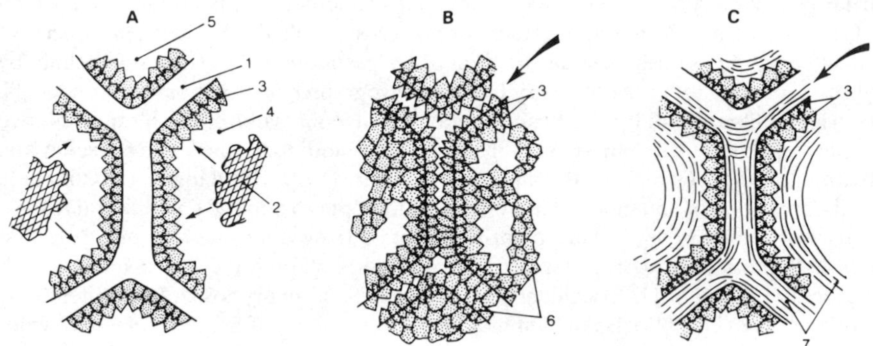

Figure 2.8. Pseudomorphoses by Septas and Late Transfers. **A, B, C** = Different stages of the formation of septa. (1) Original fissural void; (2) parental relict; (3) septa formed by two generations of oxyhydroxide crystals by relative accumulation; (4) contact void; (5) alveolar void; (6) late absolute accumulation of neoformed crystalliplasma, partially filling original fissural voids and alveolar voids; (7) late absolute accumulation of translocated and redeposited plasma (illuviation) in original fissural voids and in alveolar voids.

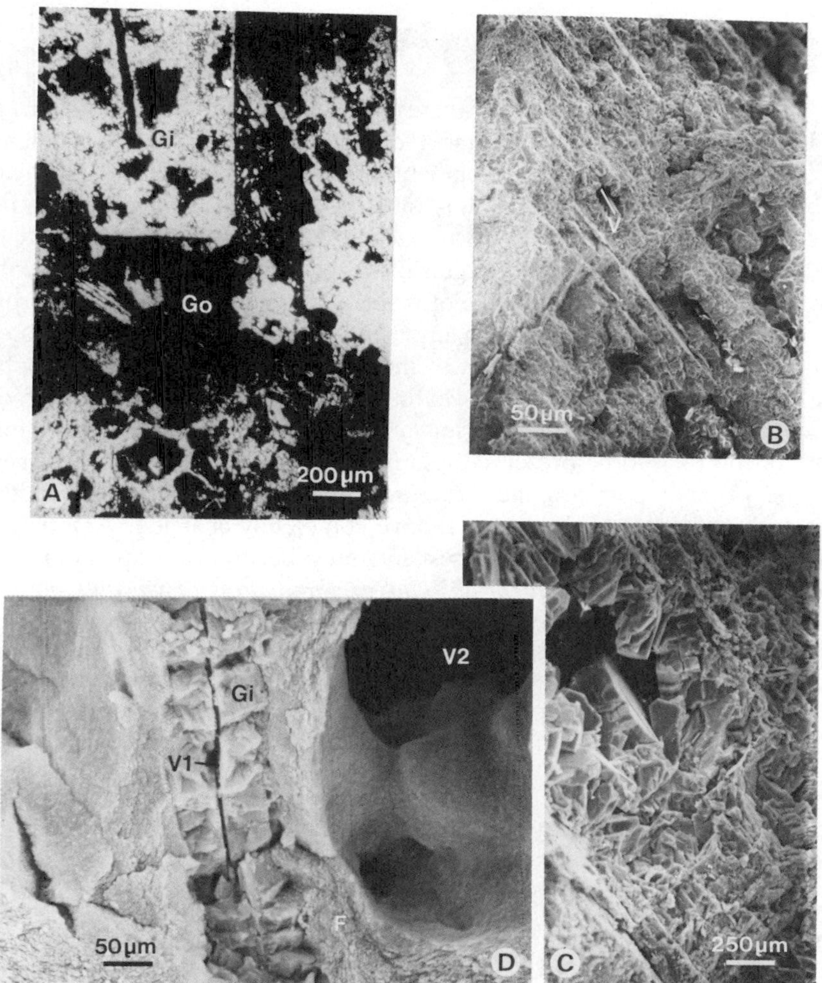

Figure 2.9. Examples of Pseudomorphoses by Septa and Secondary Transfers (Photographs courtesy B. Boulangé). **A**: Total pseudomorphosis of a plagioclase by gibbsite septa (relative accumulation) and by microtransfer of aluminum (absolute accumulation of gibbsite), Lakota granite, Ivory Coast. Petrographic microscope, crossed nicols. Notice the external shape of the original feldspar preserved as well as the relative position of feldspar crystals changed to gibbsite (Gi) and of biotite flakes changed to goethite (Go). **B**: Total pseudomorphosis by gibbsite septa (relative accumulation) and by microtransfer of aluminum (absolute accumulation of gibbsite), Lakota granite, Ivory Coast. SEM. Notice that the amount of gibbsite is volumetrically greater than the quantity that could be formed by relative accumulation only. Influx is estimated at 20% (Boulangé, 1984). V represents the relict alveolar void around which are located the better developed gibbsite crystals. **C**: Detail of gibbsite crystalliplasma from **B**. **D**: Total pseudomorphosis of a microcline by gibbsite septa (relative accumulation) and by particulated microtransfer (unconformable illuviation of ferruginous and kaolinitic constituents: ferriargillan), Lakota granite, Ivory Coast. SEM. V1 represents the original fissural void; Gi is the gibbsite septa; F is the unconformable ferriargillan; V2 is the residual alveolar void. (Fig. 9-D is from Boulangé 1984, reprinted by permission of ORSTOM Editions.)

In conclusion, within certain aggressive environments of weathering, namely where solutions are the most diluted, the least soluble elements show individualization and development of oxyhydroxides. The latter are distributed as *peripheral septas*, thus insuring the *pseudomorphosis of the parent mineral*. The examples discussed above show that this process of concentration begins along the margins of the original fissural void by means of the simple accumulation in place of residual elements, and develops from these crystalline germs by drawing upon the very same elements liberated and nanotransferred from the relict of the parent mineral toward the septa. These reticulated septas form a weathering *crystalliplasma* that, in fact, outlines the external shape of the primary minerals, thus allowing the original volume of the rock to be preserved relatively high in the profiles. However, while the original petrographic structure and volume are preserved, the nature and properties of the material have completely changed. Among the properties, porosity greatly increases and may become a trap, namely a *receptive structure for subsequent absolute accumulations of variable nature*. In this manner a transition occurs between nanotransfers of relative accumulation and macrotransfers of absolute accumulation, in other words, the transition between a system of weathering that generates relative accumulation together with porosity, and a system of transfer that generates absolute accumulation.

Pseudomorphoses by Clay Minerals: Argilliplasmas and Complex Plasmas

Under less aggressive conditions of weathering, a large proportion of elements originating from the destruction of parent minerals is reorganized in place as weathering products which are different from those forming the septa discussed above. These products are *argilliplasmas* or *complex plasmas*.

In heterogeneous environments, such as rocks, each parent mineral reacts to weathering in a different way. However, the evolution of each of the parent phases either can occur *independently* of those of the other phases and hence evolve on their own, or can be *chemically related* to those of other adjacent parent minerals.

Parent Mineral Phases Undergoing Independent Weathering

Different states of weathering coexist in rocks as a function of the differences of alterability of parent minerals (Goldich, 1938). Two cases are considered here: (1) a parent mineral that is more reactive to weathering than adjacent minerals, (2) adjacent minerals having a comparable reactivity.

Parent Mineral More Reactive Than Others

In igneous rocks plagioclases are known to be among the first minerals to destabilize and to generate argillaceous weathering plasmas (argilliplasmas).

In particular, two modes of evolution of these plagioclases occur according to whether the supergene alteration proceeds from the periphery toward the center of the crystal (centripetal), or in the opposite direction (centrifugal).

EXAMPLE OF WEATHERING OF A PLAGIOCLASE FROM ITS CENTER

Such an example was studied by Delvigne and Martin (1970) in a diorite from the southwest of the Ivory Coast. Only plagioclase (andesine with 40% An) displays an argilliplasma, whereas adjacent minerals show only fissuring. The first transformations occur in the center of the plagioclase and extend afterwards toward its periphery by means of radiating anastomosed and irregularly fingerline shapes that isolate smaller and smaller areas of feldspar as the weathering proceeds. The products of alteration are colorless, sometimes slightly yellowish, and isotropic. These are silico-aluminous amorphous materials. When the latter are well developed at the expense of plagioclase (Fig. 2.10), crystals of vermicular *kaolinite*, 5 to 50 μm in size, appear scattered within these materials. This evolution was traced by Delvigne and Martin by means of an electron microprobe. They showed that the bases, in particular sodium, are rapidly depleted. Desilicification increases gradually and continues within the gel, which shows related variations of its chemical composition. Thus, the weight ratio SiO_2/Al_2O_3, which is 2.25 in the fresh feldspar, becomes 1.59 to 1.44 in poorly evolved gels, then reaches 1.26 in evolved gels, and eventually 1.18 in kaolinite.

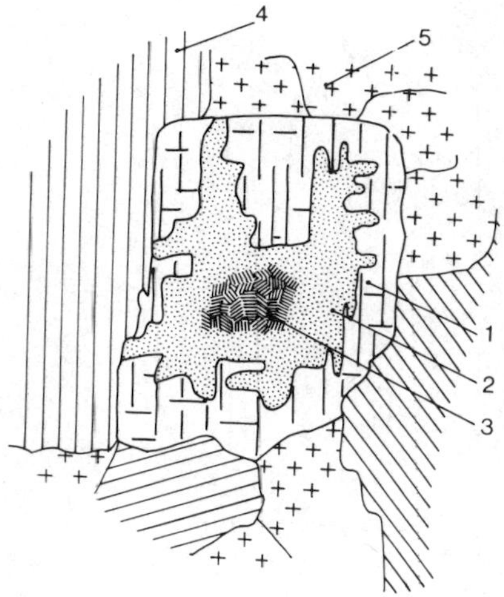

Figure 2.10. Weathering of a Plagioclase from its Center. (1) Fresh plagioclase; (2) amorphous product of transformation; (3) argillaceous product of transformation (kaolinite); (4) actinolite; (5) quartz.

EXAMPLE OF WEATHERING OF PLAGIOCLASE IN CONTACT WITH LESS REACTIVE QUARTZ AND POTASSIC FELDSPARS

Examples of this kind were described by Boulet (1974) in migmatites of the region of Garango (Burkina Faso). In this case, transformation begins along the *periphery* of the mineral and develops progressively by means of numerous branching and fingerlike *embayments* a few microns in width (Fig. 2.11). These embayments may follow cleavages and minor cracks oblique to them, or they may be independent of them. Thus, original plagioclase is corroded into a transformation plasma that corresponds either to an *amorphous silico-aluminous "gel"* close in composition to kaolinite in tropical humid countries, or to a *paragenesis of clay minerals* of smectite type in Mediterranean countries.

Adjacent Parent Minerals with the Same Reactivity

When plagioclases in rocks are associated with other minerals that react simultaneously to weathering, one can observe for each of the two types of minerals under consideration the development of two domains of argilliplasma optically distinct although contiguous. Two examples of this follow.

The first example is of simultaneous weathering of plagioclase in contact with muscovite, granite from Parthenay, Deux Sèvres, France (Meunier, 1980). Here one can observe the formation of a narrow zone of weathering for each of the two adjacent minerals investigated without noticing, under the petrographic microscope, any mixture between the argilliplasmas charac-

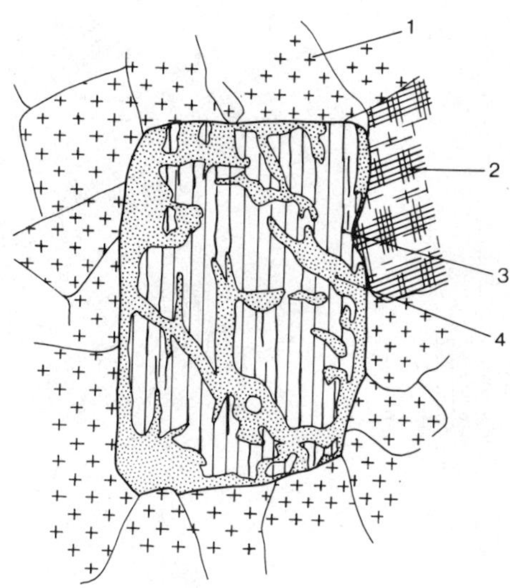

Figure 2.11. Weathering of a Plagioclase from its Margins. (1) Quartz; (2) potassic feldspar; (3) well-displayed cleavages within fresh feldspar; (4) embayments of secondary products.

teristic of each parent mineral (Fig. 2.12). Furthermore, minute relicts of the parent mineral occur in each domain of argilliplasma from which it is derived. Finally, transformation products of each mineral are identical to those that occur during an internal evolution of these same minerals. Microchemical analyses indicate the presence in these argilliplasmas of clay minerals of kaolinitic type with ratios Si/Al ranging from 1 to 2.

The second example is of simultaneous weathering of plagioclase in contact with hornblende, amphibolite from La Roche l'Abeille, Massif Central, France (Proust and Velde, 1978). In a smectitic weathering profile, plagioclase and hornblende associated within the same rock are simultaneously altered in *chemically different beidellites*. In the earliest stages of weathering, namely inside the minute cracks and immediately along the margins of the fresh relicts of each parent mineral, an intergrade aluminous mineral close to hydroxivermiculite can be identified. The chemical formulas calculated on the basis of 12 oxygens for these *intergrade minerals* (2 to 10 μm fraction) are as follows:

For hornblende: $Si_{3.46}Al_{0.54}O_{10}(Al_{1.22}Mg_{0.27}Ti_{0.01}Fe^{3+}_{0.50})(OH)_2Ca_{0.12}Na_{0.06}K_{0.05}Mg_{0.2}$

For plagioclase: $Si_{3.84}Al_{0.16}O_{10}(Al_{1.37}Mn_{0.002}Mg_{0.32}Fe^{3+}_{0.23}Ti_{0.008})(OH)_2Ca_{0.15}Na_{0.34}$.

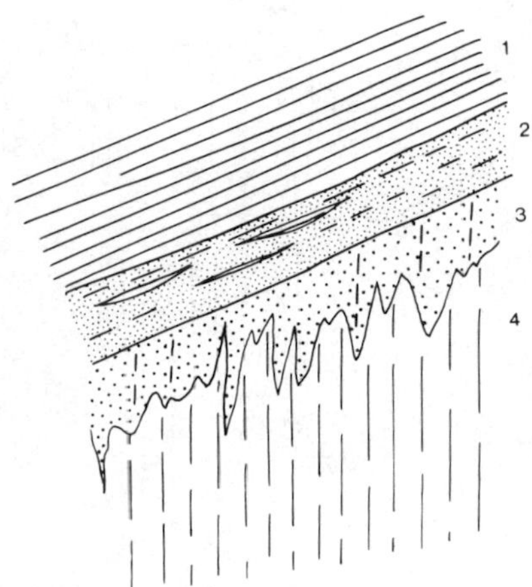

Figure 2.12. Weathering Products of Two Adjacent Reactive Minerals: Muscovite and Plagioclase. (1) Fresh muscovite; (2) transformation product of muscovite with "ghosts" of original cleavages and elongated relicts of fresh muscovite; (3) transformation products of plagioclase with traces of original cleavages; (4) fresh plagioclase.

These intergrade minerals recrystallize into a well-developed beidellitic phase and display *aluminous or ferriferous* chemical characteristics (Fig. 2.13), whether they derived from weathering of plagioclase or hornblende, respectively. The structural formulas obtained from the different mineral fractions are as follows:

For hornblende: $Si_{3.43}Al_{0.51}O_{10}(Al_{1.24}Mg_{0.21}Ti_{0.02}Fe^{3+}_{0.53})(OH)_2Ca_{0.08}Na_{0.02}K_{0.02}Mg_{0.22}$

For plagioclase: $Si_{3.56}Al_{0.44}O_{10}(Al_{1.41}Mg_{0.20}Fe^{3+}_{0.37}Ti_{0.01})(OH)_2Ca_{0.09}Na_{0.10}Mg_{0.15}$.

In conclusion, the examples investigated above show in contiguous parent minerals not only reactivity to weathering proceeding at different rates, but also *progressiveness and independence of transformations*.

Progressiveness is shown in the evolution of the gels toward kaolinite by desilification or in the evolution of intergrade minerals toward beidellitic smectites. The last two examples illustrate a petrographic and chemical independence where argilliplasmas remain not only optically but also crystallo-chemically differentiable.

A *chemical memorization* of the parent composition may, therefore, exist in the secondary products of weathering. However, the last example shows that weathering beidellitic smectites, relatively rich in iron, can develop from parent plagioclases $NaAlSi_3O_8-CaAl_2Si_2O_8$ which are devoid of it.

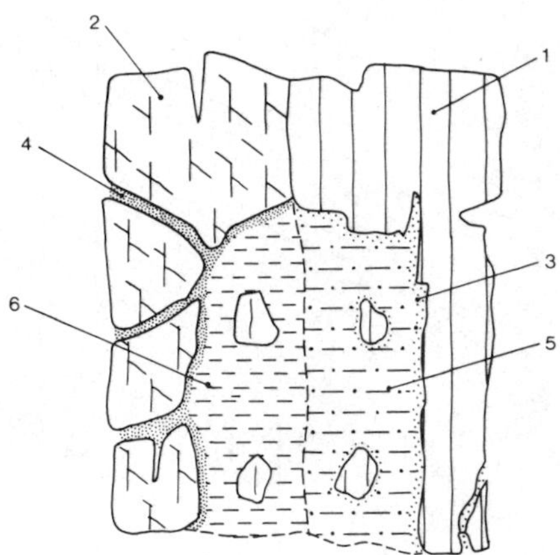

Figure 2.13. Simultaneous Weathering of a Plagioclase in Contact with Hornblende (sketch based on Proust and Velde, 1978, Fig. 3, p. 205). (1) Plagioclase; (2) hornblende; (3) intergrade alteration minerals inside cracks and around fresh relicts of plagioclase; (4) intergrade alteration minerals inside cracks and around fresh relicts of hornblende; (5) aluminous beidellites from alteration of plagioclase; (6) ferriferous beidellites from alteration of hornblende. (Reproduced by permission of the Mineralogical Society of London.)

This seems to show that, in the case of two adjacent minerals reacting simultaneously to weathering, *well-defined but limited chemical interactions* can take place. Such examples are discussed below.

Parent Mineral Phases Undergoing Simultaneous Weathering and Mutual Reactions

Weathering of *minerals of ultrabasic rocks*, such as olivine, pyroxenes, and serpentine, has been studied by many authors. An example is presented here based on the author's detailed observations on bodies of ultrabasics from the region of Moyango in the Ivory Coast (Colin et al., 1980; Paquet et al., 1981; Nahon et al., 1982; Nahon and Colin, 1982).

Parent rocks are *serpentinized dunites* (in which forsterite forms about 60% of the rock and the network of lizardite about 40%) and *pyroxenites* with enstatite and bronzite, containing from 5% to 10% forsterite. Along the margins of the bodies of ultrabasics, pyroxenites are frequent and alternate with serpentinized dunites. The example presented here is precisely a marginal profile. It displays zones of weathering in which pyroxenite facies and serpentinized dunite facies are adjacent to each other. This situation allows a comparison for the same type of weathering, of the behavior of *forsterite* with respect to other adjacent minerals, which may be either *enstatite* or *serpentine minerals*.

Forsterite, regardless of its parent facies, displays a fairly constant chemical composition that can be expressed by the average chemical formula

$$(Mg_{1.746}Fe_{0.198}Ni_{0.012}Al_{0.013}Ti_{0.001})SiO_4$$

in comparison with other adjacent parent minerals whose average chemical formulas are

$$\text{Enstatite: } [(Al_{0.029}Cr_{0.0028}Fe_{0.174}Mg_{1.714}Ni_{0.003})(Si_{1.949}Al_{0.051})O_6]$$

$$\text{Lizardite: } [(Mg_{2.540}Fe_{0.418}Ni_{0.030}\ Ca_{0.011}Mn_{0.001})(Si_2O_5)(OH_4)].$$

Moreover, no noticeable chemical heterogeneity occurs within given crystals of forsterite. For a given weathering horizon, adjacent minerals are either pyroxenes or serpentine minerals.

Forsterite Associated with Enstatite

There is a contrast in the reactivity of these two mineral phases to weathering. Forsterite may be affected by an advanced transformation of phyllosilicate type, while weathering of orthopyroxenes is expressed only by an increasing number of fissures and minute cracks along the margins of which a paragenesis of amorphous products is incipient. The margins of parent relicts of forsterite in contact with the phyllosilicate phase display embayments and protrusions (Fig. 2.14). An SEM examination of the surface condition of this forsterite shows a network of open fissures and small cracks (Fig. 2.15).

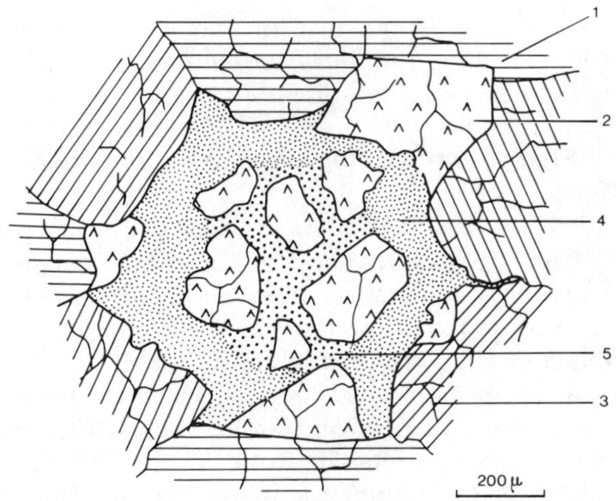

Figure 2.14. Weathering of Forsterite into Smectites (from Nahon, Colin, and Tardy, 1982). (1) Enstatite; (2) forsterite; (3) corrosion fissures and cracks; (4) marginal smectites; (5) internal smectites. (Reprinted by permission of the Mineralogical Society of London.)

Weathering phyllosilicates are *di-* and *trioctahedral smectites*; the honeycombed arrangement of these smectites has been confirmed under SEM (Fig. 2.15). Moreover, microchemical tests indicate that smectites in the center of forsterite undergoing weathering are different from those of the margins. In the center, smectites are *essentially magnesian*, tending toward the *saponite* end-member with the following average structural formula:

$$Ca_{0.046}Mg_{0.174}(Mg_{5.106}Ni_{0.050}Mn_{0.015}Fe^{3+}_{0.400}Al_{0.192})(Al_{0.558}Si_{7.442})O_{20}(OH)_4.$$

On the margins, smectites are more alumino-ferriferous, tending toward the *beidellite-nontronite* end-member with the following average structural formula:

$$Ca_{0.136}Mg_{0.111}K_{0.003}(Mg_{2.102}Ni_{0.078}Ti_{0.016}Cr^{3+}_{0.027}Fe^{3+}_{1.193}Al_{1.358})(Al_{0.640}Si_{7.360})O_{20}(OH)_4.$$

Such a zonation of the chemical composition of smectites cannot be inherited from the parent mineral. Therefore, it has to be related to the chemical interaction of these weathering products with neighboring enstatites. This is all the more obvious since forsterite, practically devoid of aluminum, displays weathering products that become increasingly enriched in aluminum as they occur nearer to enstatites undergoing alteration. This behavior is also displayed by chromium. It should be pointed out that the occurrence of

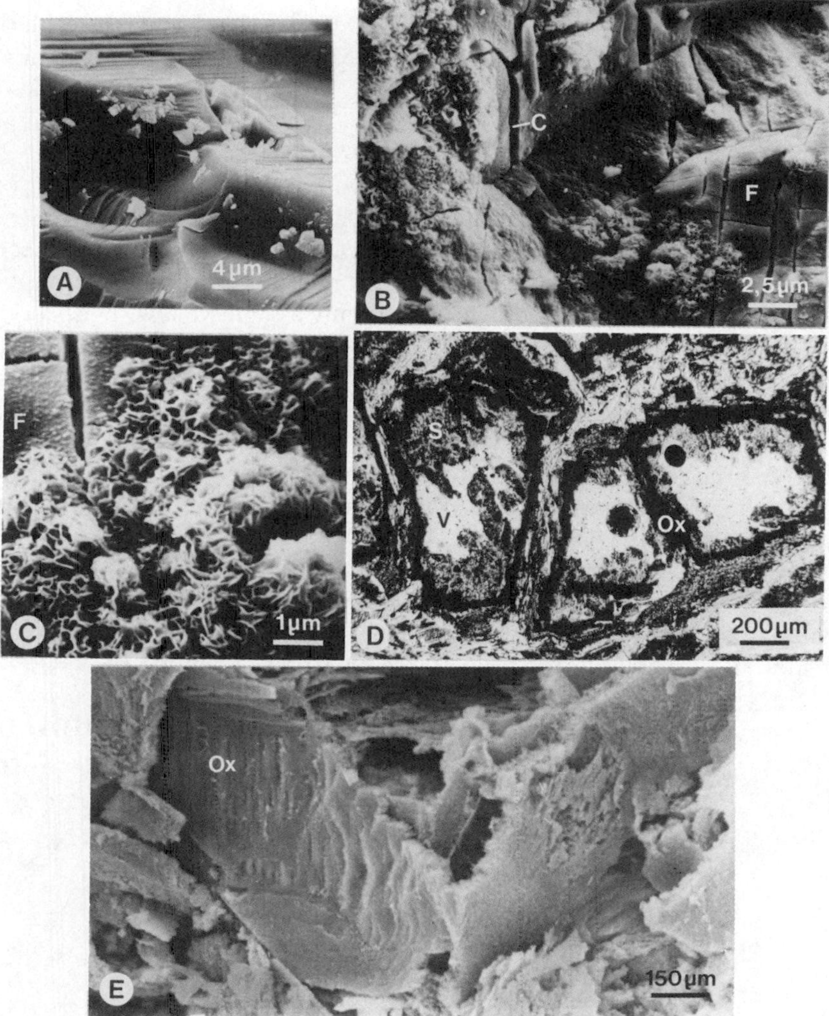

Figure 2.15. Weathering of Forsterite into Smectites and into Goethite (from Nahon, Colin, and Tardy, 1982). **A**: Surface aspect of altered forsterite (forsterite-pyroxenite of Moyango, Ivory Coast). SEM. Note that the surface displays dissolution striae. **B**: Weathering smectites from forsterite (forsterite-pyroxenite of Moyango, Ivory Coast). SEM. Forsterite (F) displays fissures and minute cracks due to dissolution (C). **C**: Enlargement of previous picture. Notice honeycombed arrangement of smectites. They appear in epitaxic growth on the surface of forsterite (F) and cover dissolution fissures. **D**: Goethite septa (Ox) developed at the expense of alteration smectites (S) from forsterite (serpentinized dunite of Moyango, Ivory Coast). Petrographic microscope, plane polarized light. Notice central void (V) resulting from dissolution of a patch of fresh forsterite or of more magnesian smectites. **E**: Goethite septa (Ox) under SEM, same sample as above. Notice the preservation of the original texture of forsterite. (Figs. 15-A, C, D, and E from Nahon et al., 1982, and reprinted by permission of the Mineralogical Society of London.)

clinopyroxene blades inside enstatites has been revealed by high-resolution electron microscopy (HREM). This would explain the amounts of aluminum provided by pyroxenes.

Forsterite Associated with a Network of Lizardite

In this case, a difference of reactivity to weathering of lizardite is also very pronounced. This difference has been shown many times in the weathering of ultrabasic rocks of New Caledonia (Trescases, 1975). Whereas a smectite phase is well developed at the expense of forsterite, lizardite, under optical observation in plane polarized light, shows only a slight yellowish aspect that expresses the initial individualization of iron oxyhydroxides (Fig. 2.16).

Analysis of forsterite-smectite-lizardite contacts shows that planar igneous contacts between forsterite and lizardite are replaced by an irregular weathering contact, with embayments and protrusions, clearly developed at the expense of forsterite and favorable to the individualization of smectites. Furthermore, SEM shows the olivine surface with dissolution striae (Fig. 2.15 A).

Mineralogical and microchemical characteristics of weathering smectites correspond to a population of di- and trioctahedral minerals, more or less magnesian or ferruginous with, nevertheless, a predominance of *dioctahedral smectites*. The average formula for these smectites is as follows:

$$Ca_{0.052}Mg_{0.114}(Mg_{1.509}Ni_{0.140}Mn_{0.009}Ti_{0.004}Fe^{3+}_{2.780})Si_8O_{20}.(OH)_4.$$

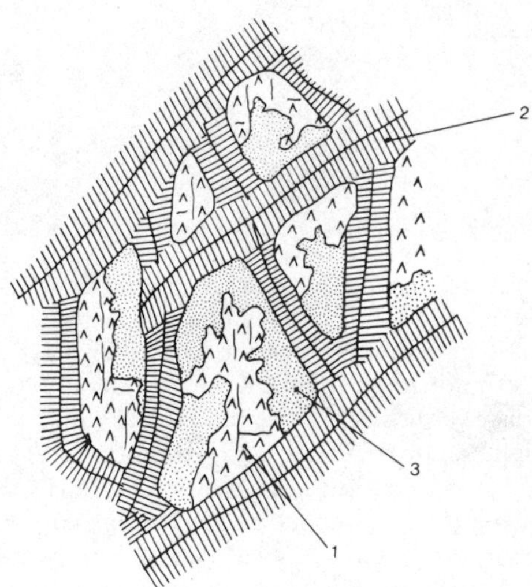

Figure 2.16. Weathering of Forsterite (1) into Smectites (3) in Contact with a Network of Unweathered Lizardite (2).

These smectites as well as parent forsterite and neighboring lizardite are *devoid of aluminum*. However, they are much *more ferruginous* than those originating from forsterite associated with enstatite. Given the constant nature of the chemical composition of parent forsterite, regardless of its location, either in facies of pyroxenite or serpentinized dunite, it becomes obvious that the lack of aluminum and the higher content of iron of smectites has to be attributed to neighboring lizardite.

Thus, for a forsterite of same chemical composition, smectitic transformation products generated in the same weathering horizon depend not only on parent minerals but also on interactions with neighboring minerals.

In the case of Moyango, this first stage of alteration of forsterite into phyllosilicates can proceed further higher up in the profiles: first, until parent relicts have completely evolved into smectites, second, until the latter lead to *iron oxyhydroxides*. These iron oxyhydroxides develop along the margins of ancient forsterite, in the vicinity of neighboring minerals where smectites are the most ferruginous. The most magnesian smectites are dissolved leaving a central void that occupies the middle of the former forsterite.

The development of these iron oxyhydroxides occurs at the expense of ferruginous smectites. They become organized into *goethitic septa* (Fig. 2.15 D and E), which allow the preservation, fairly high up in the weathering profile, of the ferruginous "ghost" of forsterite. Goethitic septa reflect, moreover, the original chemical composition of smectites with respect to the least soluble elements, namely, aluminum and iron.

Ferruginous septa originating from pyroxenite facies consist of *aluminum-bearing goethite*, while those from serpentinized dunite facies are goethite *without aluminum substitution*. In other words, oxyhydroxide septa not only generate from a transitional stage of smectites, but they also preserve the *geochemical memory* of the first stages of weathering. Thus, pseudomorphosis of forsterite by a smectite argilliplasma is followed by pseudomorphosis by a goethitic crystalliplasma. However, in the investigated case, there are no longer true septa of oxyhydroxides developed directly from the parent mineral, but rather septa expressing a second generation of weathering minerals derived from the primary mineral, the first generation being here an argilliplasma stage.

Parent Mineral Phases Undergoing Simultaneous Weathering and Whose Secondary Products Also Express an External Supply

Colin et al. (1985) reported that chemical weathering of pyroxenes from ultramafic rocks in Niquelandia (Brazil) leads either to Fe-rich amorphous products and then to goethite, or to a phyllosilicate clay. The pyroxenite parent rock consists of 69% orthopyroxenes (enstatite), 27% clinopyroxenes (diopside), and 4% chromite. The average composition of these minerals (Table 2.1) was obtained by chemical microprobe analyses.

Table 2.1. Average-Formulae of Enstatites, Diopsides and Chromites from Niquelandia (from Colin et al., 1985).

min \ el	Si	Al	Al	Fe^{2+}	Mg	Ni^{2+}	Cr^{3+}	Ti^{4+}	Ca	K	Mn	O
Ens.	1.974	0.026	0.028	0.273	1.657	0.001	0.014	0.001	0.009		0.007	6
Diop.	1.968	0.032	0.036	0.087	0.882		0.024	0.006	0.924	0.035	0.004	6
Chr.			2.340	1.102	0.790	0.004	1.726		0.003	0.003		8

Weathering of orthopyroxenes and clinopyroxenes is simultaneous. However, opening of intramineral fissures and development of corrosion embayments are more important in enstatites than in diopsides, leading us to assume that orthopyroxene weathers faster than clinopyroxene.

Weathering of these minerals (Fig. 2.17) is either peripheric and centripetal when starting from intergrain joints or in pockets and centripetal from intra- and transmineral fissures. Occurrence of unaltered chromite grains inside zones of greatest development of weathering argillaceous plasma shows that weathering was propagated first from intramineral fissures where chromite was concentrated.

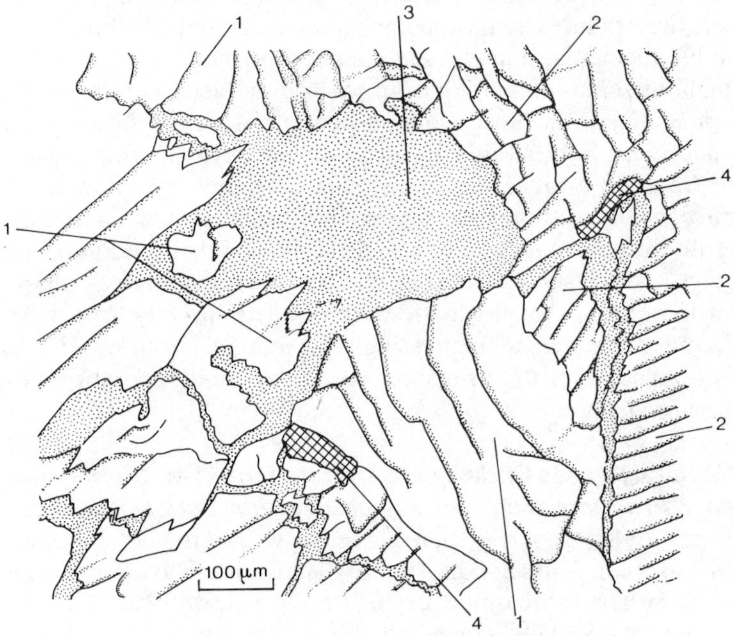

Figure 2.17. Simultaneous Weathering of Enstatite (1) and Diopside (2) in a Smectitic Argilliplasma (3); Chromite Is Shown by Crosshatching (modified from Colin et al., 1985).

Under crossed nicols, the weathering argillaceous plasma displays, an orientation parallel to the c axis of the minerals it replaces by pseudomorphosis. Therefore, preservation of the original optical orientation takes place.

Analysis by X-ray diffraction on microsamples and in situ microprobe analysis show that the weathering argillaceous plasma consists of swelling smectite with tetrahedral deficit, with the following average composition:

$$(Si_{3.222}Al_{0.752}Fe^{3+}_{0.026})(Al_{0.030}Fe^{3+}_{0.936}Mg_{0.222}Ni^{2+}_{1.529}Cr^{3+}_{0.078}Ti^{4+}_{0.007})$$
$$Ca_{0.054}K_{0.007}O_{10}(OH)_2.$$

The composition of the smectites varies according to the degree of fracturing of the initial parent rock: Al varies from 11.7% to 7.09%; Fe^{3+} from 14.02% to 19.15%; and Ni^{2+} from 22.28% to 16.04%.

Smectites derive from both diopsides and enstatites, being the result of their associated and simultaneous weathering, namely, of the geochemical reaction between these two parent minerals. Furthermore, the smectite plasma also reveals, in the case of nickel, an external interaction with adjacent parent minerals since the latter are devoid of it. Nickel must originate from the weathering of dunites occurring higher up in the sequence. This situation implies an average distance migration of nickel by solutions. These nickel-bearing smectites are discussed in detail below.

Weathering Argilliplasmas and Their Complexity

Jackson et al. (1948) and Fieldes and Swindale (1954) established the important concept of a sequence of appearance of weathering minerals and of the capacity of each primary mineral to generate a different secondary mineral. These ideas were generalized by Jackson (1968) and subsequently discussed in numerous works (Tardy et al., 1973).

Most weathering argilliplasmas were identified by means of standard X-ray diffraction methods on powders or on less than 2 micron fractions. Thermodynamic simulations were used to understand these sequences of secondary minerals in the weathering of rocks.

Complex simulations can take into account a greater number of transfer reactions (dissolution of minerals, concentration by evaporation) and reversible reactions (equilibria between aqueous species, precipitation of minerals). The evolution of the chemical composition of water and mass transfers during differentiation of a weathering profile can thus be studied by means of this approach. These simulations, which take into account simultaneously the resolution of equations of mass conservation, of electrical neutrality of solutions, as well as of all equilibrium equations, can only be attempted by means of computer programs (Helgeson, 1968, 1969, 1970; Helgeson et al., 1969, 1970; Fritz, 1975).

Fritz (1975) and Fritz and Tardy (1976) simulated under ambiant supergene conditions (initial water of percolation with average oxidizing capacity:

$pO_2 = 10^{-45}$ atm.; pH = 4.5; $pCO_2 = 10^{-2.5}$) the dissolution of an idealized granite. Weathering was simulated by dissolving ξ moles of a mixture of parent minerals by assuming that precipitated secondary minerals formed as soon as their respective saturation points were reached. These simulations reproduced an open environment (in which, during the evolution of the initial solution, precipitates do not remain in contact with the solution), then a closed environment (precipitates remain in contact with the solution and may be modified subsequently).

Qualitatively, secondary neoformed minerals, in an open system, formed the following sequence:

goethite – goethite + gibbsite, goethite + kaolinite, kaolinite + Fe smectite, Fe-Al smectites, calcite, and Mg smectite.

Two comments were made by Fritz (1975) in regard to the open system simulation (Fig. 2.18). First, kaolinite does not necessarily appear as a degradation product of smectites. Second, the stability order of primary minerals established by Goldich (1938) corresponds to the order in which these minerals reached saturation: quartz, K-feldspar, amphibole, albite, biotite, and anorthite.

In the closed system no changes occurred during the first stages of weathering with the formation of goethite, gibbsite, and kaolinite. Formation of kaolinite resulted not only from elements originating from the congruent dissolution of granite but also from elements produced by the previously formed gibbsite, which henceforth was in disequilibrium with the solution until its total dissolution. Kaolinization proceeded until precipitation of nontronite, when equilibrium with respect to the latter mineral was reached in the solution. During the formation of nontronite, goethite was dissolved. Finally magnesian smectite and calcite precipitated at equilibrium whereas destabilized kaolinite yielded its elements to solution.

Budgets for weathering as open or closed systems are therefore close. Nevertheless, in the second case, the last secondary phases formed can entirely replace earlier ones.

Such mineral sequences of weathering have often been observed in nature (for a review see Millot, 1964; Barshad, 1966; Strakhov, 1969; Tardy, 1969).

The methods used for the past ten years to identify weathering argilliplasmas – nuclear magnetic resonance (NMR), Mössbauer spectrometry, optical spectrometry, extended X-ray absorption fine structure spectroscopy (EXAFS) – showed that argillaceous phases that appeared pure when analysed by all standard methods of X-ray diffraction consisted in reality of mixtures of argillaceous particles, themselves showing variations in their chemical compositions (Duplay, 1984). For instance, the nickel-bearing smectite originating from the weathering of the pyroxenites of Niquelandia, described above, had been identified by standard methods of X-ray diffraction and of chemical analysis by microprobe. A similar sample, identified as a

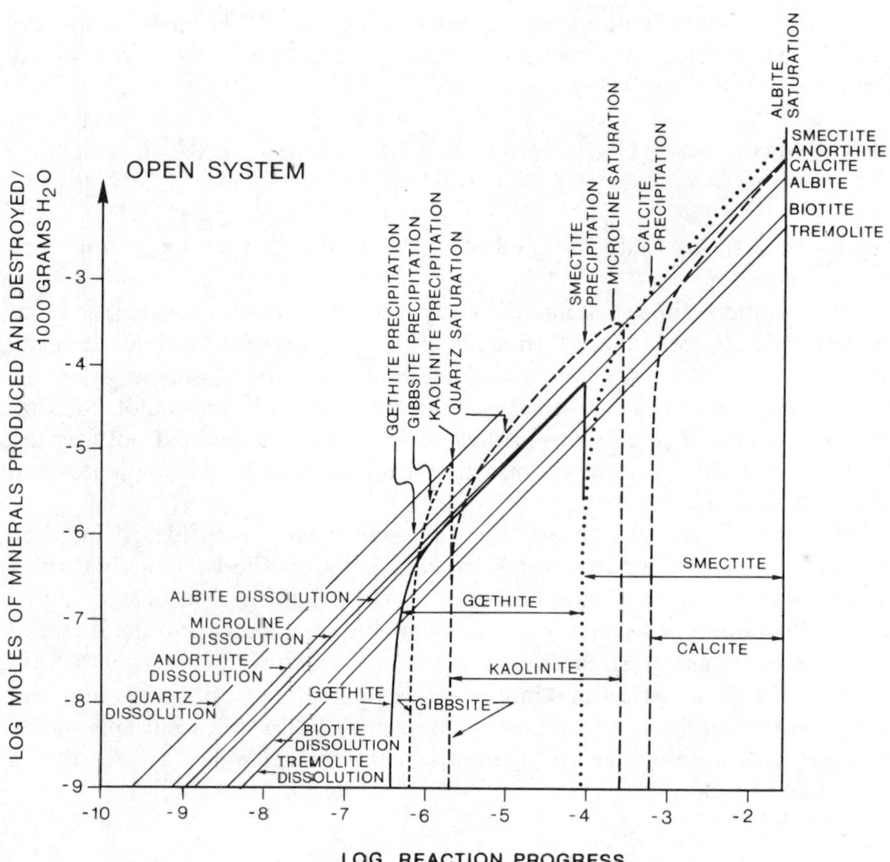

Figure 2.18. Weathering of Granite in Open System: Number of Moles of Minerals Destroyed or Produced per 1000 g H_2O as a Function of Reaction Progress (log ξ) (modified from Fritz, 1975). (Reproduced with permission of Sciences Géologiques.)

trioctahedral swelling smectite of average formula ($Si_{3.92}$ $Al_{0.08}$) ($Al_{0.17}$ $Fe^{3+}_{0.5}$ $Mg_{0.48}$ $Ni_{1.47}$ $Cr_{0.02}$) $Ca_{0.01}$ $K_{0.03}O_{10}$ $(OH)_2$, was recently studied by Decarreau et al., 1987) by means of infrared spectroscopy; diffuse reflectance spectroscopy; Mössbauer spectroscopy; EXAFS. Spectroscopic methods show atom segregations into pimelitelike domains rich in Ni and nontronitelike domains rich in Fe. From EXAFS results, Ni and Fe atoms are clustered in domains having minimum diameters of about 30–40 Å (Manceau and Calas, 1986). Infrared spectrometry results suggest that part of the Al is associated with Fe in dioctahedral nontronitelike domains, whereas small amounts of Mg(Al) occur as neighboring ions in Ni and Fe clusters (EXAFS results). The above data indicate that Decarreau et al. (1987) were able to describe more accurately the locations of these atoms in the clay structure.

The domains of segregation can be in the same layer or in alternate layers. If octahedral cations are completely segregated in separate di- and trioctahedral layers, the structural formula of the Ni-bearing smectitic clay would be:

$$2/3\ [Si_4\ Mg_{0.74}\ Ni_{2.26}\ O_{10}\ (OH)_2] + 1/3\ [(Si_{3.77}\ Al_{0.23})\ (Al_{0.5}\ Fe_{1.45}Cr_{0.07})\ O_{10}\ (OH)_2].$$

This formula shows that dioctahedral layers or domains represent one third of the sample.

Whereas the results of standard X-ray diffraction methods suggested that the smectitic clay was mainly trioctahedral, spectroscopic methods indicate that it contains mixed trioctahedral/dioctahedral sheets. However, juxtaposition of Ni-domains and Fe-domains within the same layer cannot be completely excluded. The swelling of these layers can be explained both by the small size of particles and by a disordered turbostratic lack of pimelite- and nontronitelike layers.

This example shows that argilliplasmas, even when considered pure on the basis of results from classical X-ray diffraction methods, actually display a complex composition that appears at the scale of a single particle (Duplay, 1988). This fundamental characteristic probably originates from the fact that these argilliplasmas crystallized at low temperatures, with low growth rates and low chemical diffusion kinetics expressing a formation that did not correspond to equilibrium processes (Manceau, 1989). Manceau abstracted the main structural and chemical characteristics of argilliplasmas and related properties as follows:

1. These argilliplasmas show numerous structural and chemical defects (intergrowth, nonstoichiometry); also, because of the absence of monocrystals, their precise crystallographic structure is not always known.
2. The small size of clay minerals is associated with an anisometry related to the limitation of crystal growth in certain directions; the large specific surface of these minerals generated by their microcrystallinity and anisometry, as well as by their great density of defects, provides them with a high chemical reactivity, particularly with respect to adsorption processes.
3. Because of the crystallization of clay minerals at low temperatures, solid solutions generally observed in phases formed at higher temperatures may not have any reality at the scale of a single layer.

Tardy et al. (1986) assumed that the diversity of chemical composition of the particles forming the argilliplasma depends on the variety of the layers stacked up within the same particle. Consequently, they proposed to interpret these variations by means of a model of solid solutions with multipoles in equilibrium conditions with the aqueous solutions. In this model, each particle is considered to be an ideal solid solution of several end-members consisting of individual layers. Thus, the proportions of the various end-members change from one clay particle to another. Since the theory of solid solutions requires that several ideal solid solutions cannot be in equilibrium in the same environment, these authors considered that each argillaceous particle, with the surrounding solutions, forms an independent microsystem (or nanosystem) at equilibrium. This thermodynamic modeling allows the calculation for each particle the contribution of each end-member of the ideal solid solution.

In conclusion, weathering argilliplasmas, although appearing homogeneous when investigated by standard X-ray diffraction techniques, actually consist of the juxtaposition of several populations of particles. In each population every particle displays variations of composition at the scale of stacks of layers or of micro- or nanodomains. These argilliplasmas in soils and weathering profiles endlessly interact with aqueous solutions percolating through them or filling their porosity. Equilibria at the smallest scale between particles and solution are continuously challenged. Such fugacious and localized equilibria could be modeled theoretically by means of a theory of ideal solid solution with multi end-members involving each particle and the surrounding aqueous solution at a given time.

Conclusions

Pseudomorphoses by argilliplasmas correspond either to phyllosilicate neoformations or to phyllosilicate transformations. Clay minerals that form weathering plasma can be fed in three manners: by elements released from the parent mineral only; by these same elements to which is added an influx from neighboring minerals; or by elements released from neighboring minerals to which is added an external influx from elsewhere in the profile.

Weathering Plasmation Depending on a Single Parent Mineral

Neogeneses develop from the least mobile elements of the single parent mineral involved. These elements, released by solutions during dissolution of the mineral, are reorganized in place after short transfers within the boundaries of a given crystal (*intracrystalline nanotransfers*).

Weathering Plasmation Related to the Interaction of Neighboring Parent Minerals

Neoformations develop with influxes of elements from neighboring minerals (*intercrystalline microtransfers*). The direction of these transfers is always from the least to the most weathered mineral phase.

Weathering Plasmation Related to the Interaction of Neighboring Parent Minerals and to the Interaction of External Influx

Formation of the argilliplasma depends not only on elements released by adjacent minerals undergoing weathering (intercrystalline microtransfers), but also on elements released in solution by other minerals located elsewhere in the horizon, the profile, or the sequence (transcrystalline micro- and macrotransfers).

In all instances, *neoformed or transformed argilliplasmas are mineralogically complex*. They are associated with different crystalline individuals or individuals that display variations of their chemical composition. Indeed, analysis shows the existence, from the margins of the parent relict to the center of the argilliplasmas, of an organized succession of crystalline structures (gradation from gels to kaolinite) and of chemical compositions of secondary products (gradation to intergrade minerals to smectites, and of magnesian smectites to ferriferous smectites). Finally, each population of argilliplasma itself consits of argillaceous particles and each particle varies in composition.

RELATIVE POSITIONS OF PARENT RELICTS AND OF WEATHERING PLASMAS: MINERAL PARENTAGES

Petrographic analysis of the relationship between weathering products and parent relicts is used for studying their *modes of relative distribution* and the *nature of the contact* between original mineral and secondary product. Morphology of distributions affords data on *the direction of progression of weathering*, whereas the nature of the contacts indicates the possible mechanisms.

Morphology of Distributions

Two major types of distribution may be distinguished on the basis of the relationship between weathering products and parent minerals: peripheral and patchy.

Peripheral Distribution

Whenever secondary products generated by weathering become individualized and then develop at the expense of a parent mineral (or a fragment of that mineral), from the margin toward its center, this distribution is called peripheral. The shape of the boundaries may be regular (Fig. 2.19 A) or irregular (Fig. 2.19 B), with residual patches of the original crystal that indicate a *centripetal progression* of weathering, or eventually very irregular (Fig. 19–2, C) ending as fingerlike embayments).

Patchy Distribution

Distribution is designated as patchy when weathering begins at the center of the parent mineral and develops at its expense in a *centrifugal* manner, as shown by the presence of altered zones with or without residual areas, with or without irregular boundaries, inside the original crystal that has preserved its external shape (Fig. 2.19 D and E).

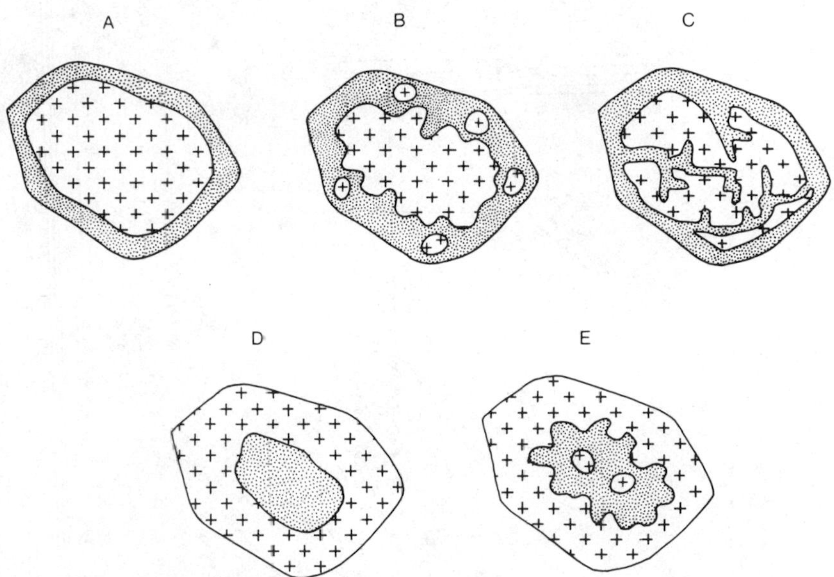

Figure 2.19. Morphology of the Relative Distribution of Parent Relicts (Cross Pattern) and Weathering Products (Stippled Pattern). **A**: Peripheral distribution of weathering products with regular shape of the boundaries. **B**: Peripheral distribution with irregular shape of the boundaries and residual patches of parent mineral. **C**: Peripheral distribution with very irregular shape of boundaries. **D**: Patchy distribution of weathering products with regular shape of the boundaries. **E**: Patchy distribution of weathering products with irregular shape of the boundaries.

Nature of Contacts

In the case of rock weathering with preservation of lithologic structures, secondary products of weathering are in contact with the parent minerals from which they originate. These are parentage contacts. The weathering phase, furthermore, preserves a certain *chemical memory of the parent mineral* and sometimes even a *structural memory* (cleavages, twinnings, fractures). Whether this transition occurs with or without the intermediate of a void, *discontinuities* or *continuities of contact* occur respectively.

There are many examples of possible continuities and discontinuities of contact (Fig. 2.20). In this figure, A represents schematically the tranformation of mica into smectitic clay mineral. It is a *crystalline transformation* caused by degradation of a primary phyllosilicate in the sense of Millot (1964) and Millot et al. (1965). This continuous transformation develops maintaining the three-layered type of phyllosilicate structure (*structural parentage*) and preserving some chemical characteristics (*chemical parentage*).

In this same figure, B represents schematically the weathering of an

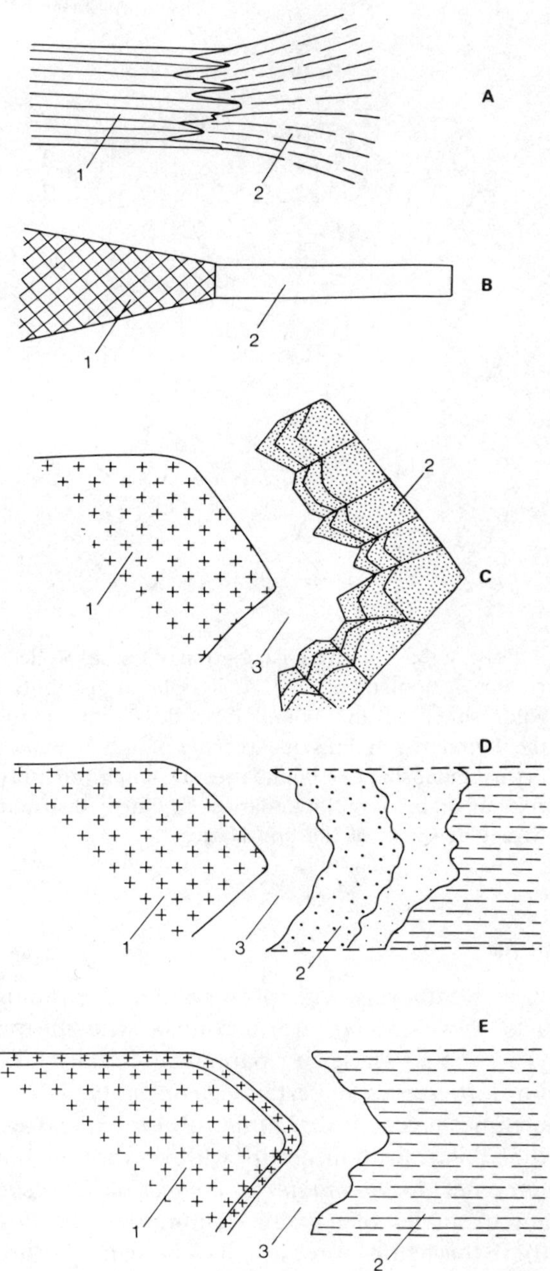

Figure 2.20. Nature of Contacts between Parent Relicts and Their Weathering Products. **A**: (1) Primary phyllosilicate; (2) weathering phyllosilicate. **B**: (1) Parent inosilicate; (2) weathering phyllosilicate as overgrowth. **C**: (1) Parent tectosilicate; (2) weathering crystalliplasma forming septa; (3) intermediate contact void. **D**: (1) Parent tectosilicate; (2) argilliplasmas with organized chemical variations; (3) intermediate contact void. **E**: (1) Parent tectosilicate with chemical surface modification; (2) weathering argilliplasma; (3) intermediate contact void.

inosilicate into a phyllosilicate mineral. The latter uses the parent relict as growth support. This is either a *neoformation with crystalline overgrowth* on the parent mineral or a transformation with high degree of inheritance and lattice coherence. This weathering occurred with either destruction of the original crystalline structures and generation of a new secondary structure from elements not removed by solutions, thus insuring a certain chemical heritage (*chemical parentage*) or with preservation of part of parent mineral lattice offering nucleation sites suitable for epitactic growth of the weathering product (structural and chemical parentage).

Contact discontinuity is represented under C, D, and E by an intermediate void (contact void) between parent mineral and secondary product. C is schematically represented neoformation of a crystalliplasma of secondary oxyhydroxides (septa) at the periphery of a parent tectosilicate.

D schematically represents a less advanced weathering of tectosilicate with *neoformation of argilliplasma* displaying organized variations of its chemical composition.

In C and D neoformation of crystalliplasma (as septa) or of argilliplasma (amorphous and phyllosilicate phases) develops from chemical elements inherited from the parent mineral not removed from the microenvironment of weathering. Such neoformations lead to destruction of the parent crystalline structure and to short-range reorganization of more or less well-crystallized secondary structures.

E differs from D only in regard to the analysis of the surfaces of parent relicts. Whereas until now examination has been restricted to structural and chemical parentages that regulate the relationship between primary minerals and weathering products, it now becomes necessary to consider the *surface conditions* of parental relicts. For this particular purpose, both shape and chemical composition must be analyzed. D is the schematic representation of a tectosilicate relict where only modifications of surface *shape* occur. However, E shows, in addition, *chemical modifications* of the surfaces of parent mineral, represented here by a tectosilicate (see Chapter 1).

The transformation of the margin of biotite into smectite (Fig. 2.21 A, after Kounestron et al., 1977) and the formation (transformation or neoformation) of blades of talc by weathering of enstatite (Fig. 2.21 B, after Nahon and Colin, 1982) illustrate continuities of contact between parent minerals and their weathering products.

The neoformation of argilliplasma by weathering of feldspar (Fig. 2.22 A, after Berner and Holdren, 1979) and of peripheral goethitic septa from hornblende weathering (Fig. 2.22 B, after Velbel, 1989) illustrate discontinuities of contact occurring between parent minerals and their weathering products.

Finally, it has been shown that in their respective relationships, parent minerals and weathering plasmas derived from them displayed *parentage contacts*. In weathered rocks, other kinds of contact may exist between parent relicts and secondary plasmas, in particular *unconformable contacts*.

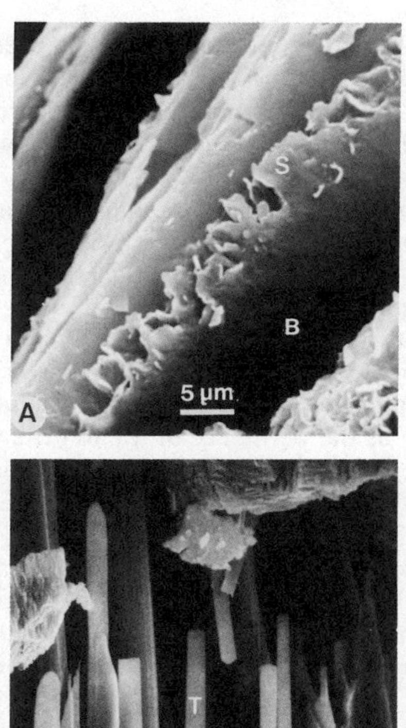

Figure 2.21. Continuity of Contact between Parent Minerals and Their Weathering Products. **A**: Transformation of biotite (B) into smectite (S) (after Kounestron et al., 1977, reprinted by permission of Académie des Sciences et Institut de France). **B**: Formation of talc (T) from enstatite weathering (E) (after Nahon and Colin, 1982, reprinted by permission of American Journal of Science).

Such contacts, as discussed later, result from transfers and accumulations over much greater distances.

RELATIVE ACCUMULATION IN STRUCTURE-PRESERVING WEATHERING

Two main weathering processes of minerals in rocks have been shown: (1) simple dissolution of minerals, which leaves no weathering product in place; and (2) incongruent dissolution, which generates secondary products (in this case relative or selective accumulation of material occurs through loss of other constituents).

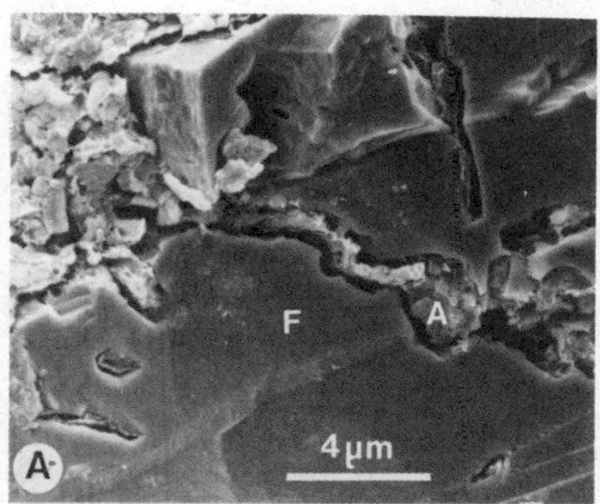

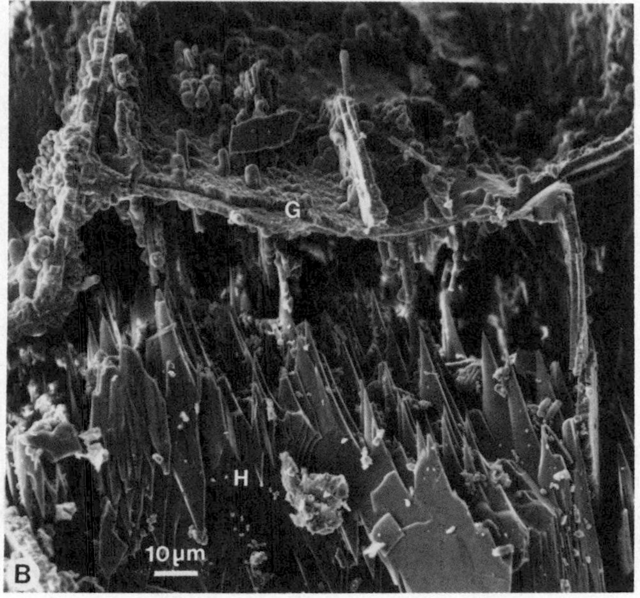

Figure 2.22. Discontinuity of Contact between Parent Minerals and Their Weathering Products. **A**: Neoformation of clay plasma (A) by weathering of feldspar (F) (reprinted with permission from Berner and Holdren 1979, by Pergamon Press plc). **B**: Neoformation of goethitic septa (G) by weathering of hornblende (H) (From Velbel, 1989, reprinted with permission of the Clay Mineralogical Society of America.)

These weathering products accumulated in place are called *alteroplasmas*. They always develop *within the boundaries of the crystal from which they were generated*. Among alteroplasmas, one can distinguish crystalliplasmas and argilliplasmas. Crystalliplasmas consist of oxyhydroxides and they lead to a *peripheral pseudomorphosis* of parent minerals simultaneously with generation of high microporosity. Argilliplasmas consist essentially of phyllitic minerals and they lead to a *complete pseudomorphosis* of the parent mineral with generation of porosity detectable only under SEM high magnification.

Alteroplasmas result from the association of chemical transfers with relative accumulation (for the residual phase) and of chemical transfers with loss (for the exported phase) (Fig. 2.23). *Chemical accumulative transfers* correspond to *nanotransfers*, namely transfers of elements that take place within the boundaries of the parent crystal. However, chemical interactions are possible between two adjacent minerals. In this case, it is a matter of *microtransfers*; the intracrystalline scale that characterizes nanotransfers is replaced by an intercrystalline transfer scale (Fig. 2.23).

Finally, to these considerations can be added elements external to the parent minerals under consideration. These elements, carried by solutions, contribute to the formation of weathering products that replace original minerals by pseudomorphosis. Such a situation corresponds to extracrystalline transfers (micro- or macrotransfers).

All short-distance transfers are chemical and provide crystalliplasmas and argilliplasmas with a memory of the chemistry of the parent mineral, and, in certain cases, with a memory also of interactions between two neighboring parent minerals. Some chemical parentage exists, therefore, between the original mineral and its weathering product. Hence, *geochemical accumulation transfers* can be defined at two scales: intracrystalline and intercrystalline, which define, respectively, geochemical *nanosystems* and *microsystems* of relative accumulation. These systems regulate the geochemistry of the least mobile elements, namely, those considered inert in the sense of Korzhinski (1965).

Relative accumulations allow *preservation of original volumes*. Indeed, accumulations of residual elements and their reorganization into newly formed crystalliplasmas or argilliplasmas operate simultaneously at intracrystalline scales with preservation of volume of the parent crystal, and at intercrystalline scales with preservation of texture; these processes reflect the association of volumes of neighboring parent crystals.

Reactions between parent minerals and solutions lead to phyllosilicate phases of weathering whose volume is larger than that of parent phases. Preservation of original volumes is possible only because subtractions of matter remain essential during weathering reactions and compensate, therefore, for the expansion forces of phyllosilicates. Nevertheless, in some cases where subtractions remain limited, forces of crystalline expansion and growth during phases of weathering lead to a new compensating fissuring. This fissuring in supergene alterations, however, remains limited in space, and textures, as a whole, are preserved. With respect to the reactivity characteristic of each parent mineral as it applies to weathering, the parent rock itself is gradually subjected to the process of relative accumulation, and its volume is thus preserved (Fig. 2.24). This modification of a volume of parent rock into a volume of weathered rock (Millot and Bonifas, 1955; Brimhall and Dietrich, 1987) through relative accumulation occurs with the appearance and development of a nano- or microporosity. The latter results from a loss of material. It is the soluble phase during weathering of parent minerals that is

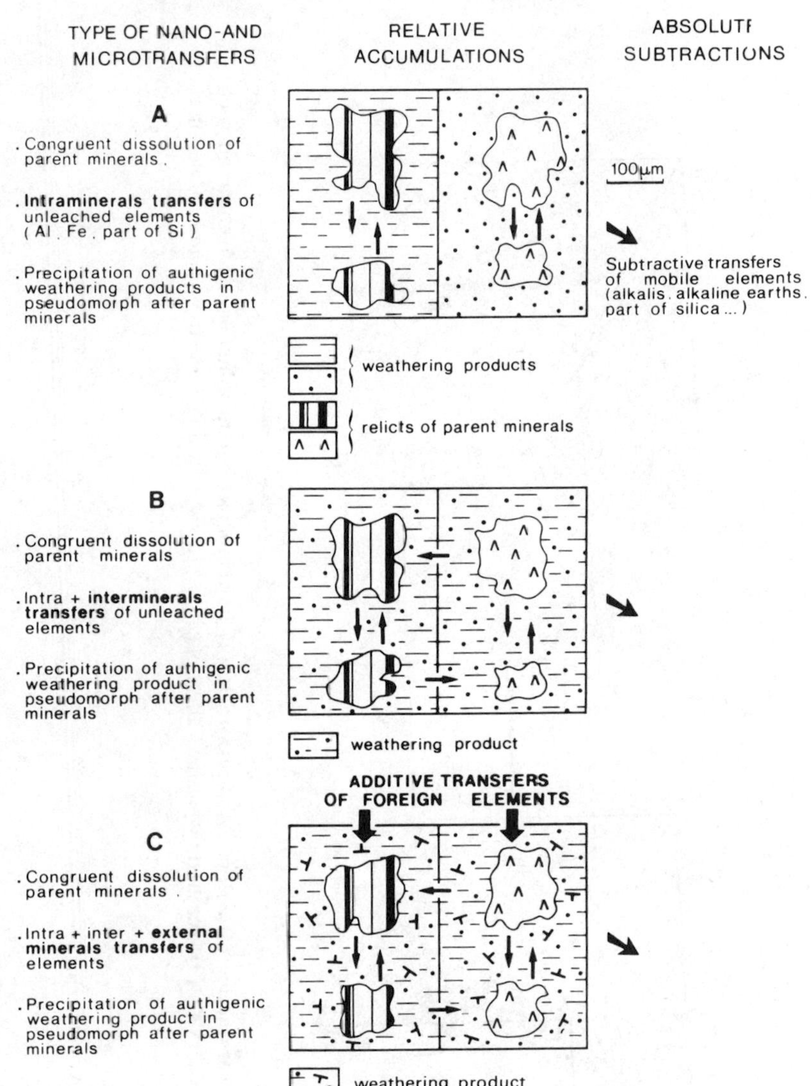

Figure 2.23. Sketch of Main Geochemical Nano- and Microsystems of Weathering (modified from Nahon, 1987). **A**: Dissolution and intraminerals transfers. **B**: Dissolution and intra- and interminerals transfers. **C**: Dissolution and intra- and inter- and external minerals transfers.

subtracted and exported out of the geochemical nano- and microsystems. There is, therefore, *absolute subtraction* of elements by *macrotransfers* that is, by transfers above the scale of several minerals. Consequently, weathering plasmation is also of a subtractive nature. Therefore, the limits of relative accumulation can be defined in structure-preserving weathering. This relative

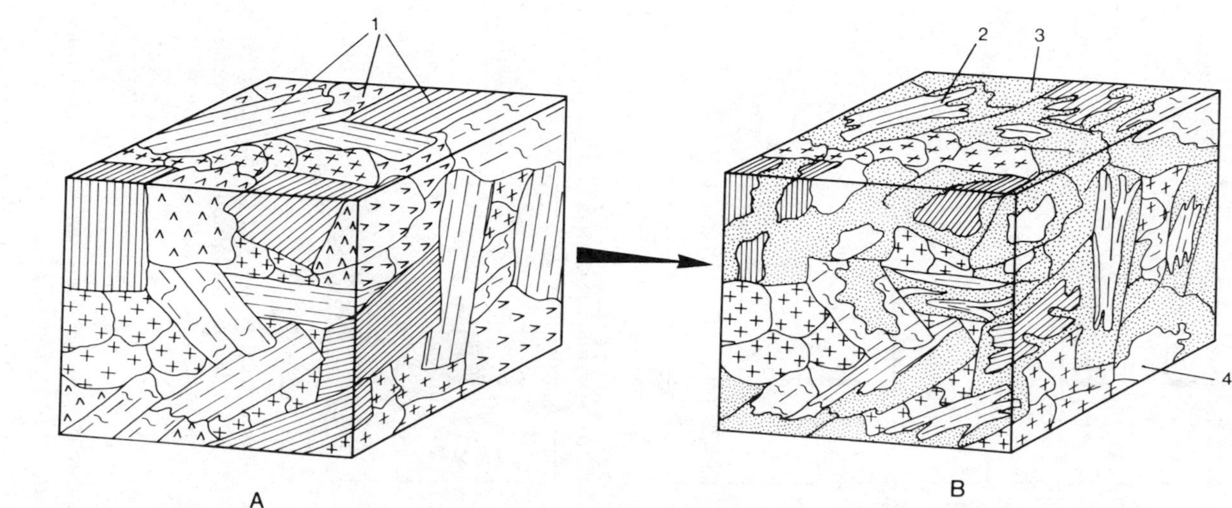

Figure 2.24. Sketch of Isovolume Weathering of a Rock. **A**: Fresh rock. (1) unweathered parent minerals. **B**: Weathered rock. (2) skeletal grains (relicts of parent minerals); (3) weathering plasma (alteroplasma) plus weathering porosity; (2 and 3) weathering s-matrix; (4) voids.

accumulation is essentially regulated by *element-accumulating nano- and microtransfers* (the residual phase) and by *element-subtracting macrotransfers* (the exported phase).

Until now, physicochemical data on weathering processes have been defined only for geochemical macrosystems. Petrologic analyses show geochemical processes, active at the scale of nano- and microsystems, with unknown physicochemical characteristics regulating them. However, petrology reveals the nature and the direction of transfers of elements that operate over very short distances. This method shows, for instance, nanotransfers of aluminum not predicted by geochemistry because this branch of science studies only equilibria with solutions at the level of macro- and megasystems. Geometric analysis of shapes and figures of weathering by means of petrographic techniques therefore affords precise data on the nature and the direction of geochemical reactions, data not provided by other methods.

CHAPTER 3

STRUCTURAL TRANSFORMATIONS OF PEDOTURBATION

In *alterites* (also called saprolites), original structures of the parent rock may be preserved during pseudomorphosis of parent minerals by the resulting weathering products (weathering plasmas called alteroplasmas). This preservation of structure is due to a framework of variable rigidity generated during the initial stages of weathering. The framework may be provided either by the crystalliplasmas themselves (septa of oxyhydroxides) or, when of granular nature, by parent relicts (skeleton grains) within the argilliplasma. The most important mineralogical and geochemical modifications of the parent rock occur during this initial phase of weathering characterized by preservation of original structures and volumes.

However, in the upper part of these alterites and in soils, materials generated by weathering generally lose their preserved original structures. Important *structural transformations* take place wherein the mineralogical and geochemical composition of weathering products may be largely preserved. All structural changes responsible for the transformation of an alterite with preserved original structure into a pedologic material are designated by the term *pedoturbation*.

Pedoturbation may appear at different positions in the upper part of weathering profiles and develop either gradually or abruptly at the expense of the structures preserved by the alterite. These structures are temporarily preserved when weathering products consist essentially of crystalliplasmas of metallic oxyhydroxides. Weathering is complete in certain arenaceous crusts, or partial when only fragments of weathered rock (lithorelicts) or isolated pseudomorphosed parent minerals (crystallorelicts) remain. But, in general, preserved structures of alterites are destroyed when *argilliplasmas predominate*. The argilliplasmas very rapidly develop new relationships with parent relicts (skeleton grains), which undergo a *redistribution*. Laboratory experiments show that these relationships are partially determined by the nature of the argilliplasmas. During this structural reorganization, which leads from an alterite to pedological horizons, transfers and accumulations of material as well as various biological actions (bioturbation) may also take place. Only the role of *argilliplasmas* in pedoturbation shall be examined in the discussion that follows. This role can be visualized through examples of petrographic analysis of natural cases as well as by means of laboratory experiments.

Microscopic studies reveal several aspects of *structural reorganizations* in weathered rocks. In one case, the rate of transitions between alterite with preserved original structure and pedoturbated material varies, in another case *new relationships* occur between weathering plasmas and skeleton grains. Each one of these transitions or new structures may be explained by the crystallochemical inheritance of weathering plasmation, or by the hydric regime or history that affected alterites and soils, or by the topographic position of the horizons and the resulting *illuviations*, and so forth. It is often the association of two or several of these factors that determines, at a variable rate, vertically or horizontally, the *new petrographic structures of pedologic horizons*.

The first analysis below of a few natural examples is of the main transition types between alterite and pedological horizons as well as of the resulting petrographic structures. Next is a discussion of which mineralogical, textural, hydric, or even organic criteria are responsible for one given structure rather than another. This problem is approached by using data from laboratory experiments.

MICROSTRUCTURAL TRANSFORMATIONS OF ISOVOLUMETRIC ALTERITES

The two examples presented here have been studied at Garango, Burkina Faso, in a tropical environment with contrasting seasons (Boulet, 1974). They pertain to alterites and pedologic horizons developed along a toposequence (catena of soils arranged along the topographic slope) over migmatites. Two types of evolution of the parent rock have been recognized in this toposequence. One leads rather gradually toward horizons with vertisolic characters, while the other leads abruptly to a ferralitic horizon.

Evolution toward Petrographic Structures of Vertisolic Horizons

To understand the evolution that leads from alterite with original structure preserved to pedolologic horizons with vertisolic characters, it is necessary to study the transition between horizons C and B in the investigated profiles.

Alterite of Migmatite (Horizon C)

Alterite develops from parent migmatite at a depth of about 2 m. Characterized by *petrographic organizations with preserved original structure*, which are called here *isostructures* or isalteric structures. In the example studied by Boulet (1974), these organizations are defined as follows.

Relicts of Parent Minerals (Skeleton Grains)

Relicts of primary minerals are either intact or weakly fractured in the case of quartz, whereas those of feldspar and hornblende are fragmented but *not displaced*.

Weathering Plasmas

Argillaceous plasmas generated by weathering in place of parent minerals are either peripheric for potassic feldspar, or patchy for plagioclase. These smectitic and kaolinitic argilliplasmas display under the petrographic microscope an asepic structure (Brewer, 1964); that is, under crossed nicols, they show a punctuated extinction. This situation indicates that the argillaceous crystallites forming these argilliplasmas have no *general relative orientation*. This lack of general orientation in argillaceous plasmas under the petrographic microscope is characteristic of weathering argilliplasmas (alteroplasmas).

Voids

Voids visible under the petrographic microscope are relatively rare because argilliplasmas have a tendency to generate complete pseudomorphoses. Only fissures preexisting in migmatite and distention voids are observed.

In summary, these *isalteric structures* correspond to the *weathering s-matrix* (Boulet, 1974), which is parent relicts plus weathering plasmas plus voids (Fig. 3.1 A).

Transition between Alterite (or Saprolite) and Pedoturbated Horizons

The transition between alterite and pedoturbated horizons corresponds to the BC horizon of the investigated example. This transition takes place over a thickness of about 0.50 m (between 2 m and 1.5 m depth).

Parent relicts (skeleton grains) are *still not displaced*. However, in this particular environment, relicts of minerals that are weakly reactive toward weathering, such as quartz, increase relatively in number and essentially preserve their original size. More reactive relicts, such as feldspar and hornblende, decrease in size and number.

Argilliplasmas are always *asepic*, but their proportion with respect to primary minerals has greatly increased. The boundaries of the original crystals from which they developed are no longer visible under the petrographic microscope. At this stage these argillaceous weathering plasmas build a *continuous phase* in which the undisplaced skeleton grains are engulfed. The association of this residual skeleton and of the weathering argilliplasma belongs to the porphyroskelic type (Brewer, 1964).

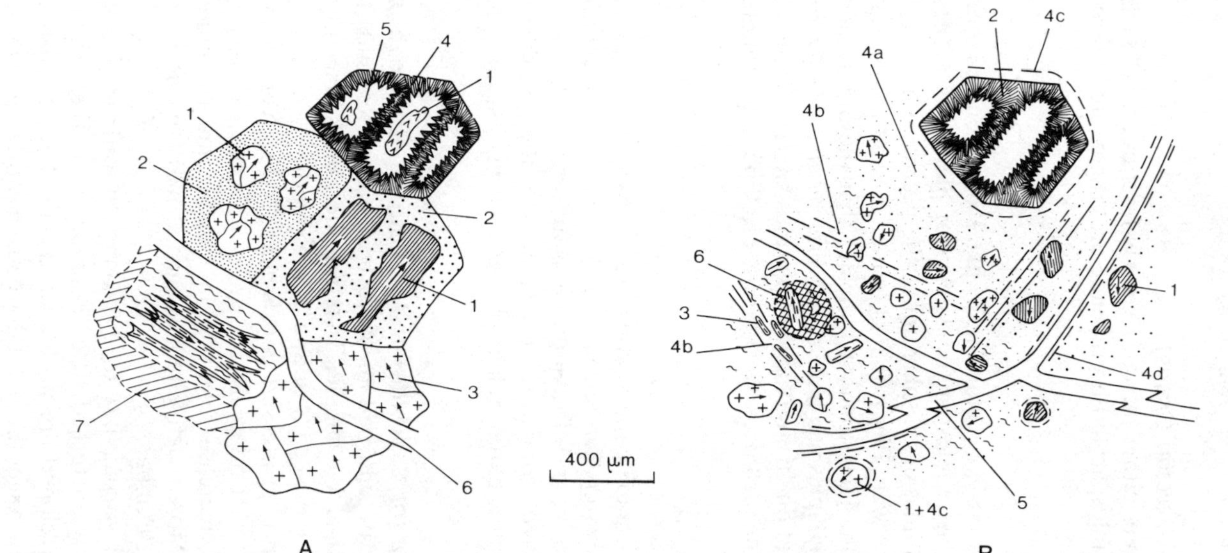

Figure 3.1. Transformation of an Alterite with Preserved Original Structure (A) into a Pedoturbated Petrographic Organization (B) in a Soil of Vertic Type. **A**: Isalteric petrographic structure: (1) Parent relict with preserved original crystallographic orientation; (2) argillaceous weathering plasma of complete pseudomorphic character; (3) fissured parent mineral; (4) crystalliplasma (oxides) of peripheral pseudomorphic character; (5) contact void; (6) original fissure; (7) ferruginous diffusion halos around weathered ferromagnesian mineral. **B**: Alloteric (pedoturbated) petrographic structure: (1) Displaced parent relict (skeleton grain); (2) crystallorelict: structure of original crystal preserved by septa of crystalliplasma (oxides); (3) very small parent relict integrated into argillaceous plasma; (4a) argillaceous pedoplasma (mixed weathering argillaceous phases); (4b) masepic orientation of pedoplasma (plasmic separations); (4c) skelsepic orientation of pedoplasma; (4d) vosepic orientation of pedoplasma; (5) planar fissure (pedologic void); (6) diffused ferruginous nodule (concentration of oxides liberated by-weathering at the same time as the argillaceous plasma).

Pedoturbated Horizon of Vertisolic Type

In the lower and middle parts of horizon B (between 1.5 m and 0.5 m depth), the association of skeleton grains and argillaceous plasmic phase remains of the porphyroskelic type, although the latter phase appears even more developed than in the underlying BC transition horizon. As a whole, the argillaceous phase remains asepic and displays *sinuous cracks*.

Along the margins of these transplasma cracks, parent relicts (the remaining quartz and feldspar) are fragmented and dissociated into a finer skeleton. The displacement or rotation of these grains of finer skeleton can be followed by observing, between crossed nicols, the extinction position of each fragment with respect to the crystallographic orientation of the original mineral, or simply by observing, when parent minerals (feldspar) allow it, the offsetting of twin planes. Furthermore, in this case the argillaceous plasma displays oriented domains ("plasmic separations" according to Brewer, 1964) located either along margins of voids when the argilliplasma takes on a *vosepic* orientation, or along margins of skeleton grains when the plasma presents a *skelsepic* orientation.

Finally, in the upper part of horizon B (between 0.5 m and 0.20 m depth), the fine skeleton is dispersed not only along the margins of cracks, but also within the argillaceous plasma. The latter displays an increasing number of oriented domains. The *plasmic structure* is said to be of the *voskel-masepic type*; that is, the domains where the plasma is oriented are located around voids (prefix *vo*), around skeleton grains (prefix *skel*), and elongated across the plasma (prefix *ma*).

It is obvious that the *gradual increase of argilliplasmas* generated by the continuous weathering of parent minerals is responsible for the disorder that appears and develops in the relationships between plasma, skeleton, and voids.

Plasmas acquire orientations at first limited to structural discontinuities (contact between plasma and relicts or between plasma and voids), then within themselves.

Skeleton grains are affected by the movements undergone by argillaceous plasmas at the same time as their chemical weathering continues. As a result, grains undergo fragmentation and physical dissociation into a finer skeleton as well as continuous dissolution. But in this case, the chemical weathering is essentially peripheral and centripetal. The final effect is an increase in the amount of argillaceous plasma, since weathering is incongruent, as well as an increase in the processes of pedoturbation; lastly, the weaker relicts are erased.

Movements of the argillaceous plasma tend to eliminate the voids that appeared during weathering plasmation. Voids generated by pedoturbation appear and develop. They depend essentially on the nature and composition of clay minerals.

In fact, a new *pedoturbated s-matrix* occurs. Weathering argilliplasmas mix, engulfing the fine skeleton, and thus change into a *pedoplasma*. Indeed, pedoplasmation, previously defined by Flach et al. (1968), results from constraints caused by movements of swelling and contraction of argillaceous phases. These phases are related, as shown by Boulet (1974), to the frequency and amplitude of wetting and drying cycles affecting the vertisolic profile (Fig. 3.1).

The transition to the new pedoturbated s-matrix is schematically represented in Fig. 3.1 (A and B). The different orientation of arrows for parent relicts of same nature indicates their displacement.

Evolution toward Ferralitic Petrographic Structures

The second example studied by Boulet (1974) shows the transition of the migmatite parent rock to pedoturbated horizons by means of a weathering cortex about 1 cm thick.

Migmatite Alterite

Two main stages can be distinguished in the weathering of migmatite. At first, plagioclase undergoes a fingerlike weathering plasmation with generation of an amorphous silico-aluminous phase. At this stage, the other parent minerals show no modification under the petrographic microscope and the rock remains coherent. A second stage appears rapidly in which the fingerlike weathering is replaced by a peripheral weathering that isolates feldspar relicts within a kaolinitic *asepic plasma*. Quartz undergoes a simple dissolution whereas biotite is ferruginized. All these parent relicts retain their original orientation within the plasma (Fig. 3.2 A), whereas fissural and dissolution voids are partially or totally filled by zoned deposits of argillo-ferruginous plasma (argillo-ferruginous coatings, also called *ferriargillans*). This argillo-ferruginous plasma is structurally discordant on the walls of the fissures and voids and also may be mineralogically different from the minerals forming these walls. These ferriargillans, therefore, result from a late and absolute accumulation of particles of minerals produced by microtransfers from upper horizons; these ferriargillans are called *illuviation plasmas*. Thus, without any displacement of the residual skeleton, illuviation plasmas are added to weathering plasmas, forming what Boulet (1974) has called an "*altero-illuvial structure*." This structure characterizes the macroscopically observed centimetric weathering cortex.

Ferralitic Pedoturbated Alterite

Overlying the weathering cortex, there develops a horizon, called "reticulated horizon (Bq)" by Boulet 1974). It consists of the juxtaposition of two very distinct s-matrixes: one of weathering corresponding to the previously described altero-illuvial structure, the other of pedoturbation consisting of an asepic, discolored, argillaceous plasma containing debris of *argilloferrans* (called *papules*), and an abundant and dispersed fine skeleton (Fig. 3.2 B). This pedoturbated s-matrix clearly intersects the altero-illuvial structures.

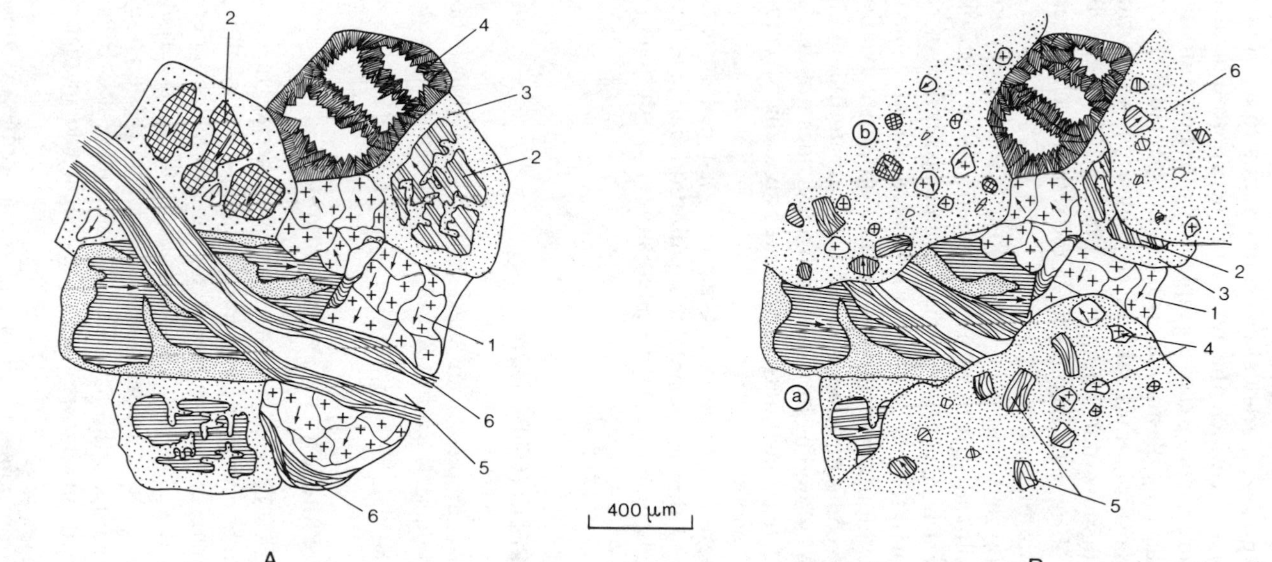

Figure 3.2. Transformation in a Soil of Ferralitic Type of an Altero-Illuvial Structure of the Weathering Cortex into a Partially Pedoturbated Petrographic Structure (Reticulated Horizon). **A**: Weathering cortex: (1) Parent mineral cracked and undergoing simple dissolution; (2) parent relict with preserved original crystallographic orientation; (3) argillaceous weathering plasma of complete pseudomorphic character; (4) crystalliplasma (oxides) of peripheral pseudomorphic character; (5) fissural and dissolution voids; (6) ferriargillans. **B**: Reticulated horizon: *Zone (a)* relict alteroilluvial structure: (1) parent mineral cracked and undergoing simple dissolution; (2) parent relict with preserved original crystallographic orientation; (3) pseudomorphic argillaceous weathering plasma. *Zone (b)* pedoturbated s-matrix: (4) displaced parent relict (skeleton grain); (5) papules (displaced ferriargillan fragments); (6) argillaceous pedoplasma.

Furthermore, the occurrence of displaced grains of a fine skeleton of the same nature as the parent relicts of the weathering s-martrix as well as the occurrence of displaced fragments of ferriargillans clearly indicate that the pedoturbated s-matrix results from pedoturbation of altero-illuvial structures.

Finally, higher up in the profile, a gradual transition occurs to an "upper red horizon," 1 to 2 m thick, in which the relicts of the altero-illuvial s-matrix with preserved structure are progressively replaced by the pedoturbated s-matrix with asepic red argillaceous plasma and with a fine dispersed skeleton.

Here, microstructural transformations take place only after weathering plasmation and argillo-ferruginous illuviation have combined their effects. Transformations are at first abrupt and incomplete in the reticulated horizon, then gradual and complete in the red horizon.

They eventually generate an s-matrix in which both fine skeleton and relicts (papules) of altero-illuvial structures are scattered. As in vertisolic horizons, pedoturbation, which achieves dissociation and dispersion of residual mineral relicts, is related to an *increase of argillaceous plasma content.* But in this case, the association of weathering plasmation and illuviation induces an increase of clay content. However, the pedoturbated kaolinitic plasma, discolored and then colored by iron oxyhydroxides, does not display oriented domains resulting from constraints; it preserves an asepic structure with the exception of relict papules.

In another example, presented later (see Chapter 5), a pedoturbated ferralitic s-matrix is followed through simple reorganization between oxyhydroxides and clay minerals by a microaggregated structure, very characteristic of ferralites (called oxisols).

CHARACTERISTICS OF PEDOTURBATION

The two above-mentioned examples from West Africa have shown how weathering argillaceous products were restructured into a pedoturbated s-matrix.

Increasing Weathering Plasmation and Pedoturbation

The amount of clay minerals in a rock undergoing weathering increases through progression of weathering itself, and also through influx of illuvial products from higher horizons. Whenever the amount of clay minerals increases with respect to the skeleton, this increase leads to a greater sensitivity to alternating periods of humidity and dryness and to a specific different behavior depending upon the nature of the clay minerals. Hence, the plasma undergoes movements and restructurings begin and increase in intensity.

During restructuring, parent mineral relicts undergo a double evolution. Those that are alterable continue their weathering into clay minerals, no longer in patches or digitations, but at their periphery, along the margins of joints or desiccation voids. They continue to feed the plasma, thus accelerating restructuring. Furthermore, the new weathering clay minerals may be different from the earlier ones if the environmental conditions become increasingly open and leaching.

Meanwhile, debris of parent minerals become smaller and begin their displacement. Quartz debris, biotite lamellae undergoing weathering, and increasingly rare feldspar fragments decrease in size, are mixed, and disperse within the plasma, which has become pedoplasma (Boulet, 1974). Larger skeleton grains (most commonly quartz) are affected by translations or rotations under the successive pressures of this pedoplasma, which orients itself along its margins in a skelsepic structure (Lafeber, 1962; Brewer, 1964; Blokhuis et al., 1970).

In this manner, pedoturbation is accomplished by means of continuous weathering and internal movements.

Pedoturbation and Isovolume

Swelling and contraction movements of the argillaceous phase have the following effects.

- to mix the respective weathering plasmas of each parent mineral phase;
- to associate the finer skeleton fragments and the pedoplasma with a common orientation;
- to displace the larger skeleton grains;
- to erase weathering porosity and replace it by pedoplasmation porosity.

These effects, as shown in the above-described examples from West Africa, are very clear under the petrographic microscope. Thus, in the same way as *isalterite* (weathering with preserved original structures) was defined in the previous chapter, pedoturbated horizons are designated as *alloterite*, since synchronously with intense restructurations, chemical weathering of parent relicts proceeds without *interruption*. However, use of the terms isalterite and alloterite should be in each particular case related to the scale of observation. Pedoturbated domains develop gradually until complete disappearance of isalterite. The isovolume concept, as defined by Millot and Bonifas (1955) for isalterite, appears inapplicable to alloterite if microstructural observations alone are considered. However, in certain cases, and particularly in some ferralites, the new pedoturbated structures, even if they clearly characterize microstructures (scale of the microscope) and even macrostructures (scale of the horizon), remain locked at the scale of the profile or of the sequence within a network of undisplaced original fractures and veins, a situation that, at this megascale of observation, demonstrates the conservation of volumes. In an appropriate example from the weathered Birrimian basement in the Ivory Coast, north of Abidjan, weathered schists gradually lose their original structure upward in the profile. In the pedoturbated horizons (mottled clays), the original schistose structure has disappeared, but undisplaced quartz veins clearly indicate that megavolumes are nevertheless preserved (Fig. 3–3). In other cases, however, and particularly when weathering argillaceous plasma consists of a swelling population, micro-, macro-, and even megastructures are disturbed. At the microscopic scale, the oriented domains of the plasma become more frequent with the increasing vertic character of the clay minerals forming it (Blokhuis et al., 1970). At the macroscopic scale, the transition between horizon C and B is more abrupt, and pedoturbation by swelling and contraction is expressed here by means of a prismatic structure of variable width. Finally, at the scale of the profile or the sequence, pedoturbation allows incorporation of solid materials between vertically and even laterally juxtaposed horizons (for instance, sands and pebbles originating from horizons A), thus disorganizing original fractures and veins. The differentiation of certain smectitic weathering covers in tropical countries may be characterized in this manner.

Pedoturbation of Nonargillaceous Horizons

This approach concerns essentially pedoturbation by gradual increase of argillaceous plasmas. Pedoturbation can be related to two other causes: namely, compaction and disturbance by fauna and flora (bioturbation), both of which shall not be discussed here. These processes either add their effects to those of argilliplasmation, or, on the contrary, constitute the essential portion of pedoturbation.

Dealing only with the effects of argilliplasmation, it is possible to understand in greater detail how plasmic structures of pedoturbation are generated. This requires examination of how clay minerals forming weathering plasmas and later pedoplasmas react to wetting and drying processes.

PLASMIC STRUCTURES OF WEATHERING AND PEDOTURBATION

The two examples discussed above and the major features of pedoturbation show that swelling and contraction movements of clays generated by weathering and possibly enriched by illuviation are responsible for restructuring which are continuously repeated by variations in the hydric state of the argillaceous plasma. It is therefore critical to understand the organization of clay minerals and their relationship with water. This study treats three levels of organization:

first-order ultrastructures that reveal the behavior of clay minerals from sheets to particles and associations of particles;

second-order microscopic structures within argillaceous plasmas;

third-order macroscopic structures of peds in soils.

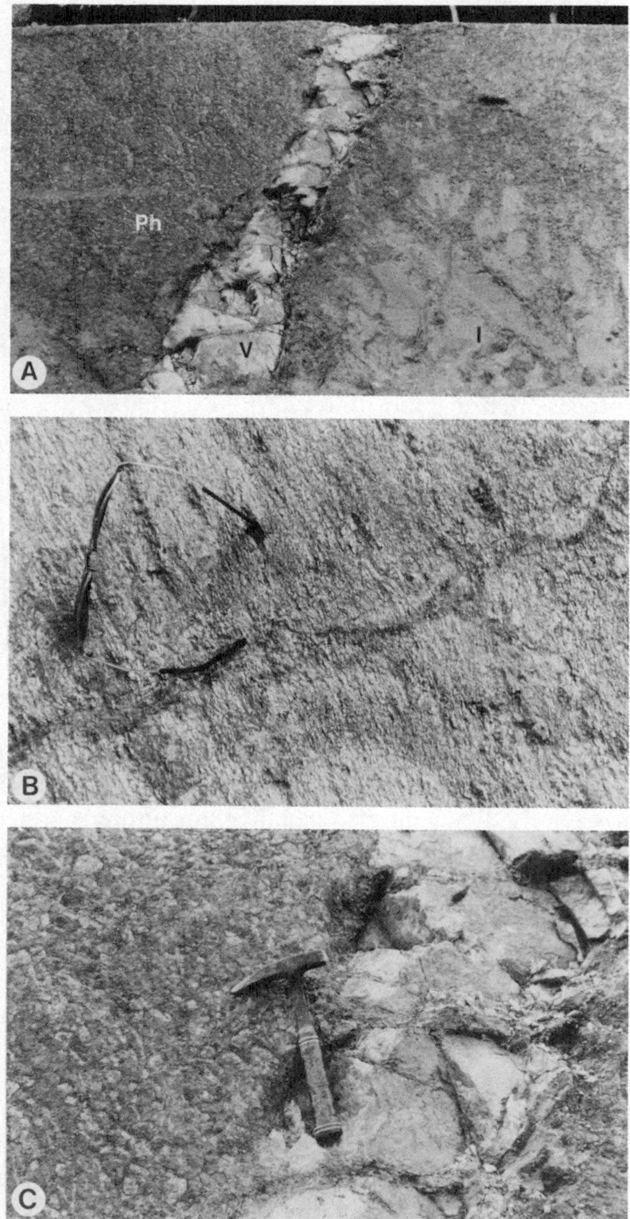

Figure 3.3. Preserved Megavolumes in Weathering Profiles Developed on Birrimian Schists in North of Abidjan (Ivory Coast). **A**: Photograph showing undisplaced vein (v) crossing a pedoturbated mottled clay horizon (ph). Relict of isalterite with preserved original schist structure occur on the right of the photograph (i). **B**: Photograph showing detail of the preserved schist structure of isalterite. **C**: Photograph showing detail of the pedoturbated structure.

First-Order Ultrastructures: Sheets, Layers, Crystallites, Particles and Association of Particles

Introduction

The organization of the argillaceous constituents of soils has been investigated by numerous mineralogists and physicochemists; (Aylmore and Quirk, 1960, 1962; Mering, 1962; Van Olphen, 1963, 1977; Quirk, 1968; Tessier, 1978a; Pédro, 1979). Only the most recent works (Tessier, 1978a; Pédro 1979, 1983; Pédro and Tessier, 1983; and Tessier, 1984) are used here to explore the smallest levels of organization of argillaceous phases. Indeed, these authors have shown that swelling of argillaceous material (property of clay minerals to change variably in volume in presence of water) is due to the introduction of a film of water at the surface of the *particles*, an exclusively interparticle process that depends only on the external surface of particles, their surface charge, and compensating ions present in the double layer.

Layers and Crystallites

Pauling's principle of charges says that any substitution at the level of a sheet is compensated by another substitution in another type of sheet, or by the addition of cations, hydrated or not, in interlayer positions. The degree of lateral extent of elementary layers and their mode of stacking are closely related to those isomorphic substitutions.

Lateral Extent of Layers

Ion substitutions that occur in tetrahedral and octahedral sheets necessarily generate lattice deformations. These deformations are distributed either regularly or irregularly, depending on whether substitutions are complete (1 or 2) and respect stoichiometry or incomplete (< 1). The elementary layer, therefore, keeps a certain crystallochemical homogeneity or conversely displays a certain heterogeneity. In the first case, the bidimensional network of the layer may display a lateral extent leading to macrocrystals (talc, serpentine, certain kaolinites). In the second case, irregular deformations in layer structure limit its bidimensional extent; phyllites become individualized into microcrystals tens of microns in size (smectite, most illites, certain kaolinites). Physicists say that these layers have a size of the order of a tenth of a micron.

Thickness Extent: Stacking Types of Layers

Whenever isomorphic substitutions are absent or complete, that is respecting the stoichiometry of the elementary layer (substitutions in the tetrahedral sheet entirely compensated by substitutions in the octahedral sheet, or by energetically fixed dry cations), interlayer spaces are not hydrated. In such a case, in addition to the bidimensional network development of the layer by lateral extent, as mentioned above, a development occurs in the third dimension in regular order along the C axis, with superposition of a great number of layers to generate large crystals and even macrocrystals (talc, serpentine, large kaolinite).

Whenever substitution remains incomplete, the charge of the layers may be smaller than 1 and must be compensated by the addition of a hydrated interlayer. The presence of these water molecules bonded to the cations ($Na(H_2O)_x$; $Mg(H_2O)_y$; $Ca(H_2O)_2$. . .) allows the latter to exchange more easily because they are less strongly bonded. The weaker the layer charge, the stronger the interlayer hydration.

This is the reason for imperfect stacking of layers. Disordered stacking may be of *translational type* (therefore limited) when the substitution is tetrahedral (vermiculites, beidellites) or of *turbostratic type* (therefore more accentuated) when the original substitution is octahedral (montmorillonite). Consequently, the disorder of stacking limits the number of stacked layers: The thickness of crystallites reaches tens of Å and rarely surpasses 100 Å (Figs. 3.4 and 3.5).

Crystallites

Vertical stacking of layers is a function of their charge, of the location of this charge in tetrahedral or octahedral sites, and of the presence or absence of compensating ions. Any regular stacking order leads to the generation of crystalline units of a size greater that 100 Å. Any disordered stacking, therefore, limits the number of stacked layers, in which case the thickness of crystalline units reaches tens of Å and rarely surpasses 100 Å (Fig. 3.5). Crystalline units resulting from the vertical stacking of layers are called crystallites.

Argillaceous Particles

Particles consisting of Crystallites

When stacking along the c axis involves electrically neutral layers, the resulting crystallites display very weakly charged or zero charged basal faces, which cannot become juxtaposed to each other in the same direction

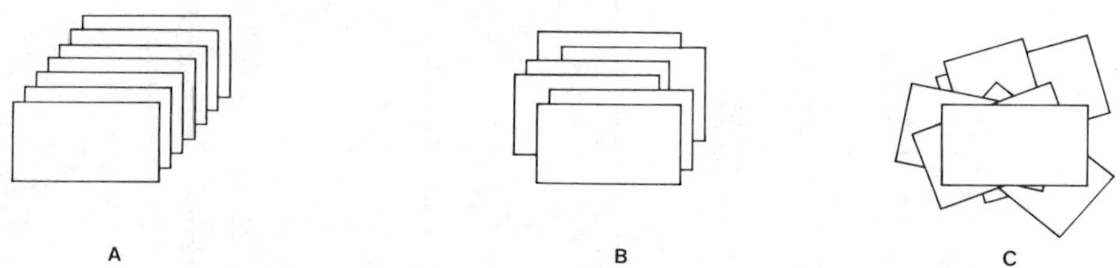

Figure 3.4. Main Types of Stacking of Layers. **A**: Regular order of stacking. **B**: Disordered stacking of translational type. **C**: Disordered stacking of turbostratic type. (From Pédro, 1979.)

Type of clay minerals	Elementary layer		Crystalline individuals	Real entities (= reactives particles)	Type of spatial arrangement of entities within materials
	Charge	Thickness	Elementary crystallites	Polycrystalline assemblages	
KAOLINITES	none	e = 7 Å	Crystallites $Ø < 2\ \mu m$ e ~ 150 - 200 Å ~ 20 to 30 layers		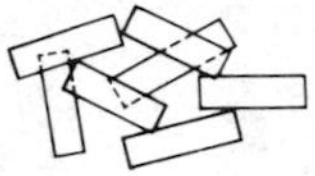 simple juxtaposition
SMECTITES (Ca)	Weak charge (z < 0.6)	e = 14 Å	Crystallites $Ø \sim 0.05\ \mu m$ e < 100 Å (3) ~ less than 10 layers	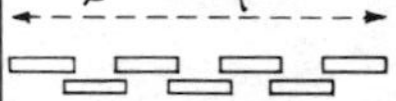 Tactoïds $Ø \sim 1 - 2\ \mu m$ e ~ 100 Å	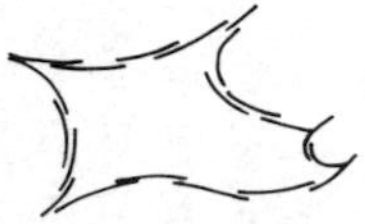 deformable network
ILLITES	Higher charge (z < 1)	e = 10 Å	Crystallites $Ø \sim 0.1\ \mu m$ e ~ 50 80 Å ~ 5 to 8 layers	Microdomains $Ø \sim 0.5\ \mu m$ 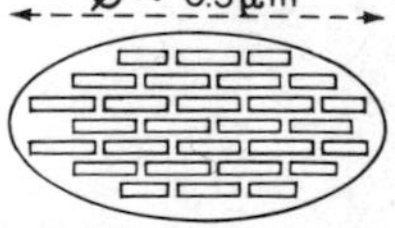	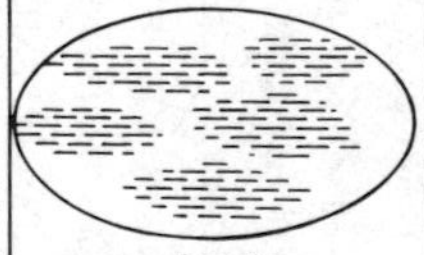 domain

Figure 3.5. Nature and Dimensions of the Various Characteristic Entities of the Major Argillaceous Materials of Soil ($\varsigma < 2\ \mu m$) (modified from Pédro, 1979).

to form larger units. In such a case, the crystallite which displays a crystallographically coherent particle is at the same time an independant and real unit (Pédro, 1979). The crystalline individual or crystallite is, therefore, the real constituent particle of the material. Such is the case of kaolinite.

Polycrystalline Assemblages

When stacked layers display a charge, the resulting crystallites may become juxtaposed on each other by means of their external basal faces and generate *polycrystalline assemblages*.

It is clear that the type of polycrystalline assemblage obtained is a function of the charges displayed by the basal faces of the crystallites. Thus, overlap of the external basal faces of two crystallites shall be better developed with an increase of their charges because bonds capable of joining them are stronger. Therefore, illite, vermiculite, and glauconite with higher basal charge display a more complete overlap and generate more compact polycrystalline assemblages (7–8 crystallites) in *weakly deformable "lenses"* called *microdomains* by Tessier and Quirk (1979). (Fig. 3.5 and Fig. 3.7). Smectites with weak basal charge form crystallites that tend to show weak overlaps, hence develop polycrystalline units of great width (1–5 μm) and of *very deformable plane type* (veils) called *tactoids* or *quasicrystals* (Quirk, 1978) (Fig. 3.6).

Summary

This section deals with the real entities of the argillaceous system (Pedro, 1979), which may also be designated as *argillaceous particles*.

In kaolinite, the basal charges of crystallites are essentially nonexisting. Therefore kaolinite crystallites behave as independent particles in the argillaceous system. Particles of illite and other micaceous clay minerals consist of polycrystalline assemblages in microdomains. Smectite particles consist of polycrystalline assemblages in tactoids.

The three great families of clay minerals occur, therefore, in materials as real figured shapes or particles of different nature. The next section, dealing with the effect of wetting and drying, shows how these real entities or particles react to alternating wetting and drying. The association or assemblage of these particles, which generate microporosity, the receptive structure for water, is examined next.

Association or Arrangement of Particles

Arrangement of Particles consisting of Crystallites

In pure kaolinitic plasmas, kaolinite crystallites are juxtaposed as a random pile of bricks without formation of polycrystalline assemblages. Hence the spatial juxtaposition of crystallites is random: margin–face, margin–margin, or even face–face.

The structure of kaolinite crystallites is easily dispersible. It provides

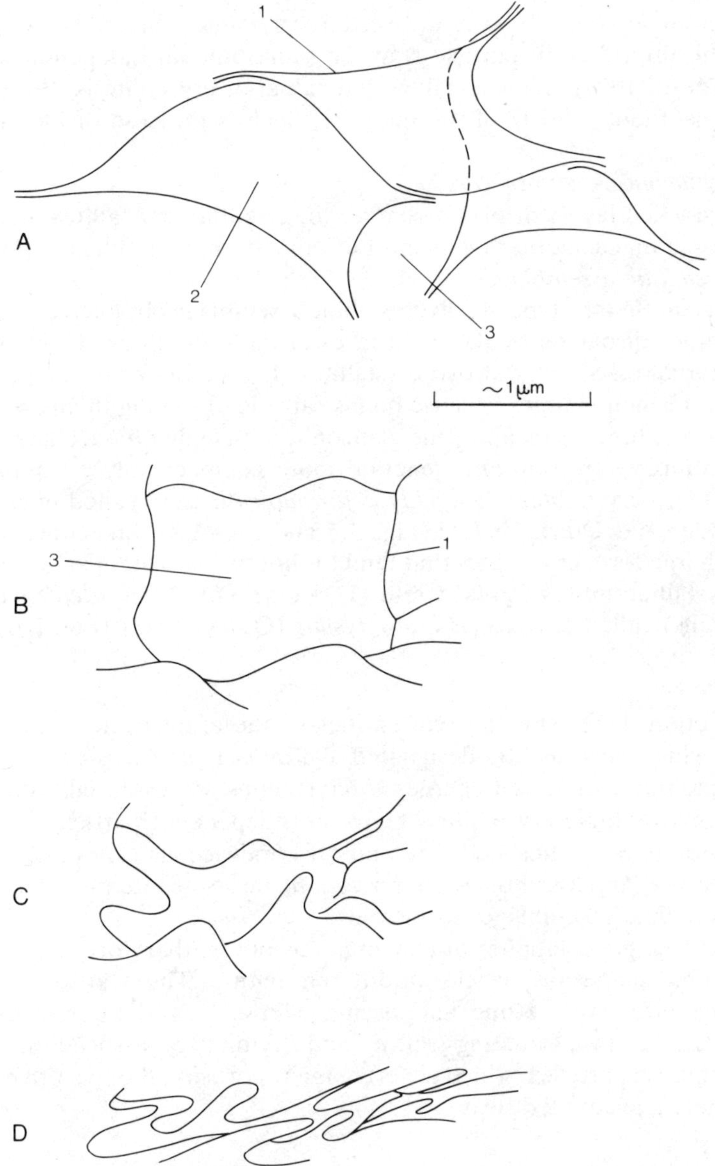

Figure 3.6. Arrangement and Deformation of Tactoid Network with Increasing Drying (modified from Tessier and Pédro, 1976). **A**: Three-dimensional arrangement of tactoids into a network: (1) tactoids; (2) tactoid veil ($\varsigma \sim 1$ μm); (3) void created by tactoid network. **B**: Plane view of tactoid network. **C** and **D**: Progressive deformation of tactoid network through dehydration. (Reproduced by permission of Science du Sol.)

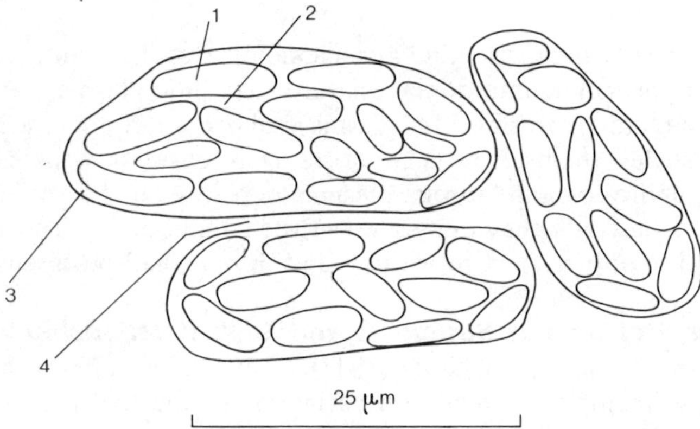

Figure 3.7. Association of Microdomains of Micaceous Clay Minerals into Domains (modified from Tessier and Pédro, 1976). (1) Microdomain (1000 Å to 0.5 μm); (2) intermicrodomain discontinuity (from 500 Å to 100 Å; (3) domain (thickness ~ 15 μm, length ~ 25 μm); (4) interdomain discontinuity = intrinsic discontinuity (< 1 μm).

numerous fine but very irregularly distributed pores. These micropores display grossly pyramidal shapes, 50–200 Å in size, at the junction of crystallites, or elongated shapes, 20–50 Å in size, between two crystallites (Bresson, 1981). These elongated micropores with parallel walls represent, in fact, the most important pore volume of these kaolinitic plasmas (Kehres, 1983; Didier, 1983; Tardy and Nahon, 1985).

Arrangement of Microdomains into Domains

Microdomains may be stacked into relatively compact packages, 15–25 μm thick, weakly deformable, and separated from one another by planar voids, called domains (Quirk, 1978). Such is the case of illitic plasmas (Fig. 3.7). Microporosity is controlled by the relationship between microdomains and domains. Thus pore size varies between 100 Å and the micron order (Bresson, 1981; Tessier, 1984). These pores are numerous, but it seems as if discontinuities between domains are the most accessible.

Arrangement of Tactoids into Networks

Tactoids are arranged into a honeycombed, three-dimensional, deformable anastomosed network. Such is the case of smectitic plasmas. The works of Tessier (1984) show the following: Contact between tactoids always occurs tangentially face to face at an angle close to 30°; alveolar porosity is more regularly distributed and limited by the walls of this network; pore size varies between 0.5 μm and 2 μm, depending upon the Na, K, or Ca nature of the smectitic plasma.

Summary

Microporosity of intercrystallite arrangements (for kaolinitic plasmas) of intra- or, in particular, interdomain nature (for illitic plasma), and of intra-network character (for smectitic plasma), allow water to reach the argillaceous particles themselves. Therefore, it is easy to visualize that the wetting capability of argillaceous plasmas depends on the size of particles and the size of voids they generate between themselves. The reaction of these particles with respect to wetting and drying shall be discussed next.

First-Order Argillaceous Structures and Their Relationship with Water

Following the works by Richards (1947), Van Olphen (1962), and Lietard (1977), experiments were made on variations of the hydric state to which argillaceous materials in soils are submitted, and a method was developed to observe materials at different stages of humidity (Tessier, 1978b; Tessier and Quirk, 1979; Tessier and Berrier, 1979). The latter stages correspond to water retained in the soil because it is submitted to tensions. The energy with which water is retained represents the one that should be applied to extract it from the soil; it is expressed by suction (h). Since Schofield (1935), the concept of pF representing the decimal logarithm of suction is generally used. Variations of the states of wetting and drying (corresponding to respective variations of suction from $pF1$ to $pF6$) affect the fundamental organization of argillaceous materials. Moreover, for certain particles, the exchangeable interlayer ion and the concentration of interstitial solutions play an important role in this organization.

Studies have been undertaken on several types of clays.

Crystallites: Kaolinite

It was previously shown that in the absence of any bonding agent, crystalline individuals (crystallites) of kaolinite play the role of fundamental particles reactive with water. The distance between the three-dimensional arrangement of these particles depends only on water content corresponding to a given pF (Tessier, 1984). The rheologic properties of kaolinite depend, therefore, on the size of particles (Lietard, 1977). Such a size is itself related to the disorder in the stacking of layers, and to the presence of trace elements in the lattice of the mineral (Lietard, 1977; Pédro, 1979). Swelling is variable, but of little importance, and represents water located between crystallites without the latter changing their stacking of layers.

Observations on soils that consist essentially of kaolinite (soils of the ferralitic domain) show very important in the reactivity to wetting. Furthermore, certain kaolinitic plasmas display domains ("microaggregates" in the sense of Chauvel, 1976) comparable to domains of micaceous clays, but of much larger size ($\sim 100\ \mu m$) and with a behavior toward water reduced in regard to crystalline individuals of pure kaolinite. In these cases, microaggregates react as elementary particles and water films penetrate into the discontinuities and spaces $< 1\ \mu m$. Ferralitic soils with predominant kaolinite (monosiallites in the sense of Pédro, 1968) have, therefore, a particular physicochemical behavior due to the role of ferric hydrates, which when combining with kaolinite at the submicroscopic or microscopic level, make the kaolinite particularly inactive to wetting. In this situation, compounds associated with kaolinite modify its physicochemical behavior (see Chapter 5). The effects on the evolution of soils in tropical countries, on their drainage, and, finally, on the agriculture of these areas are of great importance.

Domains: Micaceous Clays (Illites)

Swelling of an illitic type material (Fig 3.7) occurs between the discontinuities of 0.5–1 μm, which form the limits of domains (Aylmore and Quirk, 1971; Tessier, 1984), or between packages of crystallites, namely, microdomains (Tessier and Quirk, 1979). Maximum drying (or limit of contraction) is reached for this type of material at *pF* 4.5. Drying has a fundamental effect on the illitic material when particles tend to be oriented face to face and to increase the size of microdomains or domains, thus decreasing water content during subsequent rewetting.

Exchangeable cations and concentration of interstitial solutions here have only a reduced effect on the variation of water content because for illites, in contrast to smectites, only external surfaces are in reactional contact with the aqueous phase.

Networks of Tactoids: Smectites

In clays of smectite type (Fig. 3.6), variations of water content (hydric constraints) act at different scales. Indeed, as shown above, smectites are clays with a weakly charged layer ($Z < 0.6$); their hydration increases while the charge weakens. Such is the case of water bonded to layers and to compensating cations. The first differences of behavior of the various smectites toward variations of water content occur, therefore, at the interlayer space level. When considering the arrangement of layers into crystallites, and of crystallites into tactoids, and taking into account interlayer distances, Tessier (1984), following the works of Norrish (1954), Van Olphen (1962), Quirk (1968), and Pons (1980), showed the additional existence of a double layer of water, called *diffused layer*, in particular in Na smectites. Consequently, hydric constraints act not only on layers, but also on crystallites consisting of several layers, and on tactoids consisting of the overlap of several crystallites. This is the *amount of intratactoid water* (Tessier, 1984). But it is the structure of the tactoid network that allows stocking of the greatest amount of *intertactoid* water, which acts directly on the walls of the pores controlled by the tactoid network.

Hydric constraints are, therefore, acting in turn at different scales. This is characteristic of smectites. Indeed, an increase of hydric constraint leads to a decrease in water content, and for a given constraint the water content varies as follows.

1. As a function of the nature of the exchangeable interlayer ion. One notices a greater "crumpling" of the tactoid network in Na-saturated clay minerals, or a more rectilinear network with Ca clay minerals. In these minerals, the walls of the network are thicker, leading to an appreciable reduction of water content.

2. As a function of the concentration of interstitial solution. With increasing concentration, there occurs a simple face-to-face rearrangement of the packages of layers, also leading to a thickening of walls of the tactoid network and, hence, to a diminution of water content.

Finally, in drying and rewetting cycles, Tessier (1984) showed that certain irreversibilities may exist in the organization of tactoids. Thus, if drying is carried to high values close to that of contraction ($pF = 6$), true face-to-face bonding between layers occurs and, hence, a thickening of tactoid walls, which becomes irreversible upon rewetting up to pF 1.5.

Summary

Clearly, hydration of argillaceous materials is essentially interparticle for clay minerals of the kaolinite type (intercrystallites) and for those of the illite type (interdomain). Hydration, however, is more complex, being simultaneously intralayer, intraparticle, and interparticles for clay minerals of the smectite type. Furthermore, the very nature of smectites directly influences variations of water content.

Variations of water content related to phases of wetting and drying are, therefore, more spectacular for smectites than for illites and kaolinites. Thus, the organization of a smectitic material depends to a large extent on its hydric history and, particularly, on the maximum drying to which it has been subjected during its pedologic history.

Second-Order Microstructures: Juxtaposed Crystallites, Domains, and Networks of Tactoids

Introduction

The first-order organization of clay minerals discussed above pertains to layers, crystallites, particles, and associations of particles (particles simply juxtaposed, associated into domains, or in tactoid networks). All these objects are very small and can be investigated only by X-rays or electron microscopy: Transmission Electron Microscope (TEM) or SEM. The thickness of layers is of the order of tens of Å, and their extent is measured by fractions of microns. Crystallites have a diameter measured in tenths of microns or in microns and a thickness measured in tens of Å. Kaolinite crystallites alone reach the order of a micron. Particles have the size of a few microns (0.5–1.0 μm for microdomains, and 1–2 μm for tactoids). Finally, associations of particles that do not include kaolinite because they are only crystallites reach sizes of microns or tens of microns (1–2 μm for the lattice of tactoid networks, 15–25 μm for domains). In summary, the observation level ranges from tens of Å to the order of a micron.

Second-order structures are studied by polarizing petrographic microscope, in plane polarized light or under crossed nicols. The scale is being changed and observation no longer pertains to crystallites (except large-scale ones) or their associations, but to argillaceous plasmas and their structure or texture. Objects and structures easily observable under the microscope range in size from a few microns (5–10) to a few millimeters. Thereafter, a hand lens is used and then the naked eye in third-order structures.

A comparison of the first-order ultrastructures and second-order microstructures shows first that they are in continuity, namely, that the smallest objects or structures observable under the microscope are of the same order of magnitude as the largest organizations of the first order. It is therefore possible to observe them. Furthermore, the first order ranges from 10 Å to the order of a micron and the second order from the order of a micron to that of a millimeter. The jump in the possibilities of observation is therefore of the order of 1000. This means that a series of objects and structures smaller than a micron cannot be observed under the microscope. Conversely, a whole new field is open from sizes of a micron or of tens of microns to sizes of the order of a millimeter.

This is the field of micromorphology whose bases have been clearly formulated and codified by Brewer (1964) and have been used extensively since. They were provided 10 to 20 years before the works by Quirk (1978), Tessier (1978a, 1984), and Pédro and Tessier (1983). Today, distinctions can be made at the frontiers between the first and the second order, which pertain to the effect of ultrastructures on microstructures. The information provided by micromorphology on the structure of argillaceous plasmas is briefly examined below.

Major Types of Microstructures of Argillaceous Plasmas

The major types of microscopic structures or argillaceous plasmas in soils have been defined as "plasmic structure" (Brewer, 1960, 1964). It is necessary to describe the size, shape, and arrangement of the grains forming the plasma as well as the voids associated with such an arrangement. Brewer (1964) described observations under crossed nicols, which distinguished:

1. crystallites or plasma particles whenever possible;
2. type and degree of orientation of plasma particles;
3. type and degree of preferred orientation of "domains" which might exist, and particularly of "plasmic separations."

Brewer (1964, p. 309) proposed five major groups of plasmic structure according to their type of anisotropy and extinction under crossed nicols, namely (1) asepic, (2) sepic, (3) undulic, (4) isotic, and (5) crystic plasmic assemblages. Certain groups of asepic or sepic plasmic structures may be subdivided, particularly at the level of sepic structures with striated extinction, on the basis of the characteristics of plasmic separations. Brewer (1964) defined the following:

Insepic structures when striated extinction involves only isolated plasmic separations;

Mosepic structures when plasmic separations with striated extinction form a continuous or discontinuous mosaic;

Vosepic structures when plasmic separations with striated extinction line the margins of voids;

Skelsepic structures when plasmic separations with striated extinction immediately surround skeleton grains.

Masepic structures when, independently of voids and skeleton grains, plasmic separations display within the plasma striated extinction according to one, two (bimasepic or lattisepic), three (trimasepic), or several (omnisepic) preferred orientations, whether continuous or not.

Although this approach applied to the description of thin sections of soils is useful, it has not been related to the fundamental structures (first order) displayed by each major family of clay minerals. This difficulty has two aspects: Soil plasmas do not always consist of an homogeneous family of clay minerals, and plasmas are associated with other constituents (skeleton grains, oxides, and pedologic features) that disturb the behavior toward hydric conditions.

The following section is an attempt at relating first- and second-order structures. The question is whether or not the arrangements and associations of plasmic constituents recognizable in the first order are still identifiable under the polarizing microscope. Only cases showing a simple relationship at the boundary of the observation scales are presented.

Relationship between First-Order and Second-Order Plasmic Structures

In order to establish a scale relationship between first- and second-order structures, each recognized entity (juxtaposed crystallites, tactoid networks, domains) must be studied in regard to their type of association, hence their aspect under the polarizing microscope (Fig. 3.8 and Table 3.1).

Juxtaposed Crystallites

The example presented here is that of kaolinite. Size and type of association of crystallites define, under the polarizing microscope, the type of plasmic structures as follows.

First are crystallites of the order of a micron, with weak lateral and vertical development, randomly arranged. Although each individual crystallite is anisotropic, under crossed nicols no general anisotropy is revealed because

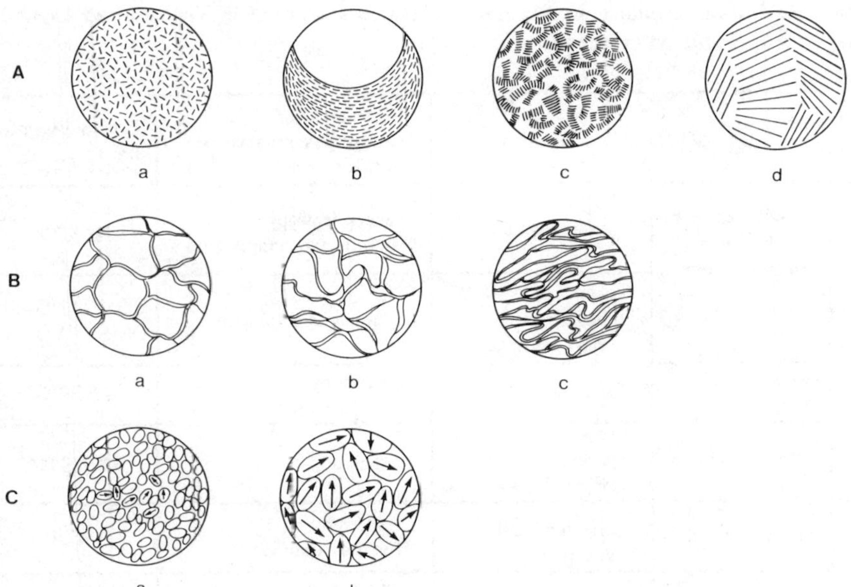

Figure 3.8. Types of Association of Argillaceous Plasmic Constituents: Second-Order Structures. **A**: Assemblage by simple juxtaposition of crystallites: (a) compensation isotropy; (b) strong and continuous extinction; (c) fleckled extinction; (d) crystic assemblage. **B**: Association of tactoids in three-dimensional deformable network. **C**: Association of microdomains and domains.

of the small size of crystallites and their random reciprocal arrangement. On the contrary, a weak anisotropy or even an isotropy occurs which is a sort of isotropy of compensation. This is the definition of an *isotic plasmic structure* (a in Fig. 3.8 A).

Second are crystallites of the order of a micron but whose c axes, perpendicular to which the stacking of layers occurs, display the same orientation. All crystallites have similar orientations and display under crossed nicols positions of equal illumination or equal extinction. Extinction figures are said to be strong and continuous (b in Fig. 3.8 A).

Third are crystallites of the order of one to several microns. Each crystallite is sufficiently developed to constitute entities observable under the petrographic microscope, but without allowing to distinguish packages of layers within them. These crystallites are tangled in all directions without general orientation. Under crossed nicols, they display a *flecked extinction*. The assemblage (or plasmic structure) is called *asepic* (c in Fig. 3.8 A).

Fourth are crystallites of the order of several tens of microns or more, namely with lateral and vertical developments of sheets sufficient to allow the optical observation of packages of layers (vermicular kaolinite).

Table 3.1. Major Argillaceous Plasmic Structures Observed in Transmitted Light under Petrographic Microscope.

Real entities = reactive particles		Reciprocal arrangement of particles	Behavior under petrographic microscope (nicols crossed)	Type of plasmic structure
CRYSTALLITES	small sizes not visible under microscope	disorganized	general isotropy = isotropy and compensation	ISOTIC
		organized : stacking of layers along c axis	continuous extinction	CONTINUOUS ORIENTATION
	small visible sizes	disorganized	anisotropy : flecked extinction	ASEPIC
	crystallites large	organized : packages of tangled layers	anisotropy	CRYSTIC
TACTOIDS	tactoid network	oriented in all directions	anisotropy : flecked extinction	ASEPIC
	collapsed tactoid network	oriented along non-dislocated planes	anisotropy : continuous extinction	SEPIC
		oriented along dislocated planes	anisotropy : striated extinction	
DOMAINS	microdomains of non-visible small size	disorganized	weak anisotropy or isotropy of compensation	ISOTIC
	microdomains of size visible under microscope	disorganized	anisotropy : flecked extinction	ASEPIC
		oriented	anisotropy : striated extinction limited to each domain	SEPIC
AMORPHOUS glomerules, filaments	poorly crystallized domains adjacent to amorphous one	disorganized	true isotropy with weak local anisotropy : weakly undulose extinction	UNDULIC
	completely amorphous particles	disorganized	true isotropy - total extinction	ISOTROPIC
OPAQUES microglaebules	opaques particles or crystals covering argillaceous particles	organized or disorganized	opaques = apparent isotropy	ISOTIC

Finally, the arrangement of crystallites corresponds to a mosaic of tangled vermicules of kaolinite where each particle is clearly recognizable. The assemblage is called *crystic* (d in Fig. 3.8 A).

Microdomains or Associated Domains

The fundamental structure of microdomains has been shown above to be an arrangement of anisotropic argillaceous crystallites with the same orientation.

How are microdomains and domains associated among themselves in order to lead to a plasmic structure observable under the polarizing microscope? Two cases may occur. Either microdomains are of very small size and the anisotropy displayed by all their crystallites is not optically visible, or microdomains and domains have observable size and display clear orientations under crossed nicols. The juxtaposition of microdomains and domains occurs without their orientation being in a preferred direction. In the first case, the small size of the entities displays only a very weak anisotropy or even a false isotropy or isotropy of compensation under the microscope. The plasmic structure is called *isotic* (a in Fig. 3.8 C). In the second case, the anisotropy of each domain is visible, but their juxtaposition in all directions gives a *flecked extinction* between crossed nicols. The plasmic structure is called *asepic* (b in Fig. 3.8 C).

Networks of Tactoids

Tactoids have been shown above to be arranged either in a network continuous in all directions, or in a network flattened along one or several preferred directions. In the first case, the network outlines a high porosity and each tactoid wall is oriented in a different direction, implying a lack of general orientation. The latter is recognizable under the polarizing microscope. Each wall is anisotropic in a given direction. Between crossed nicols, the extinction is *flecked*. In this case, the plasmic structure is also called *asepic* (a in Fig. 3.8 B).

When the network subjected to hydric constraints is flattened along one or several preferred orientations, the orientations of the walls of the tactoid network are very visible between crossed nicols. Either a *strong and continuous extinction* (c in Fig. 3.8 B) occurs if tactoids are oriented along nondislocated planes, or a *striated extinction* (b in Fig. 3.8 B) when these planes are discontinuous because of imperfect flattening or dislocation. Whether extinction is strong and continuous or striated, the anisotropy of the argillaceous crystallites forming tactoids is clearly observable under the polarizing microscope. The plasmic structure is called *sepic*.

In argillaceous plasmas, second-order structures can evidently be identical for very different first-order structures. In the case of scale relationship between first- and second-order structures, only argillaceous plasmas of the same nature have been considered for reasons of simplification. In nature, conditions can be more complex, as shown by argillaceous plasmas with different kinds of minerals, amorphous or poorly crystallized plasmas, plasmas rich in iron oxyhydroxides, nonargillaceous plasmas, and so forth. In the face of this complexity, only a few examples are presented of how some of these plasmas behave and are recognizable under the polarizing microscope.

Association of Amorphous or Weakly Crystallized Plasmic Constituents

Soils and alterites frequently display amorphous or poorly crystallized constituents whose crystallo-chemical disorder can be analyzed by X-ray diffraction and infrared spectrometry. When such constituents represent an important fraction of the weathering plasma, they generate a particular plasmic structure. These plasmas present very weakly anisotropic and isotropic domains, which, by their juxtaposition, show a *weakly undulose extinction* between crossed nicols and under high magnification. Brewer (1964) thus defined an *undulic plasmic structure*. When plasmic constituents are really amorphous, no distinction is possible within the plasma at any magnification, small or high. This is true isotropy (Table 3.1).

Association of Opaque Plasmic Constituents

Certain mineral or nonmineral (organic matter) compounds may conceal the anisotropy of the argillaceous plasmic constituents with which they are associated. Under these conditions, clay minerals can no longer be distinguished. If mineralized, these compounds (oxyhydroxides of iron or of manganese) may be anisotropic. But their small size and opacity very often prevent any observation of birefringence between crossed nicols. In all cases, plasmic constituents present, an apparent isotropy under the polarizing microscope. The plasmic structure is said to be *isotic* (Table 3.1).

Observation in reflected polarized light is necessary to establish the nature and degree of association of these opaque constituents.

Association of Crystallized and Isotropic Plasmic Constituents

Some plasmas may consist of isotropic minerals. Such is the case, for instance, in some soil horizons in arid countries where sodium chloride (halite) occurs as a mosaic of cubic crystals which between crossed nicols are at extinction. This is true isotropy.

Effects of Drying and Wetting on Microscopic Argillaceous Structures

Introduction

Experiments performed by Tessier (1984) clearly showed the influence of the size of a given sample on the water content of the argillaceous material, starting with pF 2.5. Water contents decrease with increasing size of samples. Thus, it is understandable that specific organizations exist at each scale.

As mentioned above, argillaceous plasmas are essentially characterized by asepic or sepic plasmic structures at the scale of the polarizing microscope. According to the arrangement of plasmic separations, these structures may be divided into several substructures. The generation of these different types of plasmic structures was studied by Tessier (1984). They are controlled by variations of the hydric constraints that affect the argillaceous plasma.

The types of distribution of second-order plasmic structures show that at the millimeter scale, constraints act in a differential manner within the material. The reaction of the argillaceous material during drying and then during wetting is examined next.

Effects of Drying on Plasmic Structures

Drying generates contraction cracks that occur during the reduction of apparent volume, that is, under the effect of tractions. This fissuring corresponds, therefore, to a porosity clearly observable under the petrographic microscope. It is directly induced by the reorganization of argillaceous particles at the first-order scale. As seen above, the process leads to a decrease of interparticle porosity, which reaches its peak during the maximum drying (pF 6) undergone by the argillaceous material. This condition corresponds to the void index "e," which expresses the minimum pore volume reached by each type of clay mineral. The void index during drying for kaolinite and illite, for instance, is of the order of 1 and lower for smectites ($e = 0.15$ for Na smectites and 0.08 for Ca smectites). This reduction of pore volume can therefore vary by a factor of 1 to 100 when kaolinite or Ca smectite are considered. Obviously, this property is directly related to the very structure of argillaceous particles. The more argillaceous particles contain expandable layers and the more these layers can adjust themselves through sliding during drying, the lower the void index is.

Clearly fissuring initiated and developed in argillaceous plasmas is induced by the evolution of interparticle porosity. This fissuring is better developed when porosity is reduced .

In nature, argillaceous plasmas of soils contain rigid skeleton grains (silts and sands) at the scale of the petrographic microscope, and for a given argillaceous plasma, the size of crystallites or of argillaceous particles may vary. The works of Fies (1978), Grimaldi (1981), and Tessier (1984) show that for a given suction (a given pF), the coarser pores located between skeleton grains, at contacts between skeleton grains and argillaceous plasma, or simply between rigid argillaceous particles (kaolinite crystallites of tens of microns in size or domains in the case of illite) become partially empty, while the finer pores, located at the level of the arrangements of layers continue to close. Such a situation increases traction and leads to preferential fracturing where pores became empty first, namely, at the level of the coarser pores. This is the reason contraction cracks develop for a given argillaceous material at the boundaries between skeleton grains and plasma, or for an heterogeneous argillaceous material where crystallites or particles are coarser or the most rigid. These conditions have been frequently observed in nature.

Effects of Wetting on Plasmic Structures

During wetting, argillaceous plasmas swell and change from an initial volume V_1 acquired during drying to a final volume V_2 so that $V_2 > V_1$. Therefore, the intensity of swelling of clay minerals depends on the previous state of drying to which the material has been submitted. If P_1 and P_2 are considered as the two constraints corresponding, respectively, to volume V_1 and V_2, $P_1 < P_2$. Furthermore, if Δe is the variation of the void index reached between these two values of the constraint, and if e is the initial void index, swelling can be expressed by the ratio $\Delta e/1 + e$ (Tessier, 1984). This ratio indicates that nonexpandable argillaceous material show little swelling. Thus, for a pF variation of 6 to 1, swelling expressed by the ratio $\Delta e/1 + e$ shows the following values:

For kaolinitic plasmas and according to the size of crystallites, the ratio varies approximately on the order of 0.15 to 0.3;
For illitic plasmas, the order of magnitude is the same as for kaolinite, namely, apparent volume increases to a maximum of 30%;
For smectitic plasmas, the ratio varies from about 1.30 to 20 according to the Ca or Na nature of the clay mineral.

The amount of smectite in the argillaceous material of soils therefore plays an important role in the swelling capacity of the plasma. Thus, for a plasma of pure illite to which 25% smectite is added, the ratio $\Delta e/1 + e$ rises from 0.20 to 0.46 for a pF variation of 6 to 1 (Camara, 1982; Tessier, 1984).

The constraints that develop inside argillaceous material during swelling are compressional. They are responsible for the plasmic separations with striated orientations of sepic plasmic structures (Rode et al., 1960; Brewer, 1964). Furthermore, shearings may occur at the level of such plasmic separations and may outline microaggregates still clearly adjacent to each other. Orientation and frequency of plasmic separations and of shearing planes (Tessier, 1984) depend on the nature of a given material. In other words, they depend on the behavior of crystallites and argillaceous particles that have a tendency to separate from each other. In this manner, first-order plasmic reorganizations affect reorganizations of second-order structures.

Kaolinitic plasmas with large rigid crystallites (a few tens of microns in size or more) do not display any shearing planes, regardless of the hydration constraint, whereas plasmic separations may occur in kaolinitic plasmas with small crystallites (on the order of a micron in size) for advanced stages of hydration.

Illitic plasmas display plasmic separations and shearing planes only when swelling is sufficient. The angles between these shearing planes are acute (about 30°).

Finally, smectitic plasmas may show different behaviors. In general, smectites tend to generate numerous plasmic separations because of the configuration of their particles. But, whereas Ca smectites display ruptures (shearing planes) with angles ranging from 45° to 90°, depending on the amount of hydration, Na smectites, regardless of the hydration constraint, display neither discontinuities nor rupture. This situation can be explained by the fact that certain particles, once hydrated, can easily slide one over the other. In other words, plasticity in the arrangement of the particles allows to avoid rupture observed in more rigid particles. This concept is similar to the conclusions by Baver (1948), who mentioned the greater plasticity of Na clay minerals versus Ca ones.

In summary, the number of plasmic separations with striated extinctions and rupture discontinuities (shearing planes) observable at the scale of the petrographic microscope increases during wetting. These plasmic structures are related to the intensity of swelling, itself induced by the reorganization capacity of crystallites and particles in contact with the aqueous phase. Thus, kaolinitic and illitic plasmas undergo limited reorganizations with respect to smectitic plasmas. With increasing rigidity of the particles, the angle between shearing planes, if they occur, and the direction of maximum orientation of particles becomes more acute. With increasing plasticity of the particles, such an angle becomes larger. However, if plasticity is too high, these ruptures do not occur (Tessier, 1984).

Most of the sepic plasmic structures with striated extinctions result from this process. This is the case with masepic structures, which may precede shearing planes (Sleeman, 1963; Brewer, 1964), vosepic structures, and skelsepic structures (Lafeber, 1962).

However, masepic structures, and even in part insepic structures, could, according to Brewer (1964), be inherited from altered sedimentary parent rocks. But this is contradicted by the works of Boulet (1974), which showed that insepic structures may develop in soil material. Finally, undulic and isotic structures are peculiar because they are related to the effects of opaque on oxyhydroxides or of organic compounds, which are associated with clays (Brewer, 1964); or they are simply related to the presence of crystallites or particles either amorphous or too small in size and in a disorganized arrangement to lead to structures observable under the petrographic microscope.

Summary

Plasmic structures observable under the petrographic microscope (second-order microstructures) and whose basic tabulation was made by Brewer (1964) result from several causes:

1. *Primary arrangements of crystallites and particles little or not reorganized by phases of drying and wetting*. These are certain types of first-order ultrastructures observable under the petrographic microscope. They characterize weathering plasmas of the lower part of profiles or argillaceous plasmas consisting of rigid particles, weakly reactive to pedoturbation. Asepic, crystic, and certian isotic and undulic structures belong to this group.

2. *Arrangements of crystallites and particles resulting essentially from wetting phases*. These structures characterize pedoturbation plasmas, and their features, as seen under the petrographic microscope, are a function of the amplitude of hydration constraints, themselves related to the nature of argillaceous constituents, hence to lower-order structures (first order). Sepic structures with plasmic separations and striated orientations (vosepic, skelsepic, masepic) belong to this group. However, some of these structures precede shearing ruptures leading to planar discontinuities.

3. *Arrangements of crystallites and particles associated with opaque oxy-hydroxides or with organic compounds that conceal the observation of any argillaceous plasmic structure*. Some of the isotic and undulic structures belong to this category.

Finally, drying leads mainly to contraction cracks whose amplitude is shown by macroscopic structures (third order). Contact reactions between crystallites and argillaceous particles on one side and aqueous phases on the other side are responsible for tensions (compressions and tractions) that generate:

- mixture of weathering phases;
- reorientation of plasma constituents (plasmic separation);
- appearance of fissural porosity (contraction cracks, shearing cracks).

In other nonargillaceous constituents, the following features are also generated:

- clearer separation of oxyhydroxides;
- disappearance of the more alterable skeleton grains;
- pulverization of the more resistant skeleton grains;
- redistribution of skeleton grains and integration of finer grains into the plasma.

In summary, for argillaceous plasmic constituents, first-order structures affect those of second order, either by inheritance, if they remain sufficiently rigid or little affected by hydric constraints, or by reorganizations, depending on the intensity of wetting. But the dynamics of these reorganizations is itself related to the behavior of the ultrastructures, that is, crystallites and particles in contact with the aqueous phase.

Third-Order Macroscopic Structures: S-matrix and Peds

Introduction

Field observations have shown that a certain number of macroscopic structures are capable of characterizing argillaceous soils. Pedologists have given the name of "peds" (U.S.D.A., 1951) or of aggregates (Glossaire de Pédologie, 1969) to these structural units. These three-dimensional units are easily separated from each other by less resistant surfaces.

Kubiena (1938), Brewer and Sleeman (1960), Sleeman (1963), and Brewer (1964) showed that these peds can be associated into several levels of organization, each of them with a characteristic structure. Thus, the ped or elementary aggregate (primary ped of Brewer, 1964) is itself arranged within peds or aggregates of larger size (secondary and tertiary peds of Brewer,

1964), which themselves are limited by fissures, cracks, or surfaces of least resistance including those of the groups of peds of smaller size or those of elementary peds. The larger-sized aggregates determine, by their arrangement, the soil horizon. It is wellknown that a soil consists of the superposition of several horizons.

Consequently, a complete sequence of related structures of argillaceous plasma exists at the macroscopic level. These structures are organized not only from bottom to top of the soil (horizons) but also within a given horizon from larger-size aggregates to elementary aggregates which are the smallest structure recognizable under the naked eye (third-order macrostructure). The history of a soil, and particularly its hydric history, plays an important role in the generation of the above-mentioned hierarchy of different levels of structure.

Finally, it is possible to observe at the boundary between recognizable second-order microscopic plasmic structures and third-order macroscopic plasmic ones the effects of the former on the latter. However, it is appropriate to present briefly the views of pedologist on macroscopic structures of soils.

Macroscopic Structures of Argillaceous Plasmas

The structure of argillaceous plasmas are defined by the peds or elementary aggregates and those of larger size that associate several of these elementary units. These structures are usually characterized by the measurement of the volume of argillaceous plasma at different levels of organization. These volumes are limited by faces (natural surfaces) and are defined by their size, shape, and arrangement.

An exhaustive description of the different structures and of the methods of measurement of the aggregates is not given here; only a short summary is presented of the pedologists' knowledge since the works of Kubiena (1938), Butler (1955), Sleeman (1963), Brewer (1964) and Duchaufour (1970). Several types of structures are defined to characterize the different units of plasmic organizations in aggregates. They are as follows:

1. *Polyhedral structures with peds limited by plane faces (angular volumes).* These structures can be subdivided according to whether the ped volume is equidimensional or whether one or the other of their dimensions is greater. The subdivisions are:
   polyhedric structures (shape of regular polygon);
   prismatic structures (shape of columnar prism with rounded upper face);
   cubic structures (equal faces);
   platy or lamellar structures (when two dimensions of the three-dimensional volume are greater than the third).
2. *Spheral structures with peds limited by curved faces (rounded volumes).* In this case, the subdivision is based on the shapes of the principal sections of the aggregates as follows:

polyspheral (granular) structures if the shape is irregular but rounded;
spheroidal structures if all dimensions are almost equal and faces correspond actually to the same curved surface;
lenticular structures if the shape is that of an ellipsoid;
polysphedral structures, characteristic of peds with mixed curved and plane faces.

The latter type of structure leads to the assumption that in nature some structures may be rather difficult to describe because they are of intermediate nature between several types. Therefore, many authors have had the tendency to increase the number of subtypes on the basis of the variation of dimensions in the three spatial directions, of the curvature and flattening of the faces, or even of the shape of angles between adjacent faces. In macroscopic descriptions of these structures, it is often difficult to attribute the shape of peds to a specific type. Some descriptions therefore mention a tendency toward common shapes such as polyhedric to cubic aggregates.

"Pedality," as defined by Brewer (1964), consists of all the levels of structure resulting from the individualization of elementary aggregates and from their arrangements in larger-size aggregates, namely, all the shapes and sizes of the different peds and their arrangement. The arrangement of adjacent peds for each level of structure may be defined by the degree to which faces are molds of each other (or accommodation), by the orderly or random arrangement of volumes (or packing), or by the orientation in space with reference to the vertical or horizontal line (or inclination).

Relationship between Second-Order and Third-Order Plasmic Structures

The study by Boulet (1974) of soils developed on granitic parent rocks of the region of Garango (Burkina Faso) showed very clearly the correlations that may exist from bottom to top of a profile between plasmic microstructures (second order) and plasmic macrostructures (third order). Thus, no individualization of aggregates corresponds to the second-order asepic structures of the base (C horizon); the third-order structure is said to be massive.

Upward in the profile, and in the BC and B horizon, sepic structures appear and develop. In the BC horizon, plasmic separations of skelsepic type occur simultaneously with macrostructures described as massive with prismatic tendency. At the base of the B horizon, skelsepic, vosepic, and masepic structures gradually predominate while macrostructures become clearly prismatic with smooth faces, and eventually at the top of the B horizon macrostructures are cubic in prismatic assemblage.

This example demonstrates a correlation between second- and third-order plasmic structures; namely, plasmic separations visible under the petrographic microscope increase with the accentuation of aggregate structure visible with the naked eye.

The above study of second-order plasmic structure showed that during wetting, orientation and frequency of plasmic separations increase and that shearing planes whose inclination depends on the type of argillaceous material finally occur. These rupture planes have sizes from several microns to several millimeters. When well developed, they can be observed macroscopically. Furthermore, during drying, contraction cracks may characterize second-order plasmic structures; their size may be in millimeters or in centimeters.

Therefore, the effects of second-order structures on third-order ones is obvious. Cracks generated by wetting (shearing) and by drying (contraction) eventually isolate plasmic assemblages that are aggregates.

It is known and confirmed by the example given by Boulet (1974) that the macroscopic structure of argillaceous soils increases higher up in a profile. Consequently, hydric constraints to which the argillaceous material is submitted vary with depth. It is appropriate at this stage to understand how these constraints act on the argillaceous soil, how they evolve with depth, and, finally, how the argillaceous plasma undergoing these constraints acquires a structure at the macroscopic scale.

Effects of Drying and Wetting on Macroscopic Argillaceous Materials

Pore volume of samples of macroscopic argillaceous material is always higher than that of smaller samples. It is therefore understandable that specific third-order structures are generated when episodes of wetting and drying occur in soils during their history.

However, conditions of rewetting and redrying can vary considerably during such a history and have important consequences on the macrostructure of argillaceous plasma. The works of Tessier (1984) again have to be consulted to understand the effects of these conditions on contraction and swelling.

Effects of Drying on Macroscopic Plasmic Structures

The works by Haines (1923) and Tessier (1975, 1980, and 1984) showed that contraction is subdivided into three successive phases:

> A first phase in which water elimination is accompanied by equal decrease of void volume. Since samples are originally water saturated, they remain the same;
>
> A second phase during which contraction is smaller than water loss but volume continues to decrease slightly, consequently air enters the sample (air point of entry);
>
> Finally, a third phase during which volume remains constant; this situation corresponds to the lower volume limit called contraction limit. Subsequently any loss of water leads no longer to volume decrease. (Tessier, 1984, p. 279).

If Vw represents the volume occupied by water and Vs the solid volume, the values of water index equal Vw/Vs express the contraction reached by argillaceous materials.

Thus, for kaolinitic plasmas with large crystallites, the contraction limit appears for a value of about 1, whereas for kaolinite with small crystallites, this value decreases to 0.75. In illitic plasmas, contraction occurs at a value close to 0.8, whereas in Na smectitic plasmas, the values reached are close to 0.3.

In other words, this situation indicates a greater diminution of void volume for smectitic plasmas than for kaolinitic ones. This is easily explained by the configuration of the finer structures of the argillaceous material. When crystallites or particles are rigid (kaolinite, illite), contraction occurs when the latter come in contact with each other, whereas for smectites, expandable layers allow a more intimate readjustment that becomes better developed when smectite contains an increasing number of expandable layers.

In this case, first-order and second-order structures again affect third-order ones. For a given argillaceous material, tensions generated by the reduction of fine porosity induce, as shown above, contraction cracks visible under the microscope and with the naked eye. Furthermore, this contraction fissuring (and cracks) increases in importance when the argillaceous material reaches the smallest minimum volume. In other words, macroscopic contraction cracks are more abundant in smectitic plasmas than in kaolinitic ones. This situation is frequently displayed in nature.

Therefore, during contraction, the closure of microporosity, characteristic of first-order structures, leads to the development of macroporosity of fissures and cracks. This process, already visible at the level of second-order structures, increases in magnitude and becomes characteristic of third-order ones.

Effects of Wetting on Macroscopic Plasmic Structures

When considering argillaceous plasma as a whole, it is well-known that its reactivity to the aqueous phase depends to a great extent on its position in the soil profile, that is, to its accessibility to wetting. Thus, certain plasma particles can be rewetted more rapidly than others; water circulating in fracture porosity does not have the same properties as the one penetrating smaller pores; and a highly dried material reacts differently from one having undergone a lesser contraction. These three variants of accessibility to wetting are actually interrelated when a soil profile is considered.

Argillaceous plasma is increasingly drier (hence fissured) and cracked as it is nearer to the surface. Furthermore, it gets much more water than at depth, and this water is free in comparison to the one variably bonded to crystallites and argillaceous particles in the mass of the material. All these conditions were experimentally investigated by Tessier (1984) and the major results are briefly described below.

Between two given pF (pF6 to pF1), the effect of wetting on the water content of identical argillaceous materials varies, depending on whether wetting occurs gradually or instantly. In the first case, the argillaceous plasma observed with the naked eye keeps its general coherence, whereas in the second case it breaks up into small parallelepipedic volumes a few millimeters by 1 cm in size.

These differences in the type of wetting are encountered in nature. When argillaceous plasma is near the surface, rewetting is more rapid than for plasma at depth. Indeed, contraction fissuring being greater near the surface, the amount of free water penetrating this macroporosity is also larger. Consequently, argillaceous material in contact with these fissures rehydrates more rapidly, leading to a new fissuring, but this time under constraints generated by wetting.

Clearly the structure of argillaceous material is better developed near the surface than at depth.

Therefore, accessibility of argillaceous material and its degree of macrostructure are functions of several closely related parameters:

initial contraction fissuring, namely the state of drying to which the argillaceous material has been previously submitted;

potential of the water reaching the argillaceous material (free water in macroporosity, variably bonded water inside the mass of the material);

pedostatic pressure undergone by the material, namely vertical constraints (Towner, 1981), and lateral ones due to the weight of the soil material.

Other parameters more directly related to the nature of argillaceous constituents are also active. Their important role played in first-order and second-order structures has been discussed above. Unquestionably, these structures affect those of third order. This situation was noticed by Peterson (1944), who showed the change of shape of aggregates according to the kaolinitic or montmorillonitic nature of clay minerals. Similarly, Wolkewitz (1958) showed that for a given argillaceous material, the type of contraction fissuring varies when considering a Ca or a Na smectite.

It is nevertheless certain that for an argillaceous material consisting of the same type of minerals, the third-order macrostructure varies considerably according to its position in the soil profile. It can easily be predicted that if the nature of crystallites or argillaceous particles plays an important role in second- and third-order structures, the latter depends also to a great extent on other specific parameters and particularly on the type of wetting, namely, the gradient of the water potential affecting the argillaceous material.

Summary

Analysis of a weathering or soil profile indicates that the structures of argillaceous materials, as observed with the naked eye (third-order structure), is increasingly differentiated upward in the profile and for a given material.

It has been shown how, under the effect of hydric variations (drying and wetting), first-order structures of crystallites and argillaceous particles preceded the organization of the argillaceous plasma at the scale of the petrographic microscope (second order). Similarly, ceaselessly repeated oriented readjustments of argillaceous plasma into plasmic separations themselves foreshadow, by the shearings (plane and curved fissures) they produce, the third-order structures. But the gradient of the water potential seems to play a major role in the macroscopic structure of argillaceous materials (Tessier, 1984). This gradient increases as water that reaches the argillaceous material is closer to free water. Because contraction fissuring sets limits to large aggregates and hence occurs in highly dried up materials near the surface, water circulates more abundantly and wetting becomes rapid, leading to greater swelling of argillaceous material and, finally, to a greater fragmentation into finer aggregates.

This statement is confirmed by the observations of Boulet (1974), who, under the petrographic microscope, showed, that rotation and crumbling of skeleton quartz grains under the effect of the strongest hydric constraints occur first along the margins of fissure porosity.

CONCLUSIONS

Pedoplasmation is essentially controlled by the behavior of argillaceous plasma during phases of wetting and drying in the history of soil. The effects of pedoplasmation on the entire spectrum of soil structures are observable only when the argillaceous phase of the s-matrix is sufficiently abundant.

Therefore, it is critical to understand the structure of clay minerals and their relations with the aqueous phase. This organization exists at three main levels: (1) from layers to crystallites or particles and their associations (first-order structure); (2) from associations of particles to plasmic separations, even to microaggregates (second-order structure); (3) from plasmic separations to aggregates of the soils themselves (third-order structure).

The first-order structure is the base of all the others and depends on the nature of the argillaceous constituent. Argillaceous entities or particles really reacting with water depend on the nature of the elementary layer, whether it is charged or not. Thus, for kaolinite, the crystallites themselves play the role of an argillaceous particle. For the other clays, the association of crystallites by overlapping constitutes the basic particle: microdomains or domains for illite, tactoids or tactoid network for smectites.

Crystallites and domains react as rigid particles; most of the reactions with water during phases of wetting and drying take place in intercrystallite and interparticle porosity. Smectites that have expandable layers readjust and relax not only in intertactoid porosity, but also at the level of layers. The constraints thus developed are much stronger in smectitic plasmas than in others and affect all higher-order structures.

Thus, in second-order structures, at the scale of the petrographic microscope, plasmic separations, that is, groups of particles with the same orientation, develop during wetting phases. These separations are more abundant and more complex within plasmas where constraints are stronger, namely, in smectitic plasmas. Ruptures by shearing occur in relation to these plasma separations and result from compressions generated by wetting. Drying acts in a tensional manner on the argillaceous phase and leads to contraction fissuring.

In third-order structures, increase and association of ruptures by shearing (plane or curved fissures) during wetting with contraction fissures lead to the individualization of pedoturbated units of s-matrix that are visible with the naked eye and are easily separated from one another: the aggregates. Although this is a clear case of the effect of second-order structures, and hence of first-order structures, on the generation of third-order ones, the latter preserve specific features at their scale. Indeed, it is the quality (water potential gradient) and the quantity of water that allow increasing reorganizations along the macroporosity of fissures and cracks. This explains the better developed structure in aggregates of argillaceous soils near the surface than at depth. Furthermore, aggregates may be subdivided according to this type of wetting into smaller ones, which are often erroneously called elementary aggregates.

Thus, the argillaceous plasma displays different levels of structures expressed by the shape and size of crystallites and particles and of macroscopic aggregates in alterite and soils. These different levels were generated by constraints developed by the reaction of this plasma, at different scales, to wetting and drying.

If these plasma structures remain moderate, particularly in kaolinitic alterite and soils, they allow a preservation of structures in the parent rock at the scale of the horizon, although often structures are disturbed at the microscopic and ultramicroscopic scales.

However, plasmic reorganizations in smectitic alterites and soils are early, intense, and rapid. They quickly destroy structures of the original parent rock preserved at the scale of the horizon and replace them by an aggregate structure at the centimeter or decimeter scale. This organization of the s-matrix into aggregates of variable size constitutes, therefore, the *macroscopic expression of pedoturbation*.

CHAPTER 4

TRANSFERS AND ACCUMULATIONS

Differentiation and transformation of weathering s-matrixes, as outlined in the previous chapters, result from transfers of elements and particles by percolating solutions, or simply from redistribution of plasma and skeleton under the effect of wetting and drying.

Once differentiated, s-matrixes continue to be affected by solutions that percolate through the pore network. These structures undergo numerous changes to the extent of being replaced by new ones compatible with the new environmental conditions. These changes consist of losses of material leading to complete redistributions of skeleton grains, and of accumulations of material occurring in the s-matrix or in the pore system.

CHARACTERISTICS OF LOSS OF MATERIAL

Leached Horizons

Initial differentiations due to weathering and to related transformation are characterized by intrinsic s-matrixes (weathering s-matrix, pedoturbated s-matrix), each one displaying its own relative arrangement in space of plasma, skeleton grains, and pore system. These structural types may be affected by separation plasma–skeleton, resulting from partial or total transfer of plasma. This transfer can be ionic or particulate. Consequently, new structural types appear at the expense of earlier ones, which essentially involve skeleton grains when plasma becomes very rare or is absent. Hence, these new structures with high porosity belong to a pedologic differentiation and characterize leached A2 horizons.

Microscopic analysis of assemblages of skeleton grains, regardless of their mineralogical composition, therefore enables us to establish criteria of recognition of the microstructural units of leached horizons generated by partial or total loss of plasma. The advantages of such an approach are obvious: we can, with certainty, (1) recognition of a leached structure derived from any kind of previous structure, (2) classify these various superposed structures, and (3) characterize and define the history of paleosoils.

All these changes, which lead from an initial structure (an association of weathering products and relicts) to another, pertain to the crystallo-chemical nature of constituents and their redistribution. In other words, these changes are regulated at different scales by transfers, either of relative accumulation or of absolute loss of elements and particles.

Elsewhere in the s-matrix or in the pore system of the weathering profile, these transfers of material can produce absolute accumulations, either simultaneous or successive, corresponding to new structures. These can themselves lead to structural discontinuities through autodevelopment (or self-organization).

In summary, ionic and particulate transfers and related accumulations are characterized both by structures of residual products and skeleton resulting from losses of material, and by generation of new structures typical of influx of material. These various structures are specific to alterites and to soils; they correspond to pedologic differentiations.

Differentiation of Microstructures of Leached Horizons

The example chosen here was described by Bocquier (1973) in leached ferruginous soils on granite upstream of the Kosselili sequence (profile KB) in Chad.

With the exception of the surficial horizon, which is 20 cm thick, the entire profile K 13 (125 cm thick) is a vertical sequence of leached A2 horizons resting on a coherent parent–rock granite. These brown to reddish-brown horizons consist essentially of a high porosity silty to gravelly skeleton. Plasma is scarce, yet always present as coatings ("cutans" according to Brewer, 1964) and, together with the skeleton, forms an s-matrix only between 45 and 65 cm depth.

The most significant microstructural features of these leached A2 horizons consist of the distribution and arrangement of skeleton grains in structures that Bocquier (1973) calls *"caps on gravel," "striae,"* and *"lamellae."* In reality, these structures are closely related to each other and differ only in size and composition of skeleton grains.

According to Bocquier's sketches and descriptions (1973), an attempt was made to illustrate these petrographic structures of leached horizons (Fig. 4.1). In each case, skeleton grains display a "clustered distribution."

The case illustrated in Figure 4.1 is a *cap structure*. It is characteristic of A2 horizons of the upper part of the profile: a cap made of fine-skeleton grains (silt to fine sand), associated with a complex plasma, on top of a gravel clast. The cap is composed of argillaceous particles, fragments of biotite, and concentration of organic matter; it provides a bond between fine skeleton grains. The *assemblage of plasma and fine-grain skeleton* of the cap is of *porphyroskelic type* (according to Brewer, 1964). Skeleton grains adjacent to capped gravel clasts are always of smaller size than the clasts and generate the largest intergranular cavities underneath them. Skeleton grains located between gravel clasts are associated with a small amount of plasma, which either surrounds them, in which case the *assemblage is of granular type*, or which connects them by bridges, in which case the *skeleton plasma grain assemblage is of intertextic type* (Brewer, 1964).

The case illustrated in Figure 4.1 B is a *striated structure*. It occurs at the base of a leached A2 horizon that displays cap structures. The striations are characterized by the occurrence of a single cap that covers two or more gravel clasts located approximately in the same plane.

Caps have a composition identical to that of the above-described one. Skeleton grains of intermediate size can have a more diversified composition toward the bottom of the profile. Intergranular cavities generated below gravel clasts are the largest; they display an elongated shape in the sub-horizontal plane and are called *cavities beneath striae* (Bocquier, 1973, named them "under laminar cavities"). The assemblage of the small amount of plasma and skeleton grains of intermediate size ranges from intertextic to granular type.

The case illustrated in Figure 4.1 C is merely a more complex variant of the two preceding types. It is called *lamellar structure* and is characteristic of the lower part of profiles with leached A2 horizons. Skeleton grains show a more diversified mineralogical composition and their size increases toward the lower part of the profile, until fragments of granitic rocks sometimes appear ferruginized in fissures and along intergranular contacts. The arrangement of the coarsest skeleton grains follows approximately subhorizontal planes. Each plane determines a lamella overlaid by the same cap of fine skeleton grains, but coarser than those of the caps described in Figure 4.1 A and B. Furthermore, in the example described by Bocquier (1973), the caps of lamellae of the A2 horizon, located at a depth of −80 and −100 cm, do not seem to display plasma with porphyroskelic assemblage. Nevertheless, it was illustrated as such in the figure because it was observed in other examples.

A large cavity or fissure always occurs beneath each lamella. The lower part of this *under-lamellar cavity* is sometimes covered by an oriented argillaceous coating (cutan), unconformable over the skeleton grains forming the wall of the cavity. Obviously, it is a late argillaceous accumulation.

The leached A2 horizons studied in the Kosselili example represent eluvial environments in which structures consist mainly of the distribution of skeleton grains that generate a high porosity because of variable intergranular contacts. This distribution leads, particularly through vertical transfer of the finest fractions, to the generation of microstructural features specific to this type of environment, called, by Bocquier (1973), "laminar features," namely,

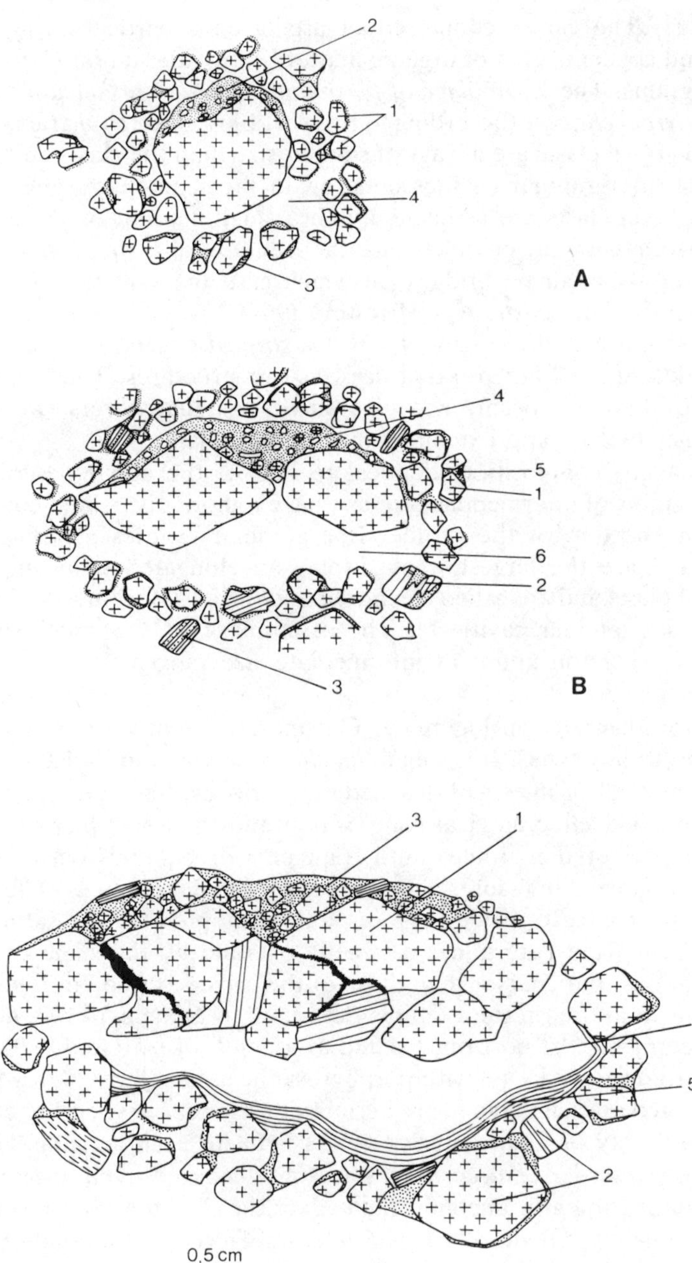
2
1
4
3
A
4
5
1
6
2
3
B
3
1
4
5
2
0,5 cm
C

caps, striae, and lamellae. Because they consist of a caplike coating of fine skeleton, associated or not to an argillo-organic plasma, these microstructures represent a precise criterion of recognition and of polarity of leached horizons.

High porosity of eluvial environments thus generated becomes a receptive structure for subsequent accumulations of material superposed on the laminar features. During the evolution of the pedologic cover, therefore, microstructural differentiations developed in leached horizons can either be submitted to transformations or, inversely, keep their characteristics and become relicts of a previous history.

Two main situations may occur: Either the *superpositions by new organizations* are part of the eluvial environment and remain limited, or they represent a continuous accumulation and lead to the development of an *illuvial environment* followed by a *transformation environment*.

Thus, *microstructural features of accumulation* can intersect and even destroy microstructural features of leaching, a problem that has an impact on the survival of laminar features during the succession of chronological differentiations that occur during the history of soils.

Survival of Microstructures of Leached Horizons

Limited Superpositions

Examples studied by Bocquier (1973) in Chad and by Boulet (1974) in Burkina Faso show that groups of leached A2 horizons display *discontinuous and limited illuvial accumulations* appearing as argillaceous coatings accompanied by iron oxyhydroxides adsorbed at the surface of argillaceous particles. These coatings are *illuviation cutans*, named according to their composition: argillo-ferruginous cutans or argilloferrans (in the sense of Brewer, 1964).

These cutans are superposed on the previously described leaching structures and are located, in particular, in cavities beneath striae and laminae, where they overlie unconformably skeleton grains forming the

Figure 4.1. Microstructures of Leached A2 Horizons (modified after Bocquier, 1973). **A**: Cap structure: (1) Skeleton grains consisting of quartz grains; (2) cap on gravel-size clast consisting of an association of plasma and fine skeleton grains; (3) plasma that overlies in small amount skeleton grains or joins them by means of intergranular bridges; (4) the largest intergranular cavities always located underneath capped gravel-size clasts. **B**: Striated structure: (1) Skeleton grains that consist essentially of quartz grains (1) but which may be grains of feldspar (2) and biotite (3); (4) caps shared by two gravel-size clasts. (5) plasma associated with skeleton grains of intermediate size. (6) cavity beneath striae. **C**: Lamellar structure: Skeleton grains consisting of granitic rock fragments (1) ferruginized along internal fissures, and of quartz grains or feldspar grains (2); (3) cap shared by several fragments and grains; (4) under lamellar cavity; (5) argillo-ferruginous coating over the bottom of the under lamellar cavity.

walls of the cavity. These cutans are always thicker on the bottom of the cavity. Furthermore, they can also overlie caps (Fig. 4.2 A).

It was shown that structures resulting from leaching or laminar features result from the vertical displacement of the finest particles. Caps consisting of fine skeleton and plasma and discontinuous illuviations of plasma (cutans) lead to the clogging in A2 horizons of vertical porosity, thus favoring subhorizontal porosity, and to the increase of the anisotropy of water circulation by decreasing vertical resultants to the benefit of horizontal resultants. Furthermore, within these plasma accumulations, localized hydromorphism can occur in sites of better development of these accumulations, namely, at the level of laminar features; this leads to a separation of adsorbed iron oxyhydroxides from argillaceous particles. Iron thus put into solution within these reducing microenvironments migrates over short distances and is redistributed in intergranular spaces of laminar features. Consequently, ferruginous concentrations become gradually differentiated in leached A2 horizons at the level of laminar features thus fossilized by cementation (Fig. 4.2 B). Resulting ferruginous nodules, specific to leached horizons, can, in turn, play the role of skeleton grains and display a cap of fine skeleton (Fig. 4.2 B). This alternation can be repeated several times, leading

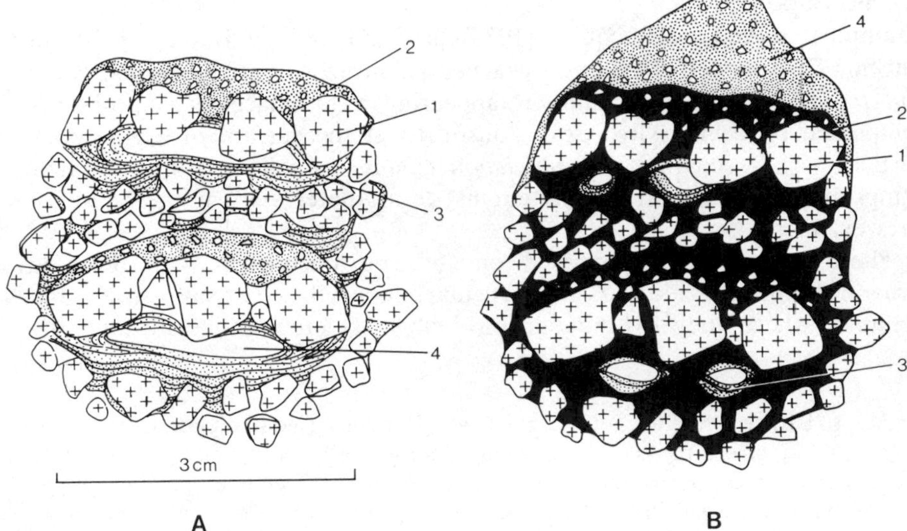

Figure 4.2. Schematic Illustration of Discontinuous Argillaceous Illuviation and of Ferruginous Nodulization in Microstructures of Leached Horizons (modified after Bocquier, 1973). **A**: Discontinuous accumulation of illuvial plasma: (1) Coarse skeleton grains; (2) cap consisting of fine skeleton associated with an argillaceous organic plasma; (3) coatings of illuvial argillo-ferruginous plasma (microbedded argilloferrans); (4) residual under laminar cavity. **B**: Intergranular ferruginous nodulization: (1) Skeleton grains; (2) highly ferruginous cement; (3) illuviation argilloferrans; (4) nonferruginized late cap covering nodule.

to "ferruginous nodules with vertical development by apical growth" (Bocquier, 1973, p. 170). The most spectacular examples were described by Bocquier (1973) in the groups of leached horizons of the pedological sequence of Mindera in Chad. Similar nodules were also reported by Gavaud (1968).

Limited superpositions of argillaceous plasma or of iron oxyhydroxides are characteristic of leached horizons. Not only do they modify the structures in laminar features of skeleton grains, but also, in the case of cementation by oxyhydroxides, they can also fossilize them.

Continuous Superpositions and Related Microstructural Transformations

Discontinuous illuvial accumulations can continue in the form of argillans within leached A2 horizons. Under these conditions, they are taken over by *continuous illuvial accumulations*, which are superposed on the skeleton of laminar features without disturbing the structure. It is the transition of leached horizons of A2 type to illuvial accumulation B horizons (Bocquier, 1973; Boulet, 1974).

When these illuvial argillaceous accumulations continue, the amount of argillaceous particles becomes sufficient for reactions with the aqueous phase by wetting-swelling to generate internal tensions capable of restructuring plasma and skeleton. Zoned structures, characteristic of illuviation plasmas, are destroyed through disorganization and reorientation of particles under new constraints. The arrangement of the skeleton acquired in leached horizons is also disturbed: *Redistribution by dispersion* takes place as an effect of the new dynamics. It is the transition of illuvial accumulation B horizons to structured accumulation B horizons.

In short, continuous superpositions of illuvial argillaceous plasmas lead gradually to the destruction of microstructures of leaching type and their replacement by new structures typical of accumulation.

Continuous Superpositions with Little or No Transformation of Microstructures of Leaching Type

In certain cases, continuous accumulation, superposed on laminar features of leached horizons, does not destroy initial structures but, on the contrary, fossilizes them. Three examples illustrate this situation:

1. Microstructures previously acquire within the leaching environment itself a limited superposition of iron oxyhydroxides, which protects them from restructuring due to continuous argillaceous accumulation. Such is the case of ferruginous nodules of leached horizons; the nodules congeal the microstructures in striae and lamellae described by Bocquier (1973) in the soil sequences of Mindera in Chad.

2. Microstructures of leaching type are submitted to a continuous argillaceous accumulation whose phyllosilicate particles do not react sufficiently with the aqueous phase for the generated tensions to rework the caps over

gravel clasts. This is the example described in nodular horizons under ferruginous crust of Ndias Massif in Senegal (Nahon, 1976).

3. Microstructures of leaching type are submitted to a nonargillaceous continuous accumulation that cements them, and hence fossilizes them. A typical example consists of the paleoprofiles with siliceous accumulation (paleosilcretes) described by Thiry (1977, 1981) in the Paris Basin of France.

FERRUGINOUS NODULES OF LEACHED HORIZONS AND CONTINUOUS ARGILLACEOUS ACCUMULATION

In the sequence of Mindera in Chad (Bocquier, 1973), the previously described ferruginous nodules formed in leached horizons (Fig. 4.2) can be submitted, at the same time as the leached horizons enclosing them, to a continuous illuvial argillaceous accumulation. Such is the case of the lateral gradation to B horizons in the sequence of soils studied by Bocquier (1973). With increasing accumulation, cavities under nodules are first invaded by illuvial argillo-ferruginous coatings, which subsequently take on a more general importance and surround the nodule and its cap. Local hydromorphism can develop at the level of argillo-ferruginous plasmas, particularly in the B horizons, located downstream of the sequence; in can also cause a separation between argillaceous paticles and ferruginous particles adsorbed at the surface of the former (Brinkman, 1970; Bocquier and Nalovic, 1972). In the illuvial plasma, which surrounds the ferruginous nodules of leaching type (Fig. 4.3 A and B), this separation is centripetal and leads to a ferruginization of the cap and to a ferruginous concentration on the periphery of the nodule.

Thus, the nodule gradually acquires a peripheral, ferruginous, and zoned cortex. Contraction phenomena that affect the argillaceous, locally deferritized plasma can subsequently generate a discontinuous cavity between plasma and ferruginous cortex (Fig. 4.3 C).

Consequently, eluvial leached beds, in which skeleton grains largely dominate over a scarce and discontinuous argillaceous phase consisting of kaolinite and micaceous phyllites, grade into illuvial horizons in which the continuous argillaceous phase consisting of predominant montmorillonite and kaolinite is abundant. During this transition, microstructures of leached horizons, represented by ferruginous nodules with caps, become surrounded in the accumulation horizon by a peripheral ferruginous cortex that does not destroy the structures of the central part of the nodules. This central part, whose microstructures indicate an initial eluvial stage, form a *pedorelict* (following Brewer, 1964).

RELICT FERRUGINOUS NODULES UNDER CRUST AND CONTINUOUS ACCUMULATION OF FERRITIZED KAOLINITE

A ferruginous nodular horizons, some tens of centimeters thick, develops at the base of massive arenaceous-ferruginous crusts (iron crust of "simple arenaceous facies" of the profiles of the Ndias Massif of Senegal). Nahon (1976) showed that this nodular horizon results from an initial leaching of

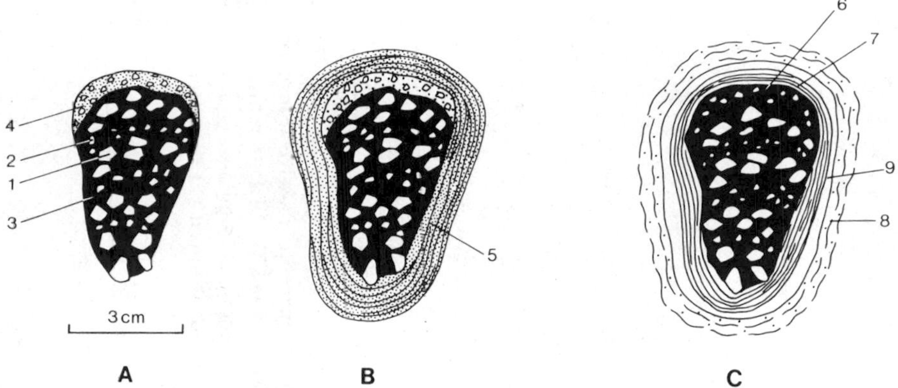

Figure 4.3. Schematic Evolution of a Capped Ferruginous Nodule During Argillo-Ferruginous Illuvial Accumulation. **A**: Ferruginous nodule of leached horizon: (1) Gravel-size skeleton clasts with (2) their initial ferruginized caps of fine skeleton grains; (3) ferruginous cementation acquired in leached horizon; (4) most recent non-ferruginized cap. **B**: Ferruginous nodule of leached horizon surrounded by argilloferruginous plasma of illuviation (5) **C**: Late differentiations acquired by the ferruginous nodule in the horizon of illuvial accumulation: (6) Ferruginization of cap over nodule; (7) peripheral ferruginous cortex generated by separation of iron-clay minerals from surrounding plasma; (8) deferritized argillaceous plasma; (9) contraction crack appearing at the discontinuity ferruginous cortex-argillaceous plasma.

the lower part of the ferruginous crust. The vertical or horizontal transition between crust and nodules displays several main stages (Fig. 4.4, A to D). Most of the arenaceous-ferruginous nodules thus generated display the same simple arenaceous facies as the crust from which they derive. Therefore, these nodules appear as *pedorelicts*; they constitute the coarse grains of the skeleton. The finest skeleton grains (silt), which consist of quartz grains and grains of iron oxyhydroxides (fragments of ferruginous plasma), are associated in caps at the apex of the coarsest grains, namely, they are relict ferruginous nodules with a simple arenaceous facies (Fig. 4.5).

The nodular horizon generated by leaching was submitted to a continuous illuviation of argillo-ferruginous plasma forming an s-matrix with a structure of agglomeroplasmic to porphyroskelic type. The argillo-ferruginous plasma is red and consists of kaolinite and of fine particles of oxyhydroxides adsorbed at the surface of the argillaceous particles. This ferritization of kaolinite (see chapters 3 and 5) considerably reduces the swelling capacity of kaolinite in contact with the aqueous phase (Chauvel, 1976; Pédro et al., 1976). Thus, the movements of the plasma, due to the dynamics generated by phases of swelling and contraction, remain restricted to local integration of argillaceous cutans to the remaining portion of the plasma, as shown by the presence of papules (fragments of zoned argillaceous cutans). In all instances, this dynamics of the ferritized kaolinitic plasma does not allow a

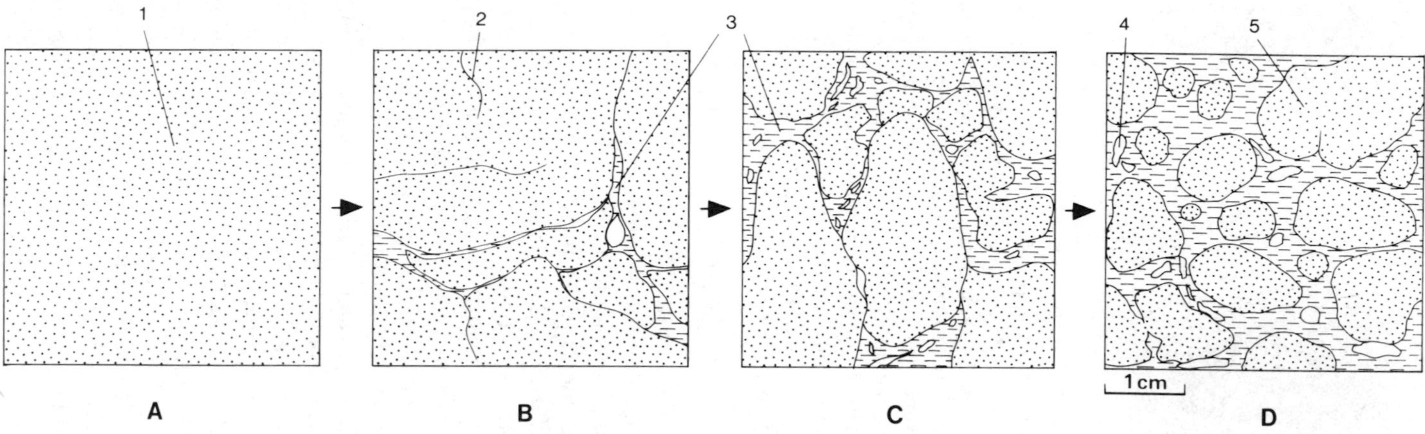

Figure 4.4. Transition between Iron Crust and Relict Ferruginous Nodules (from Nahon, 1976). **A**: Arenaceous massive iron crust (1). **B**: Development of voids as fissures, tubules (2) and alveoles and their filling up by argillo-silty plasma (3). **C**: Differentiation of coarse relict ferruginous nodules. **D**: Differentiation of coarse and fine relict ferruginous nodules (5) among an argillo-silty plasma. Number 4 is alveolar void. Relict ferruginous nodules present the same facies than the original arenaceous iron crust. (Reprinted by permission of Sciences Géologiques.)

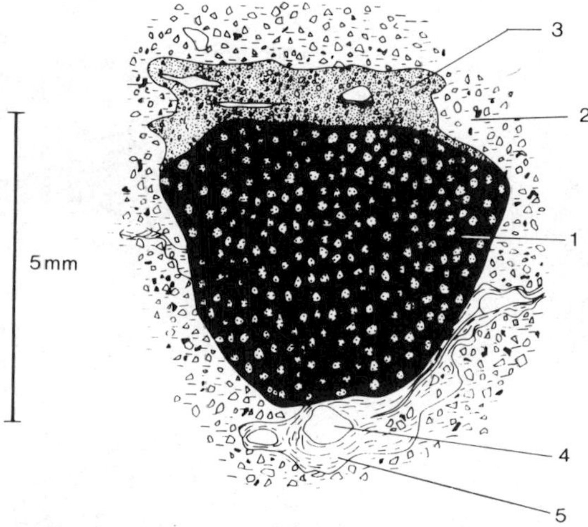

Figure 4.5. Detail of a Fine Relict Ferruginous Nodule from Figure 4.4 (from Nahon, 1976). (1) Relict ferruginous nodule; (2) argillo-silty plasma; (3) cap on relict ferruginous nodule consisting of an association of plasma, and fine skeleton grains; (4) under laminar cavity; (5) argillane coating (cutan). (Reprinted by permission of Sciences Géologiques.)

wiping out of the microstructures in caps generated during an initial pedologic stage of leaching type.

CAP STRUCTURES ON PEBBLES AND CONTINUOUS SILICEOUS ACCUMULATIONS

The Nemours conglomerate of the southern part of the Paris Basin, attributed to the Lower Eocene, is a detrital formation with rounded chert pebbles. This formation has undergone pedologic processes, particularly an important silicification at the lower Eocene-middle Eocene boundary (Thiry, 1977, 1978, 1981; Thiry and Schmitt, 1983). Investigations by these authors on paleo-profiles with siliceous accumulation (paleosilcretes), particularly in the type section of the Nemours conglomerate, can be summarized as follows. At the base of the section, and overlying the Cretaceous chalk, is a conglomerate with rounded pebbles of cherts that are more or less in reciprocal contact within an unconsolidated argillaceous sand (Fig. 4.6 A, horizon I). Higher up in the section, chert pebbles display a silicified cap, whereas the interstitial matrix is not silicified (Fig. 4.6 A, horizon II). Some cherts show embayments and small cavities due to dissolution. When these cavities are located at the apex of the cherts, they also act as receptive structures for caps (Fig. 4.6 B). However, cavities that possess only a lateral opening, or that are located in the lower part of pebbles, display no filling structures in microbeds. These cavities and this type of filling, therefore, demonstrate that dissolution (hence, leaching) of cherts preceded the deposition of caps and that leaching and the resulting deposition of fine grains and particles was downward. Furthermore, these caps are often polyphased and connect

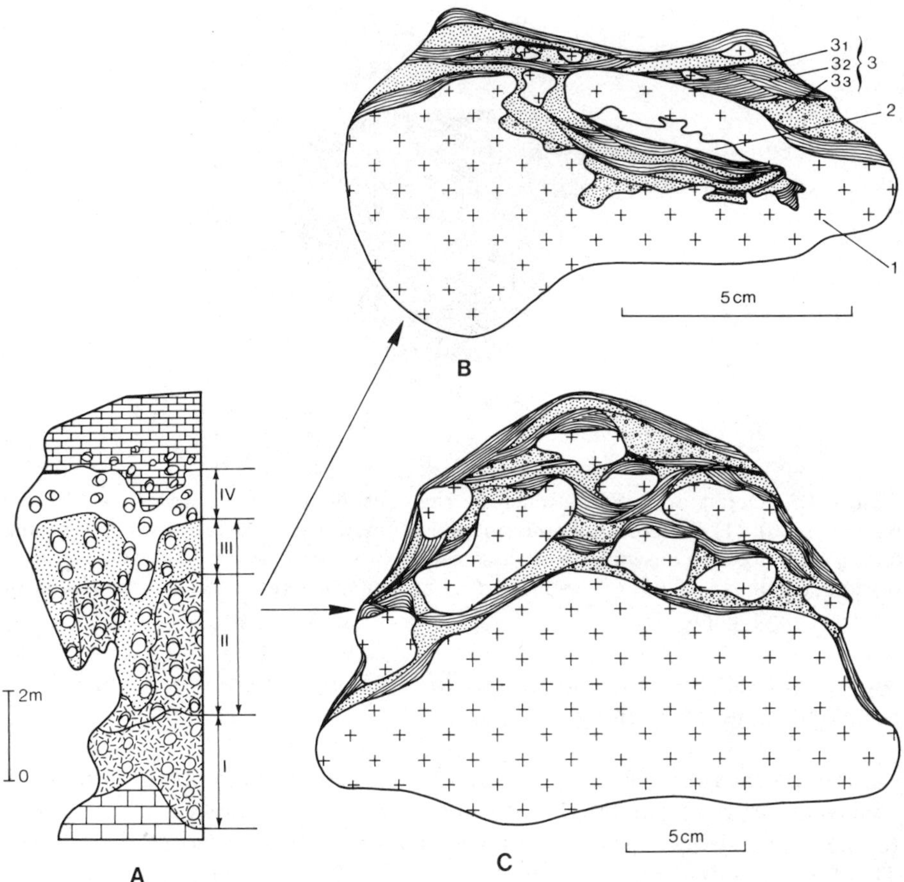

Figure 4.6. Macrostructures and Microstructures in Paleosilcretes Developed in the Eocene Nemours Conglomerate (modified after Thiry, 1977, pp. 108–109). **A**: Field section of paleosilcrete. I, II, III, and IV are the main recognized horizons. **B** and **C**: Complex caps over chert pebbles. (1) Chert pebbles forming the conglomerate; (2) solution cavity inside pebble generated during leaching of the profile; (3) silicified alternation of deposits of fine sand (3_1), of laminae (3_2), probably originally argillaceous, and of medium sand (3_3). Notice microstratification and crossbedding of these deposits. (Reproduced with permission of Bulletin du Bureau de Recherches Géologiques et minières.)

several adjacent cherts (Fig. 4.6 C). The size of the caps is proportional to the size of the host pebbles.

Microstructures of caps (Fig. 4.6 B and C) consist of repeated alternations of fine sands or silts and yellow to ochre cross-bedded laminae, perhaps originally argillaceous. These caps always closely follow the shape of the supporting pebbles. Microbeds of fine sand with a thickness ranging from 1 mm to 1 cm consist of angular grains of quartz, small splinters of chert, rare grains of feldspar, and, very rarely, grains of zircon and tourmaline. Average grain size ranges between 0.1 and 1 mm. The siliceous interstitial plasma consists of cryptocrystalline opal and quartz finely speckled with rutile. The laminae, which alternate with the microbeds of fine sand, consist of opal; their orientation is emphasized by pigments of titanium oxides.

Finally, the walls of intergranular cavities occurring within the coarsest beds are coated with fine laminae of opal. The central part of these cavities can be filled by geode-type microcrystalline quartz.

Farther up in the profile (Fig. 4.6 A, horizon III), silicification affects not only caps, but also the argillo-sandy matrix between pebbles. This matrix is entirely replaced by opal and microcrystalline quartz and appears as an ochre-beige lustrous sandstone.

At the top of the type section of the conglomerate, the silicified zone (Fig. 4.6 A, horizon IV) displays a coarser porosity, which often follows the margins of the chert pebbles, giving the horizon a cavernous aspect. The cement of the conglomerate becomes granular. This is the effect of restructuring by loss of the silicified zone with rotation of some capped pebbles, as demonstrated by caps that are no longer in apical position. In the newly created pore system, this restructuring is associated with recrystallization of geodelike quartz.

In summary, the section described by Thiry (1977, 1978, 1981) reveals the existence of a succession of structures of pedologic origin, which display among themselves a historical relationship.

Vertical leaching of chert conglomerates and the resulting deposition of fine particles leads to the generation of *caps on pebbles*, consisting of alternations of microbeds of fine sands and of fine, probably argillaceous, materials.

Silicification of caps represents a discontinuous and selective accumulation of silica. It affects only the caps in replacing their argillaceous matrix by opal and cryptocrystalline quartz, thus *fossilizing* the caps on pebbles.

Generalized silicification that extends to the argillo-sandy matrix between pebbles represents *continuous accumulation of silica*; it leads to the generation of silcrete.

The above example, as well as those of Chad (Bocquier, 1973) and of Senegal (Nahon, 1976), shows the three successive stages of leaching, of limited discontinuous accumulation, and of generalized continuous accumulation, which lead from the differentiation of microstructures in caps over pebbles to their fossilization by accumulation of cryptocrystalline quartz and opal.

Succession and Hierarchy of Microstructures Typical of Loss and Accumulation of Material

Transfers by loss of material, and particularly of the finest argillaceous particles, lead, when sufficiently developed, to new distributions and new arrangements of residual products, essentially represented by the skeleton. Skeleton grains display, therefore, arrangement discontinuities that are specific to the so-called leached environments. Their distribution becomes clustered and characterized by the granulometric contrast between the finest and the coarsest grains. As a consequence of the vertical leaching, the finest grains concentrate as caps on top of the coarsest host grains. Several comparable microstructures that are only variants with respect to size and complexity of arrangement are thus generated. They are *caps on gravels or pebbles, striae, and lamellae*, designated as "laminar features" by Bocquier (1973).

These microstructures appear, therefore, as very reliable indicators of leached horizons.

Porosity, which is characteristic of these leached environments and which results from losses of material and from new arrangements of skeleton grains, acts, in turn, as a privileged receptive structure for subsequent accumulations. Thus, several other microstructures resulting from the discontinuous and subsequently continuous accumulation are superposed on the initial differentiation of microstructures of leached type. A hierarchical succession of several microstructures takes place. These accumulations can have several effects on the original microstructures of leaching type.

Accumulations can be *nonindurating and dispersive*. They are argillaceous with predominant swelling minerals, which lead, by reaction with the aqueous phase, to a gradual destruction of initial structures by redistribution of the skeleton.

Accumulations can also be *nonindurating and temporarily preserving*. They are argillaceous by weakly active or inactive in the aqueous phase and intersect only initial structures.

Accumulations can be *indurating and permanently preserving*. They correspond to localized cementation by iron oxyhydroxides (which appears earlier in leached environments undergoing discontinuous accumulation) or to generalized cementation by silicon oxides, crystallized or not. These accumulations fossilize microstructures of leaching type, which then correspond to *pedorelicts*.

In summary, petrographic microstructures permit us to characterize losses of material. These losses not only generate accumulations elsewhere in the horizon or profile, but also create, by means of porosity developed in situ, structures capable of receiving accumulations originating from other places of the horizon, either from the profile or from the sequence of soils. These accumulations, which were described above in connection with laminar features, generate typical microstructures, much more varied and complex than those characterizing leached environments.

CHARACTERISTICS OF ACCUMULATION OF MATERIAL

In weathering and in soils, accumulations of material appear as concentrations of plasma developed either within the s-matrix of the rock or in the system of anastomosed cavities that occur throughout the rock. These different types of accumulation correspond to pedologic features that is "recognizable units within a soil material which are distinguishable from the associated material for any reason such as differences in concentrations of some fraction of the plasma" (Brewer, 1976, p. 142).

Accumulations Developed in S-Matrix: Glaebulization

Concentrations of plasma developed in s-matrix correspond to three-dimensional accumulations. They allow, within the s-matrix, differentiation of three-dimensional plasmic units, subspherical to spherical, called *glaebules* (Brewer, 1964). Relationships between glaebules and surrounding s-matrix provide data on the nature and direction of transfers of material. As proposed by Brewer (1964, 1976), it is essential to analyse the following features of glaebules: internal structure, degree of relative concentration and mineralogical composition of their constitutive material, and shape and nature of their boundaries with the s-matrix. Consequently, glaebular accumulation displays several types of differentiation.

Weakly Differentiated Glaebular Accumulations: Initial Concentrations

Weakly differentiated glaebular accumulations appear, under the naked eye, as small spots that generally do not exceed 1 cm in size and are weakly indurated. Even though this degree of induration is difficult to appreciate, even under the microscope, it is possible to characterize the nature of the structural and mineralogical relationships between glaebule and enclosing material.

In general, *initial glaebular concentrations* are characterized by the existence of structural, geochemical, and mineralogical relationships between them and the surrounding s-matrix. Boundaries between these two plasmic structures are gradual, hazy, and very irregularly interfingering. Only one of the plasmic constituents shows an increasing and centripetal concentration when grading from s-matrix to glaebule. Distribution and shape of skeleton grains demonstrate that glaebular differentiation occurs without significant disturbance of microstructures and host constituents (Fig. 4.7).

Such is the common case at the base of weathering profiles, within the weathering plasmas themselves. In such places, ferruginous and manganese-rich diffused glaebular concentrations are differentiated and intersect the s-matrix without appreciably affecting mineral constituents. When these concentrations are ferruginous, they are enriched in goethite (FeOOH), associated or not with amorphous oxyhydroxides; when they are manganese-rich, they are enriched in birnessite ($Mn_5^{4+}Mn_2^{3+}O_{13}\cdot 5H_2O$), in cryptomelane ($K_xMn_{8-x}^{4+}Mn_x^{3+}O_{16}$), or in manganite ($Mn^{3+}OOH$).

In short, initial glaebular concentrations develop without significant modifications of mineral constituents and structures of the host s-matrix. They result from the simple centripetal accumulation of a fraction of the plasma. The *plasmic concentration is displacive.*

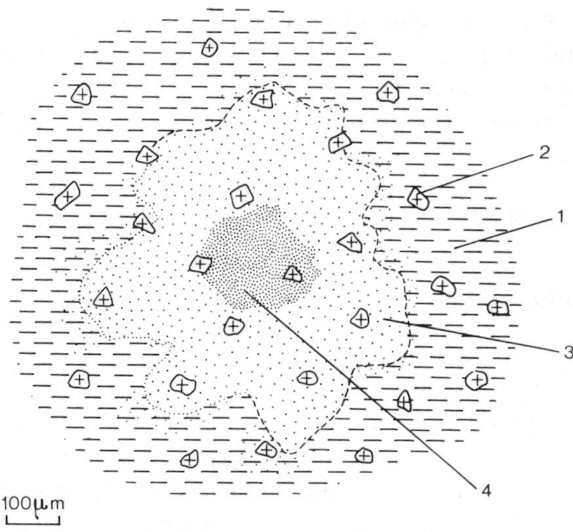

Figure 4.7. Schematic Representation of an Initial Glaebular Concentration. (1) Argillo-ferruginous plasma (kaolinite plus goethite) of s-matrix.; (2) skeleton grains (quartz) of s-matrix.; (3) reddish glaebular zone: weak concentration of ferruginous plasma (goethite); kaolinite of plasma and skeleton grains are not affected. Boundary between (3) and surrounding s-matrix is hazy and gradual; (4) red to brownish glaebular zone: more important ferruginous concentration (goethite), but kaolinite still abundant and skeleton grains unaffected. Boundary between (4) and (3) is hazy and gradual.

Moderately Differentiated Glaebular Accumulations: Nodules

In this case, centripetal plasmic accumulation within the s-matrix always displays transitional external boundaries. However, in the center of the glaebule, where accumulation is the strongest, a profound modification of the host mineral constituents takes place. Thus, instead of simple plasmic superposition observed in initial concentrations, a partial or total replacement of the host plasma by the concentrating plasma occurs. During this replacement, either some of the chemical elements of host minerals participate in the crystallo-chemical "construction" of the crystallites of the concentrating plasma, or no element of the replaced minerals takes part in the precipitation of the crystallites of the concentrating plasma. Between the enclosing s-matrix and the glaebule, the structural and mineralogical parentage occurs only in the transitional zone (periphery of the glaebule). The central part of the glaebule forms a structurally and mineralogically different plasmic entity, resulting from the epigenetic replacement of host minerals by crystallites of the concentrated plasma. Thus, for this type of glaebular accumulation, *plasmic concentration is displacive and reconstructive*. The resulting three-dimensional structures are *nodules*. Two examples are presented. The first deals with the formation of hematitic nodules in lateritic weathering profiles of the Ndias Massif of western Senegal; the second pertains to calcitic nodules differentiated in a granitic gruss of the Tasiast Massif in Mauritania.

Ferruginous Nodules in Weathering Profiles of the Ndias Massif

In lateritic weathering profiles developed over Maestrichtian glauconitic sandstones, Nahon (1976), Nahon et al. (1977), and Didier et al. (1983) studied the generation of ferruginous nodules within an essentially kaolinitic plasma associated with a few quartz grains and with goethite. The centripetal accumulation of iron is expressed by plasmic concentration of aluminous hematite at the same time as the amount of kaolinite diminishes until its final disappearance. An electron microprobe traverse of a ferruginous nodule (Fig. 4.8) clearly shows the behavior of the three main elements. In the

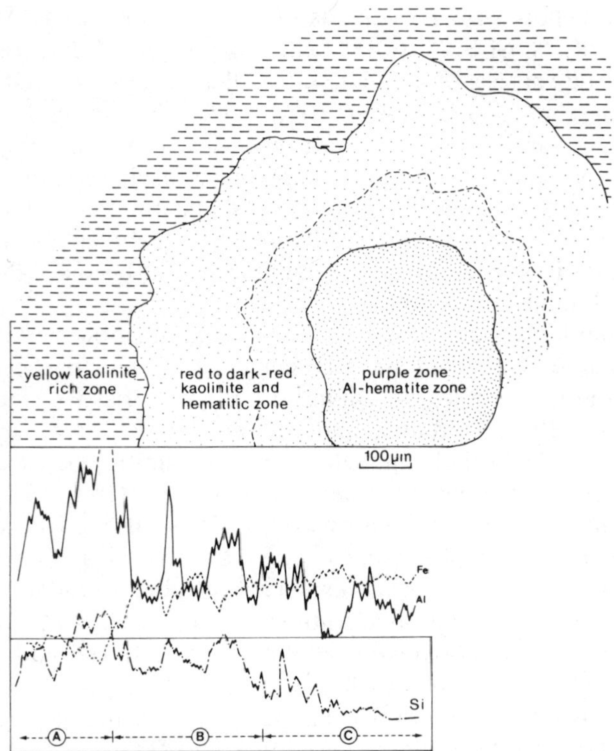

Figure 4.8. Moderately Differentiated Ferruginous Plasmic Concentration: Formation of Hematitic Nodule. Distribution of Al, Fe, and Si along Microprobe of Section of Nodule. **A**: Yellow s-matrix rich in kaolinite. This kaolinite is an iron-bearing clay with 1 mole % $Si_2Fe_2O_5(OH)_4$. **B**: Red to dark red external zone of glaebule consisting of iron-bearing kaolinite with 3 mole % $Si_2Fe_2O_5(OH)_4$ and of aluminous hematite with 4% mole of Al_2O_3. **C**: Purplish red internal zone of glaebule consisting only of aluminous hematite with 10–14 mole % of Al_2O_3.

s-matrix, rich in kaolinite (zone A of the traverse), Al and Si vary in a parallel manner; in the external part of the glaebule (transitional zone B), Fe increases whereas Al and Si decrease, but the last two elements are not always parallel, and Al combines locally with Fe. Finally, in the center of the glaebular concentration, Si is eliminated whereas Al and Fe are concentrated. Epigenetic replacement of kaolinite by iron oxide occurs gradually with leaching of Si from kaolinite, followed by its dissolution, elimination of a small portion of alumina liberated from kaolinite, and participation of the remaining alumina in the formation of aluminous hematite, which contains in its network 10–14% mole of Al_2O_3 (Nahon, 1976).

Furthermore, Didier et al. (1983) show with the same example, using microsamples of the various stages of ferruginization and Electron Spin Resonance (ESR), that kaolinite plasma, during the concentration of hematite and its replacement by hematite, displays the following mineralogical modifications. The yellow s-matrix consists of ferriferous kaolinite containing in its lattice 1 mole % $Si_4Fe_2O_5(OH)_4$ whereas the red to dark red glaebular transitional zone consists of kaolinite of smaller size, not so well-crystallized and containing 3 mole % $Si_4Fe_2O_5(OH)_4$. Obviously, during dissolution and reprecipitation of iron over very short distances, which are responsible for hematite accumulation, the original kaolinite is itself replaced by a disorganized and more ferruginous kaolinite. At first this latter kaolinite is in equilibrium with the conditions regulating the glaebular concentration of iron oxide; later it is in complete desequilibrium and becomes dissolved and replaced by aluminous hematite.

Many examples of hematitic glaebular accumulation within kaolinitic plasmas have been described since Didier et al. (1983), particularly in the lateritic weathering profiles of Diouga in Burkina Faso (Ambrosi, 1984; Ambrosi et al., 1986; Herbillon and Nahon, 1988): There, vermicules of kaolinite, which form by their entanglement the kaolinitic plasma, are replaced in the center of the nodules by aluminous hematite. However, the original structure of the vermicules frozen by iron oxide is still recognizable demonstrating the epigenetic replacement of the clay mineral by hematite.

It has recently been shown that manganese glaebular accumulation within kaolinitic plasmas lead to the same features (Nahon et al., 1983): as hematitic glaebular accumulation booklets of kaolinite are dissolved and replaced by lithiophorite, which is an Al-bearing manganese hydroxide (Nahon et al., 1989b). Here also, the shape of original booklets of kaolinite is preserved in the core of lithiophorite nodules (Fig. 4.9 A and B).

Calcitic Nodules in Weathering Gruss of Migmatites of the Tasiast Massif (Western Mauritania)

The migmatites of Inkebdene in the Tasiast Massif display weakly developed weathering profiles, reaching about a 1 m thickness. They consist of weakly altered massive gruss with preserved original structure. The weathering s-matrix consists of a porphyroskelic assemblage of greenish smectitic

Figure 4.9. Manganese Nodule Formation within Kaolinitic Plasma (after Nahon et al., 1989b). **A**: Contact between manganese nodule consisting of lithiophorite (L) and kaolinitic plasma consisting of vermicules of kaolinite (K). Packing of automorphic platelets of kaolinite are replaced by honeycomb texture of lithiophorite crystallites. **B**: Core of manganese nodule consisting of lithiophorite (L). Original booklets structure is preserved by lithiophorite. (Reprinted by permission of Elsevier Science Publishers B.V.)

argillaceous plasma and skeleton grains of mica, feldspar, and quartz. Within this s-matrix are differentiated calcitic nodules with indurated center and powdery, hazy margins intimately related to the encasing s-matrix (Fig. 4.10). From the margins of a given calcitic nodule toward its center, the following features can be observed: rapid disappearance of montmorillonitic plasma and decrease in size and number of parent relicts until total disappearance. These relicts are feldspar and quartz in the powdery margins of the nodule, but quartz only toward the center. Moreover, all these relicts display corrosion shapes with salients and embayments and may even be separated into fragments. In the latter case, each fragment within the calcitic plasma keeps the same crystallographic orientation as that of the original grain. This is demonstrated by similar positions of extinction of the fragments and by correspondence of twinning planes between a given fragment of feldspar and a neighboring one.

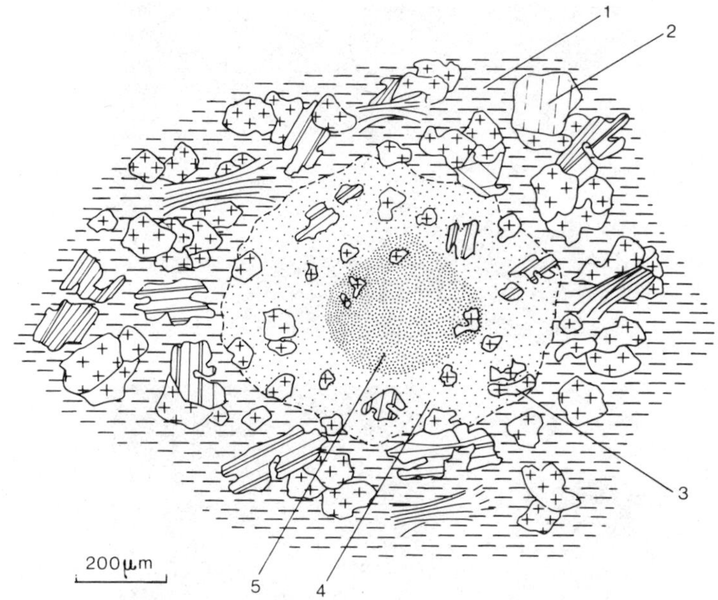

Figure 4.10. Moderately Differentiated Calcitic Plasmic Concentration: Formation of Calcitic Nodule. (1) Smectitic weathering plasma of s-matrix; (2) skeleton grains (lithic fragments, grains of quartz, feldspar, and mica) of s-matrix; (3) skeleton grains within calcitic glaebular plasma; note their smaller size and their local wide separation into several pieces; (4) calcitic glaebular plasma of external zone; (5) calcitic glaebular plasma of central part.

This situation proves that the calcitic glaebular orientation developed through radial epigenetic replacement in situ of the constituents of the weathering s-matrix.

In tropical soils of northern Burkina Faso, Boulet (1974) showed that inside their powdery margins such calcitic nodules displayed an impregnation of iron and manganese oxyhydroxides subsequent to the calcitic glaebular concentration. At Thies, in western Senegal, Nahon (1976) described in vertisols, glaebular calcitic concentrations accompanied by a very weak iron concentration leading to the generation of siderite crystals inside the calcitic nodules.

Strongly Differentiated Glaebular Accumulations: Well-defined Nodules and Individualization of Cortex

As shown above, generation of a three-dimensional plasmic concentration by epigenetic replacement of a host plasma leads to a differentiation between the nature of the constituents of the nodular structure and those of enclosing plasmic structure. Differences in the nature of constituents leads to differences in their respective reactivity to wetting and drying cycles. Differences of reactivity, weakly recognizable petrographically when plasmic concentration is weak, is expressed by the appearance of two essential structural features: a constraint cutan and a peripheral cavity.

The encasing argillaceous plasma displays an orientation of its constitutive particles along the margins of the nodule; this orientation is parallel to the external boundary of the nodule. It is a plasmic separation resulting from constraints existing within the argillaceous s-matrix and acting in three dimensions perpendicularly to the external surface of the nodule. These plasmic separations are therefore simple in situ reorganizations; they do not represent real coatings by transfer. Brewer (1964) calls them *constraint cutans*.

The argillaceous s-matrix encasing the nodule allows, by contraction during drying phases, the appearance of a *cavity peripheral* to the nodule.

These two structural characteristics develop gradually. Intermediate cases between glaebular concentrations with hazy margins and glaebular concentrations with peripheral cavities are common. A typical example of this situation, described by Boulet (1974), consists of calcitic nodules generated in bottom-slope vertisols of the Garango sequence in Burkina Faso (Fig. 4.11). The calcitic nodule indeed results from an in situ concentration, as shown by skeleton grains (feldspar and quartz) strongly epigenetically replaced by calcite in comparison with skeleton grains of the enclosing s-matrix (Fig. 4.12 A). This calcitic nodule is partially surrounded by constraint cutans in its upper portion where a sharp boundary with the enclosing plasma exists; in its lower portion, on the other hand, a hazy boundary with

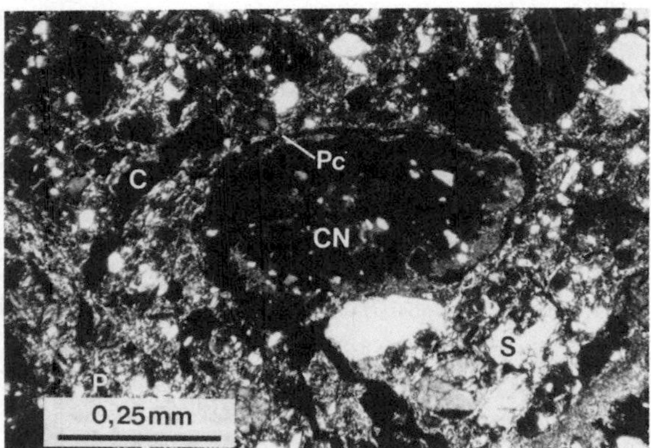

Figure 4.11. Calcitic Glaebular Concentration with Peripheral Cavity at the Top and with Hazy Margins at the Base (after Boulet, 1974). P. voskelmasepic plasma; S is skeleton grain; CN is calcitic nodule with secondary difuse ferruginous impregnation; C is cavity; PC is peripheral cavity at the top of calcitic nodule; hm is hazy margin between calcite and smectitic plasma. (Reprinted by permission of ORSTOM Editions.)

Figure 4.12. Feldspar Grains within Calcitic Plasma of a Calcitic Nodule (after Boulet, 1974). Feldspar (F) is penetrated and separated into several fragments by calcitic plasma (C). Each fragment of feldspar within the calcitic plasma keeps the same crystallographic orientation as that of the original grain. This is demonstrated by similar extinction and by correspondence of twinning planes between fragments. (Reprinted by permission of ORSTOM Editions.)

the same plasma, devoid of plasmic separation, occurs. This allows Boulet (1974) to consider the nodule with hazy boundary as the precursor stage of nodules with sharp boundaries. In other words, the intensity of the glaebular concentration, expressed macroscopically by an increase of induration, has an orientation. The intensity is stronger at the top of the nodule and leads gradually to its individualization from the enclosing s-matrix.

Thus, the *circum-nodular peripheral cavity* appears as a structural characteristic generated by the centripetal glaebular accumulation, that is, as a structural discontinuity acquired during the continuous geochemical evolution of glaebular accretion. This structural discontinuity leads to a new evolution of the nodule. Cut off from its supply by centripetal diffusion across the network of cryptocavities of the argillaceous s-matrix, the evolution of the nodule continues by internal restructuration from its margins. The presence of the peripheral cavity modifies local geochemical conditions at the surface of the nodule and the latter acquires a banded *cortex*. The development of the cortex is centripetal and occurs by means of reduction of the size of the nodule, by modification of the mineralogical constituents of the nodule, by reorientation of the plasma of the cortex (banded aspect), and by greater induration of the cortex as observed megascopically.

Differentiation of the cortex occurs at the expense of the previously acquired plasmic concentration. It may be complete, so that the entire nodule is transformed. The nodule is then called *pisolite* or *pisolitic concretion*. It is the ultimate stage of centripetal three-dimensional glaebulization.

An example was chosen from the formation of ferruginous pisolites generated in kaolinitic plasmas of weathering profiles in the Ndias Massif. The formation of ferruginous hematitic nodules within kaolinitic plasma was examined above; a discussion of the development of banded cortex at the expense of hematitic nodules follows. A brown halo appears in ferruginous nodules of aluminous goethite with 10–14 mole % Al_2O_3 at the contact between their margins and the surrounding peripheral cavities. These halos become thicker while developing at the expense of the reddish purple hematitic core of the nodules. Thus, the latter display a concentrically banded brown cortex. It is possible to observe at Ndias all intermediate stages between the largest pisolites with thin cortex and large hematitic core (Fig. 4.13 A) and the smallest pisolites with reduced core (Fig. 4.13 B), or devoid of core (Fig. 4.13 C).

The transition of pisolites of type "A" to those of type "B" (Fig. 4.13) occurs in two ways. In the first, the cortex develops at the expense of the red purple hematitic core by isolating scales of hematitic plasma, by decreasing gradually the importance of these scales when the cortex develops, and by destroying the existing microporosity of the hematitic nodule. In the second, the cortex consists in natural light of banded zones, which are alternately light (orange red under the microscope) and dark (brownish red). X-ray diffraction and Mössbauer spectrometry analyses show that this cortex consists of aluminous goethite with 16–22 mole % AlOOH. The electron microprobe indicates that the lighter bands have a higher content of alumina.

No polarity in the development of cortexes was observed at Ndias. However, Muller et al. (1980) showed in other examples in the Congo that ferruginous cortexes were preferentially developed at the apex of nodules.

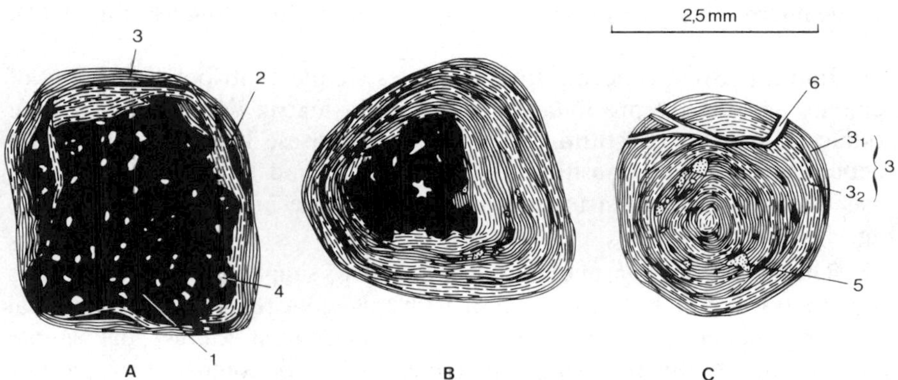

Figure 4.13. Differentiation of Goethitic Banded Cortex at the Expense of Hematitic Nodules. Formation of Ferruginous Pisolites (modified after Nahon, 1976). **A**: Large pisolite with well developed hematitic core (1) and thin cortex (3). Number 2 is relict of hematitic core in the cortex; 4 is microvoids. **B**: Intermediate pisolite between A and C. **C**: Small pisolite devoid of hematitic core. Cortex (3) consists of alternatively dark bands (3_1) and light bands (3_2). Number 5 is relict of skeleton quartz grains; 6 is a secondary fissure. (Reprinted by permission of Sciences Géologiques.)

These authors also noticed the complex aspect of banded ferruginous cortexes, in which certain bands are interrupted by still others, and so forth. This indicates that the development of the cortex is not always isotropic in three dimensions, but that it can display a preferred direction at one instant or another of its history. This complexity in the orientation of cortexes has often been reported, particularly by Jones (1965).

The example presented here is a simple case of a glaebular concentration evolving into a pisolitic concretion. The profiles of Ndias studied below show in particular that glaebular concentrations may be so important within the kaolinitic s-matrix that they become reciprocally anastomosed and generate continuous hematitic accumulations. In such a case, pisolitization develops under the effect of each discontinuity such as fissures, cracks, or alveolar cavities which occur in these continuous hematitic accumulations.

The Different Stages of Glaebular Microstructure

Glaebulization as described above presents several stages characterized by the structural and mineralogical parentages or by discontinuities displayed by plasmic concentration with respect to the host s-matrix. These various stages are expressed by the intensity of plasmic concentration, namely, by their degree of differentiation within the s-matrix. Thus, a transition exists from weakly differentiated concentrations (domain of *initial concentration*), through moderately differentiated concentrations (domain of *nodules*), to strongly differentiated concentrations (domain of *well-defined nodules* and *pisolitic concretions*).

Increasing plasmic concentration is expressed macroscopically by a gradual increase of induration. Glaebular evolution, as observed under the petrographic microscope, can be schematically summarized as follows (Fig. 4.14):

1. Initial plasmic concentration expresses simple centripetal transfers of elements and their reprecipitation within the s-matrix without appreciable modifications of its constitutive mineralogical phases. The boundary of the glaebule is hazy and transitional. Mineralogical and structural agreement exists between the two structures, namely glaebule and enclosing s-matrix (Fig. 4.14 A).

2. The next type of plasmic concentration shows simple centripetal transfers of elements followed by reprecipitation with epigenetic replacement of original host constituents. Thus, a mineralogical replacement follows the simple superposition of the first stage. A mineralogical disconformity separates both structures. A structural and mineralogical agreement is maintained only at the boundary, which can still remain gradual and hazy, demonstrating that the differentiation of the nodule occurred in situ (Fig. 4.14 B).

3. The intensity of concentration is responsible for mineralogical individualization of each structure, which by its difference in reactivity to the water phase induces a structural disconformity. This disconformity is expressed

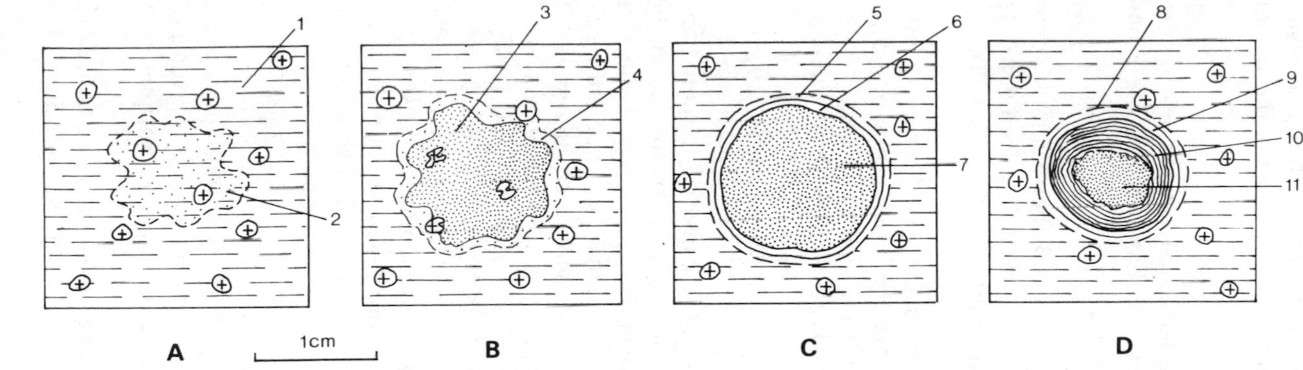

Figure 4.14. Stages of Glaebular Microstructure. **A**: Initial glaebular concentration: (1) S-matrix of argillaceous plasma and figured skeleton grains; (2) glaebular plasmic concentration respecting the constituents of the s-matrix. **B**: Nodule: (3) Glaebular plasmic concentration having replaced, by epigenesis, the constituents of the smatrix, (4) hazy external boundary of nodule. **C**: Well-defined nodule: (5) Reorientation under constraint of argillaceous particles of the s-matrix (constraint cutan); (6) peripheral contraction cavity; (7) indurated nodule. **D**: Pisolitic concretion (or pisolite): (8) Constraint cutan; (9) peripheral contraction cavity; (10) banded cortex developed at the expense of central part of nodule); (11) pisolite core (residual central part of nodule).

by reorientation under constraint of the argillaceous particles of the s-matrix and by generation of a peripheral cavity (Fig. 4.14 C).

4. The nodule thus outlined undergoes its own evolution, which consists of centripetal generation of a banded cortex through reorganization and reorientation of the glaebular plasma with decrease of the size of the nodule (Fig. 4.14 D).

Generation and development of glaebular structures belong, therefore, to a pedologic differentiation. Its geochemical mechanisms are examined below.

Processes of Glaebulization

Glaebular plasmic concentration in soils and alterites, most commonly called "nodulization" or "concretioning," was often described when its individualization was sufficiently clear and emphasized by its color, structure, or mineralogical composition.

Examples dealing with ferruginous, manganese-bearing, carbonate, sulfate, and siliceous plasmic concretions are the most common. Among these are those investigated by Kubiena (1938), Drosdoff and Nikiforoff (1940), Humbert (1948), Bryan (1952), Kovda et al. (1958), Yamasaki and Yoshizawa (1961), Brewer and Sleeman (1964), Brewer (1964, 1976), Flach et al. (1968), Bocquier (1973), Boulet (1974), Nahon (1971, 1976), Leprun and Nahon (1973), and Leprun (1979).

Whereas most of these authors accept the in situ differentiation of glaebular concentrations, they disagree on the processes of formation. For instance, the disappearance of the host s-matrix at the place where the glaebule developed was often explained by its gradual expulsion during the growth of the nodule, an expulsion due to the pressure of crystallization of the constituents of the nodule. This point of view is encountered in the works of Boulet and Nahon (1970), Boulet (1974), Nahon et al. (1975), and Nahon (1976) on calcareous nodules, and of Nahon (1976), Nahon and Millot (1977), Didier et al. (1983), and Ambrosi (1984) on ferruginous nodules; all these authors clearly show that the original mineral constituents of the host s-matrix are dissolved and replaced during centripetal plasmic accumulation. This replacement occurs with conservation of initial volumes and hence is epigenetic. It is therefore possible to explain generation in situ of nodules without any mineralogical agreement of the latter with their immediately adjacent matrix environment.

Nevertheless, the very process of centripetal plasmic concentration (the cause of transfers of elements, their precipitation at a given place of the s-matrix rather than another, and the nature of the precipitated mineral) has received little attention.

Drosdoff and Nikiforoff (1940) suggested that during desiccation, solutions concentrate in small pores, inducing the initial nucleation of sesquioxides from which the nodule develops.

Tardy and Monnin (1983) noticed that nodules in soils and alterites develop preferentially in argillaceous s-matrixes, that is, with fine porosity; they presented a geochemical interpretation of siliceous nodulization. These authors calculated the differences in solubility of minerals as a function of the size of the pores in which they precipitate. Thus, an aqueous solution saturated with respect to a given mineral in a large cavity can become supersaturated with respect to the same mineral in a smaller pore. After drainage or desiccation, the smallest pores retain the solution as small-radius concave menisci. Under these conditions, the solution becomes supersaturated with the mineral, which then precipitates and tends to fill partially the fine porosity, and the process continues.

The best-known cases (Nahon, 1976) are those of ferruginous glaebular structures generated in soils and alterites of humid tropical countries (laterites). This study explains the formation of such nodules, particularly as a function of s-matrix porosity. It was clearly demonstrated, petrographically and mineralogically, that any indurating ferruginous nodulization within a kaolinitic plasma occurs by epigenetic replacement of kaolinite crystals by aluminous hematite (whose substitution rate in Al_2O_3 moles ranges from 4% to 14%). To grasp the process of ferruginous glaebulization is therefore to understand the movements of iron, its precipitation as a hematite, and the dissolution of kaolinite.

Migrations of iron are known to occur under a reduced state because of differences in the oxido-reduction potential (Eh) that exist continuously within soils and alterites. Furthermore, in a reducing environment, concentrations of ferrous ions other than Fe^{2+} are negligible (Nahon, 1976). In this case, concentration in total ferrous iron of percolating or diffusing solutions can be considered as being that of the simple Fe^{2+} ion. This situation can be schematically expressed by the following equation:

$$2Fe^{2+} \Leftrightarrow 2Fe^{3+} + 2e^-.$$

Hematite precipitates from the soluble Fe^{3+} cation according to the reaction:

$$2Fe^{3+} + 3H_2O \Leftrightarrow Fe_2O_3 + 6H^+.$$

Indeed, hematite precipitates in kaolinite-rich zones. These two equations express the oxidation-hydrolysis stage called "ferrolysis" (Brinkman, 1970, 1979; Brinkman et al., 1973). Ferrolysis is responsible for the dissolution of clay minerals, and particularly of kaolinite, by means of protons liberated during hydrolysis of the Fe^{3+} ion in the second equation. This dissolution of kaolinite can be schematically represented by the following equation.

$$Si_2Al_2O_5(OH)_4 + 6H^+ \Leftrightarrow 2Al^{3+} + 2H_4SiO_4 + H_2O.$$

However, precipitation of hematite or of its hydrated precursor ferrihydrite (Fischer and Schwertmann, 1975; Schwertmann and Murad, 1983), instead of goethite, as well as its precipitation in kaolinitic zones of fine porosity rather than in zones of coarser porosity, can be explained by the works of Didier et al (1983) and Tardy and Nahon (1985). These authors suggested for iron compounds the same behavior proposed in a hypothesis by Tardy and Monnin (1983) for siliceous concretions. Indeed, the direction of iron migration and its accumulation as hematite appear to be controlled by the pore size of the s-matrix. The direction of transfers is always from larger pores (in which atmospheric pressure is higher and solubility of oxyhydroxides greater) toward smaller pores (in which pressure is lower than atmospheric pressure). Moreover, the activity of the water that percolates or diffuses is related to the size of the pores containing it. This condition was demonstrated by Bourrie and Pedro (1979) and applied to alterites and lateritic soils by Didier et al. (1983) and Tardy and Nahon (1985). Thus, in the pore system between crystallites of kaolinite, the size of the pores is of the order of 30–50 Å, activity of water in them is weak, and dehydration of goethite to hematite is the rule, favoring precipitation of the latter. This case is shown on the thermodynamic diagram of Tardy and Nahon (1985), which outlines in the $Al_2O_3-Fe_2O_3-H_2O$ system the stability fields of aluminous goethite and aluminous hematite as a function of water activity and concentration in Al_2O_3 (Fig. 4.15). Considering aluminous goethite and aluminous hematite, respectively, as solid solutions of goethite-diaspore and hematite-corundum, dehydration of goethite into hematite occurs at higher degrees of water activity than dehydration of diaspore into corundum. A decrease of water activity is expressed by an increase in the content of aluminum in aluminous hematite when the two solid solutions are in equilibrium.

In summary, these processes explain the ferruginous nodular differentiation so characteristic of the mottled clay horizons of tropical landscapes. This nodulization can reach a major development and lead gradually to highly indurated horizons located in the upper part of lateritic profiles. It is the domain of continuous glaebular structures.

Continuous Glaebular Structures: Duricrusts

When glaebular accumulations are moderately differentiated, they appear in the field as indurated pedologic features scattered within soil or alterite. In other words, their occurrence remains local and limited to *discontinuous indurations*. Certain horizons may be the site of important accumulations of elements that may lead to highly indurated zones. In such a case, the accumulation is *continuous*, and, according to the nature of the dominant precipitated mineral, these accumulations are called ferricrete or ferruginous crusts (for iron oxyhydroxides), calcretes or calcareous crusts (for calcium carbonates), mangancrete or manganesiferous crusts (for manganese oxy-hydroxides), and silcretes or siliceous crusts (for SiO_2 compounds). All these accumulations result from the multiplication of glaebules until their reciprocal welding into a continuous indurated and anastomosed network.

Whereas the isolated glaebule appears as a pedologic feature, glaebules

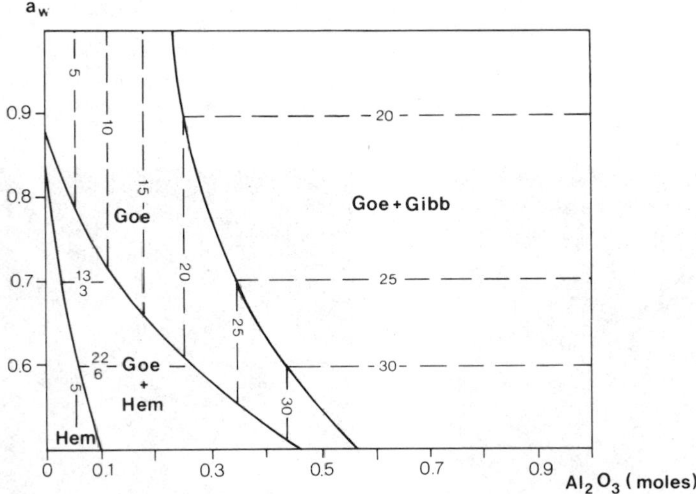

Figure 4.15. Goethite and Hematite Al Substitutions Depending on Water Activity a_w and Al_2O_3 Content for 1 Mole of Fe_2O_3, t=25°C (modified from Didier et al., 1983, and Tardy and Nahon, 1985). Iso-Al lines in goethite or hematite in mole percent. In the presence of an excess of Al (goethite plus gibbsite field) the Al content of goethite is dependent only on a_w. In goethite and hematite fields, all the Al is incorporated into one solid phase and hence the Al content of goethite or hematite is directly related to the bulk composition of the system. In the goethite plus hematite field, both a_w and the bulk content of Al in the system play a part in the fractionation of Al between goethite and hematite. (Reprinted by permission of Sciences Géologiques.)

associated in continuous indurated horizons become responsible for characteristic geomorphological features of landscapes. Such is the case of the silcretes of Australia (see Langford-Smith, 1978, for a review), the ferricretes of Africa (King, 1962; Michel, 1973), and the calcretes of the semiarid areas of North America and North Africa (Gile et al., 1966; Ruellan, 1971a).

In short, the differentiation of these indurated horizons by gradual glaebulization occurs by means of epigenetic replacement of argillaceous or sandy-argillaceous s-matrix by one or several of the plasmic constituents that accumulate secondarily and selectively. This explains the monomineralic character of these glaebular accumulations, which are first discontinuous and then continuous. They express a geochemical differentiation, of long duration, which favors the most stable mineral in a given environment.

Numerous examples of crusts have been studied in the past ten years. Only calcretes are discussed below because they are the most spectacular and perhaps the easiest to understand.

With a few exceptions, due to the structure of the host rock, calcretes macroscopically display profiles with the following succession, from bottom

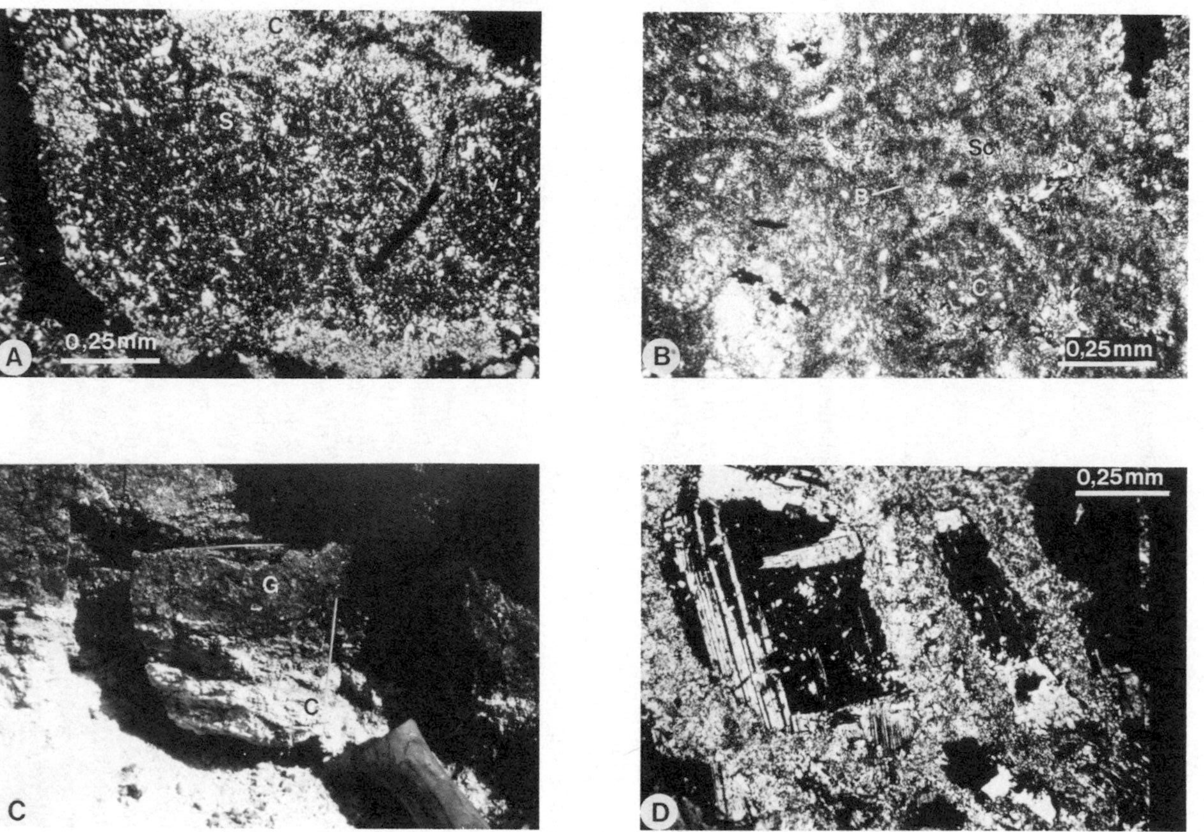
C
S
A
0,25mm
Sc
B
C
B
0,25mm
G
C
C
0,25mm
D

to top of of calcitic plasmic concentrations: spots, nodules, sheets of crust, indurated slab with or without banded film (Ruellan, 1968, 1971a and b).

Microscopic studies completed on several examples in Congo, Burkina Faso, Senegal, Mauritania, Morocco, and Spain (Stoops, 1968; Boulet, 1974; Nahon et al., 1975; Nahon, 1976; Millot et al., 1977; Bech et al., 1980) show that spots and nodules are glaebular plasmic concentrations of calcite of a size ranging from a few microns to 2 cm. Their roughly spherical shape shows that calcitic concentration is centripetal and isotropic. Furthermore, in argillaceous soils, the size of these glaebules seems to be controlled by the size of the aggregates (Nahon, 1976), namely, calcitization began in the fissural cavities limiting argillaceous aggregates and developing at their expense. This situation was verified in other examples, particularly in Morocco (Millot et al., 1977), where calcitization developed from fractures or fissures, but at the expense of the enclosing rock by epigenetic replacement of its constitutive minerals (Fig. 4.16). Clearly, calcite does not precipitate in large pores (fissures limiting aggregates), but from solutions that diffuse from these fissures until they reach small pores of the plasma of the enclosing rock.

Calcitic glaebulization implies the transfer by solutions of Ca^{2+} ions. The latter can originate from the surrounding s-matrix itself by means of intraplasma transfers. However, if Ca^{2+} ions are missing in the s-matrix, long-distance transfers have to be considered. In such a case, the origin of calcium is often controversial. However, an elegant geochemical tool can help the investigator, namely, the value of the isotopic rate Sr^{87}/Sr^{86}. The work of Clauer and Tardy (1971) showed that this ratio is characteristic for each type of rock: basic igneous rocks, acid igneous rocks, freshwater carbonate rocks, and marine carbonate rocks. Establishing the ratio Sr^{87}/Sr^{86} for soil glaebules allows us to determine the origin of calcium, which accompanies the strontium in source rocks, whether marine, freshwater,

Figure 4.16. Epigenetic Calcitization Developed from Fractures and Fissures at the Expense of the Enclosing Rock. Example of Soils and Calcretes Developed on Granite in Southern Morocco (after Millot et al., 1977). **A**: Smectitic microaggregate (microped) partly calcitized in its external boundaries. S is smectitic plasma of the core of the microaggregate. C is calcite replacing smectite. Observation by means of petrographic microscope (crossed nicols). **B**: Aggregates entirely replaced by calcite (micrite C). Boundaries of original smectitic aggregates can be recognized (B). Sparite and microsparite crystallarias fill original fissures, cracks, and alveoles (SC), which previously limited clay aggregates. Observation by means of petrographic microscope (crossed nicols). **C**: Calcite (C) is replacing enclosing fresh granite (G) at the base of profile. Structure of granite is preserved by calcitization. **D**: Detail of C: Feldspar replaced by calcite. Note correspondence of twinning planes between feldspar fragments, indicating in situ epigenetic replacement without displacement. Observation by means of petrographic microscope (crossed nicols). (Reprinted by permission of Sciences Géologiques.)

acid, or basic intrusive, and, hence, to find the sources of the Ca^{2+} supply. This method was successfully applied by Nahon (1976) in collaboration with Clauer in the case of calcretes of the Achouil and of Tasiast in Mauritania.

The gradual and permanent accumulation of calcium in a given horizon is expressed by the multiplication of glaebules, their induration, their coalescence, leading to the formation of decimetric and discontinuous subhorizontal sheets, in French called "crust sheets" (*feuillets de croûte*) (Ruellan, 1968, 1971a). These sheets can become thicker and then display inside planar subhorizontal fissures ranging in width from a few microns to a few centimeters. At this stage, sheets indurate and can grade toward the top of the incrusted horizon into a resistant slab (Fig. 4.17 A and B). These fissures are zones of dissolution and the simultaneous induration of the sheets clearly indicates that dissolution–precipitation of calcite is controlled by vertical or horizontal circulations of solutions. In essence, these anisotropic transfers are responsible for the sheet structure of the upper part of the calcrete. This process is also revealed by dissolution within the sheets of residual quartz grains of the original s-matrix. Some quartz grains may be cut into several fragments by planar and horizontal fissures of dissolution involving a given

Figure 4.17. Calcareous Duricrusts (Calcretes) and Their Main Successive Macrostructures. **A**: Calcrete profile from Achouil Valley (Mauritania) and developed on weathered schists. (1) Weathered schists; (2) elongated calcitic glaebules preserving the original structure of the schist; (3) coalescence of elongated calcitic nodules leading to indurated nodular horizon; (4) highly indurated calcitic slab. **B**: Calcrete profile from Thies (Senegal) developed on Eocene marls. (1) Weathered marls; (2) calcitic nodular horizon; (3) subhorizontal indurated calcitic sheets of crust; (4) highly indurated calcitic slab.

sheet; these fissures can be filled up by calcite (Fig. 4.18). Each quartz fragment displays, under crossed nicols, the same position of extinction as that of the original grains. This texture demonstrates that fissures cutting across quartz grains are in situ dissolution cavities.

These dissolution–precipitation couplets, clearly visible in the upper parts of calcretes, can also affect glaebules, although less commonly. In such a case, fine fissures or discontinuous, curved to concentric alveoles affect the calcareous nodule. The shape of these fissures shows that ion transfers by solution are three dimensional and isotropic. These nodules are often called septaria.

In summary, continuous accumulations lead to essentially monomineralic and highly indurated horizons in which the original s-matrix is almost entirely replaced epigenetically and remains only locally in the state of relicts. This replacement requires dissolution of alumino-silicates and silicates, and loss from the profile of the elements put into solution. Such a process is close to the one called alkalinolysis (Pédro, 1983). The fundamental role played by these large-scale epigenetic replacements in the geochemical evolution of landscapes is examined later (see chapter 5). For the time being, the ubiquitous structure of calcretes, mainly regulated by couplets of dissolution–precipitation of calcite in the upper part of the profile and by epigenetic replacement of host material at the base of the calcrete, should be kept in mind.

Evolutionary Mineralogical Sequences in Glaebular Accumulation

Accumulations of material that take place in the s-matrix replace the constituents of the latter. This epigenetic replacement can become complete

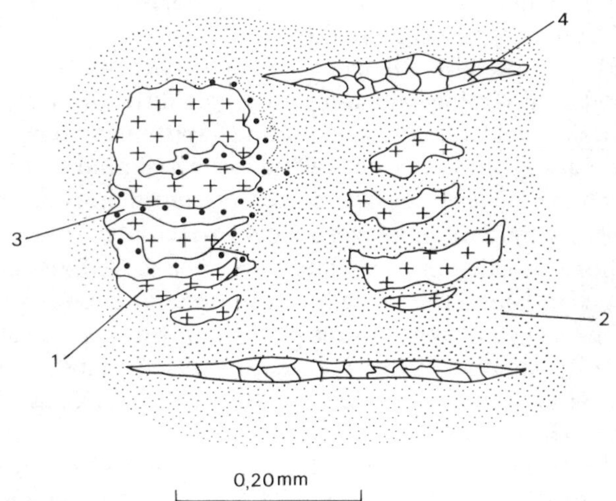

Figure 4.18. Replacement of Quartz Grains by Calcite within a Calcareous Crust Sheet. (1) Quartz grains cut into several fragments; (2) micritic plasma; (3) microsparitic plasma; (4) sparite filling horizontal fissures and voids.

during the gradation from discontinuous to continuous accumulations. From bottom to top of profiles, and from upstream to downstream of sequences, differentiation of these accumulations is concurrent with individualization of a succession of minerals forming them. Indeed, each mineral originates from the previous phase by solution followed by recrystallization in situ with changes of shapes and orientations of crystallites. Thus, from differentiation of glaebules to generation of continuous and indurated horizons, accumulating minerals form a real evolutionary mineralogical sequence. This sequence can be as follows:

1. During glaebular differentiation, mineralogical phase thermodynamically metastable in the considered environment precede, for kinetic reasons, the more stable phases.
2. Physicochemical characteristics, particularly the concentration of solutions, vary during glaebular accumulation and among the deposited mineralogical phases. Earlier ones become destabilized to the benefit of others in equilibrium with the new solutions.
3. By its own differentiation, glaebular accumulation automodifies the microenvironment (particularly the pore system), thus allowing development of new physicochemical conditions favorable to a new mineralogical phase.

In all instances during glaebular accumulation, a sequence of minerals is generated in which the more recent replace the earlier ones. Consequently, it is sometimes difficult to establish for a given glaebule these mineralogical evolutionary sequences, in particular in the earliest accumulations. However, the study of vertical or lateral 'glaebular parentage,' that is, of differentiation of glaebular structures in time and space in a given profile or sequence, allows detection of even the most ephemeral stages in these evolutionary mineralogical sequences of accumulation.

Recent studies show the different stages of accumulation of calcium carbonate in soils. Mineralogical studies by X-ray diffraction of the different units characterizing calcareous accumulation (spots with hazy boundaries, nodules, sheets) always indicate that calcite is the essential constituent of the mineralogical phase. Clearly, calcite represents the major and stable phase of this type of accumulation. Nevertheless, systematic study of the most recent generations of calcium carbonate that appear in the s-matrix, where accumulation occurs along the margin of the functional pore system, shows the ephemeral existence of other forms of calcium carbonate: calcite needles (Bocquier, (1973); Nahon, (1976); Durand, (1979); Bruand, (1985); Pouget and Rambaud, (1980); Butel, (1982); Verges et al., 1982) as well as other crystalline forms (Butel, 1982).

However, the works of Nahon et al. (1980), Dupuis et al. (1984), and Ducloux et al. (1984) show the existence of different mineralogical phases in the calcretes of Vouillé, France, summarized below. The investigated microsequences of plasmic carbonate concentration reach 1 cm thickness. They consist of four microhorizons, numbered from 1 to 4. The carbonate plasma of each microhorizon was sampled and studied by infrared spectrometry between 4000 and 250 cm^{-1} following the technique of KBr tablets (0.7 mg of sample powder for 300 mg of KBr). For the field of spectra pertaining to the vibration of the OH radicals, powders distributed in "nujol" were placed between NaCl lamellae. The powder used as standard was from the calcite of Cumberland and obtained spectra were compared to published ones (Baron et al., 1959; White, 1974). Sampling in situ was done during winter and summer and showed the following results:

WINTER SAMPLING Analyses reveal the general occurrence of calcite in all four microhorizons. Furthermore, horizon 1 contains *amorphous calcium carbonate* (characteristic bands at 1480 and 1400 cm^{-1} for the valence vibration of CO_3^{--} groups). In horizons 2 and 3, amorphous calcium carbonate is accompanied by calcite, but the infrared spectrum of the latter displays the main band at 1445 cm^{-1} instead of 1420 cm^{-1} for regular calcite. This 1445 cm^{-1} calcite contains OH^- groups in its structure as disclosed in the powders diluted in nujol. These OH^- groupings disappear with differential thermal analysis (DTA) between 300° and 550°C, leaving a residue of real calcite. Finally, horizon 4 consists of regular calcite.

SUMMER SAMPLING Analyses show that microhorizon 1 contains calcite and small amounts of amorphous carbonate associated with 1445 cm^{-1} calcite. In horizons 2 and 3, amorphous carbonate occurs in an even smaller quantity, but in addition to 1445 cm^{-1} calcite, a small amount of aragonite is present (vibration bands at 1490, 1520, and 1565 cm^{-1}). Furthermore, aragonite-bearing samples are rich in aliphatic organic matter (strong vibration bands of CH_2 and CH_3 groups between 2800 and 2900 cm^{-1}). This situation is particularly clear in microhorizon 2. Finally, horizon 4 consists entirely of regular calcite.

Analysis of microprofiles of calcretes under pebbles at Vouillé (Vienne, France) and experiments performed under conditions similar to natural ones afford data on the processes of precipitation and the succession of types of calcium carbonate in pedologic environments under temperate climate. Amorphous calcium carbonate appears as a precocious, very hydrated, and metastable form. This amorphous carbonate becomes organized very rapidly into disordered calcite and aragonite, or only into aragonite. Disordered calcite, called here 1445 cm^{-1} calcite, contains OH^- groups in its structure. Eventually, aragonite and disordered calcite recrystallize in turn into regular calcite, the latter being the major constituent of calcretes.

The above-described behavior expresses the differentiated evolution of carbonate during the various seasons. In winter, under cool and humid conditions, amorphous carbonate changes only into disordered calcite, and then into regular calcite. In summer, a period of strong biological activity, amorphous carbonate changes into an association of disordered calcite and aragonite, both of which recrystallize into regular calcite. Although aragonite was described by Goudie (1973) in calcretes of South Africa and the Mediterranean regions, this is the first report of its seasonal development and of its significance in the succession of dissolutions and recrystallizations of the evolutionary mineralogical sequence characterizing the pedologic accumulation of calcium carbonate.

The number of analyzed samples and performed experimentations are sufficient to reach the conclusion that the proposed evolution of carbonates can be generalized to the generation of calcretes under temperate climate.

In summary, calcretes show that seasonal alternations of wetting and drying accumulate new carbonate generations that evolve through several crystallo-chemical stages, from amorphous calcium carbonate to disordered calcite and aragonite, and finally to true calcite. Thus, the mineralogical sequence of carbonate accumulation is established and can be added to the other sequences previously recognized in the major types of meteoric mineralogical accumulation, namely ferruginous, aluminous, siliceous, and argillaceous.

Accumulations of Material in Micropore and Macropore Systems: Coatings

Accumulations of material taking place in the pore system correspond to *plasmic concentrations* directly related to the network of cavities visible under the naked eye or under the petrographic microscope (micro-, macro-, and megacavities) and which involve materials of soils and alterites. In other words, these concentrations result from an illuviation inside cavities. They include cutans, neocutans, quasicutans, and crystallaria described by Brewer (1964), with the exception of constraint cutans, which appear as oriented *plasmic separations*, that is, simple modifications in situ of pedoplasma or weathering plasma under the effect of constraints generated by wetting–drying phases.

In these accumulations, the characterization of their types of distribution and their genetic significance can be understood only if they are related first to the wall of the cavity formed by the s-matrix, and second to the nature, structure, and arrangement of the constituents of the accumulation.

The Different Types of Cutanic Accumulation

Plasmic concentrations differentiated within cavities display very variable structural and mineralogical characteristics. They are ubiquitous in weathering profiles and soils. Hence, their identification indirectly provides data on the nature and the dynamics of percolating solutions.

Plasmic concentrations coating cavities can be either deposits of detrital particles, or products precipitated from solutions, or transformation products of both.

Detrital Plasmic Accumulations

Detrital cutans result from the deposition, on the walls of cavities, of fine particles transported by traction or in suspension by solutions entering these cavities. Coatings consist of detrital particles derived from crystallized or noncrystallized bodies such as debris of organic matter, grains of quartz or of opaque minerals, and argillaceous particles. All of them are temporarily insoluble or weakly soluble in that particular environment.

Deposition of this type of cutans can be considered as a sedimentation in an aqueous microenvironment. Hence, bedforms are those of sedimentary deposits. Indeed, study of detrital cutans shows a microstratification of particles (real varves), which, as for sediments, are controlled by the principle of superposition. In short, two fundamental structural features characterize detrital cutans: alternating microdeposits and their constant geopetal character.

ALTERNATIONS OF MICRODEPOSITS

This situation is called *true microstratification*. All particles are arranged in such a manner that their great axis is either parallel or subparallel to the floor of the cavity underlying the deposit. Microstrata overlie each other either in a regular and rhythmic manner, without discontinuities (a common case in B horizons of argillaceous accumulation), or with internal discontinuities. In such a case, the first deposits are partially eroded and unconformably overlaid by later microstrata (a common example of cutans in leached horizons).

This microstratification may be emphasized by color hues. Studies showed that these hues may result either from a change in the packing of the deposited particles or from the alternation of deposits whose constituents vary in composition, or simply in amount, particularly for oxides.

CONSTANT GEOPETAL CHARACTER OF MICRODEPOSITS

The action of gravity on percolating solutions leads to thicker deposition at the bottom of cavities. However, under the effect of capillarity, which may exist in pores, particles can also be deposited over other parts of the walls of cavities. Nevertheless, in all instances, a geopetal structure develops since detrital cutans are thicker in the lower part of pores.

A typical example is represented by the detrital cutans of the upper horizons of the weathering profiles of Lam-Lam (Senegal). These profiles, investigated by Flicoteaux et al. (1977), developed on a sedimentary bedrock consisting of alternations of smectitic clays, calcium phosphate, and arenaceous-silty beds. Lateritic weathering leads to the generation of horizons of aluminum phosphate with a pore system containing numerous examples of detrital cutans (Fig. 4.19). Many other examples of detrital cutans have been described, in particular by Brewer (1960, 1964), Bocquier and Nalovic (1972), Bocquier (1973), and Boulet (1974). In the lateritic soils of tropical

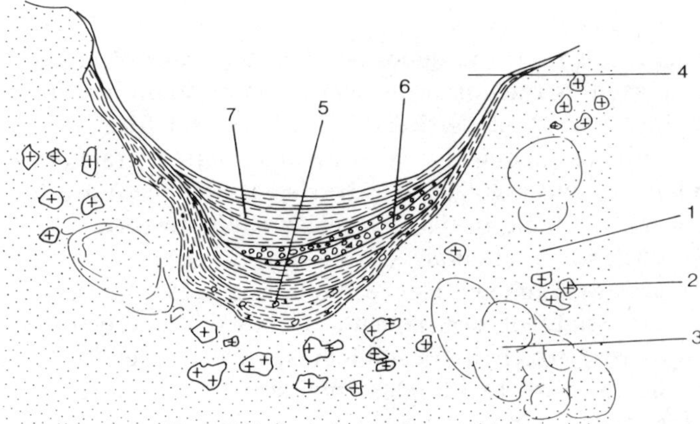

Figure 4.19. Detrital Argillo-Quartzose Coating in an Alumino-Phosphatic Weathering Horizon at Lam Lam, Senegal. (1) Alumino-phosphatic weathering plasma with dominant crandallite; (2) grains of detrital quartz with strongly corroded margins; (3) internal mold of rotaliform foraminifer; (4) alveolar cavity; (5, 6, 7) argillo-quartzose coating with alternations of reddish microbeds consisting of a mixture of kaolinite plus quartzose fine sand or silt plus a small amount of pigmentary goethite on the clay mineral (5), of microbeds of fine sand and predominant quartz silt with reverse graded bedding (6), of pale yellow to gray microbeds of kaolinite (7).

countries, detrital cutans of soils and alterites consist essentially of argilloferrans (kaolinite plus goethite) or of silty coatings of quartz and iron oxides. Illuviation of kaolinite particles on the surfaces of which fine particles, or crystallites of goethite or of amorphous iron hydroxides, are adsorbed, were described by Maignien (1958) and, particularly, by Carroll (1958).

Neoformed Plasmic Accumulations

These plasmic accumulations result from crystallization or simply precipitation on the walls of cavities from elements in solution. Neoformed cutans display different arrangements of their constitutive crystals, crystallites, or amorphous materials. When crystals are well individualized and clearly observable under the petrographic-microscope, they display an oriented growth; namely, crystals grow perpendicularly to the walls of cavities or, more rarely, parallel to them. "Crystallaria" described by Brewer (1964) and "crystal chambers" of Kubiena (1938) belong to this group of neoformed cutans. The latter occur in all horizons of a given profile and their nature affords, indirectly, data on the composition of circulating solutions at a given time of the history of the profile.

Two simple cases can occur: Either plasmic concentration is different mineralogically from the surrounding s-matrix, or the cutan is identical or close in mineralogical nature to the s-matrix.

Two examples illustrate the first case (Fig. 4.20 A and B). First, cutans of

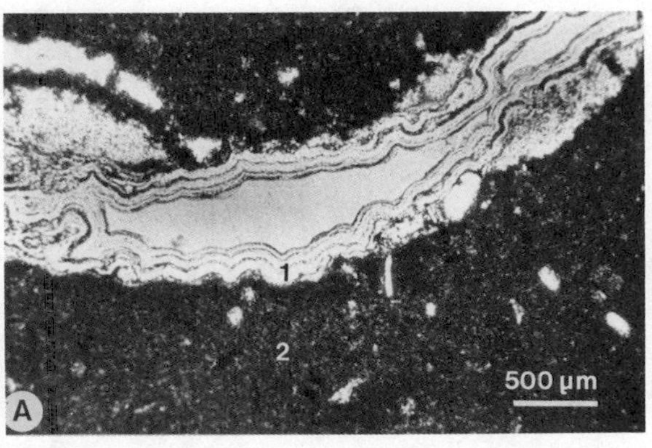

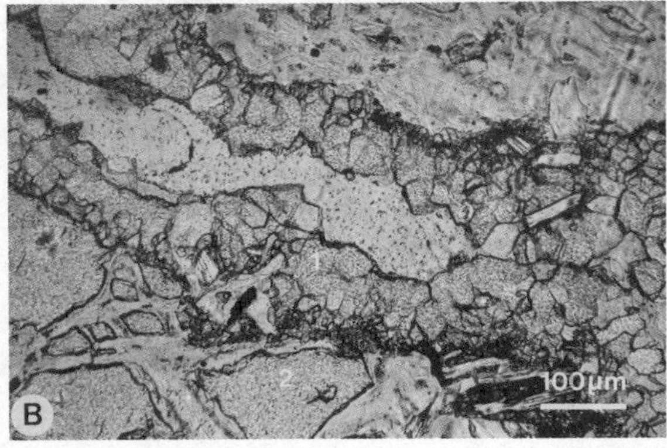

Figure 4.20. Neoformed Plasmic Concentrations, Cutans, Different Mineralogically from the Surrounding S-Matrix. **A**: Cutan of amorphous silica (1), on the walls of alveolar cavity within a calcitic nodule (2). **B**: Cutan of Mg-calcite (1) in fissure of a weathered serpentinized dunite (2).

amorphous silica at the base of a calcrete horizon developed on a migmatitic gruss of Tasiast in Mauritania (Nahon, 1976). Weathered migmatites underwent carbonate accumulation. In alveolar cavities of calcareous nodules at the base of the profile, coatings of well-banded amorphous silica, black under crossed nicols, occur. Silica might have originated from dissolution of quartz grains of the gruss at the top of the calcrete. In a second example, Nahon et al. (1982) described cutans of magnesian calcite at the base of a weathering profile of ultrabasic rocks (serpentinized dunite) at Sipilou (Ivory Coast, West Africa). There, magnesian calcite precipitated only in the transition zone to the fresh rock.

Two examples illustrate the second case (Fig. 4.21 A and B). In the first example, Nahon (1976) described a coating of sparitic calcite on the walls of an alveole within a sheet of crust of a calcrete at Thies (Senegal). The s-

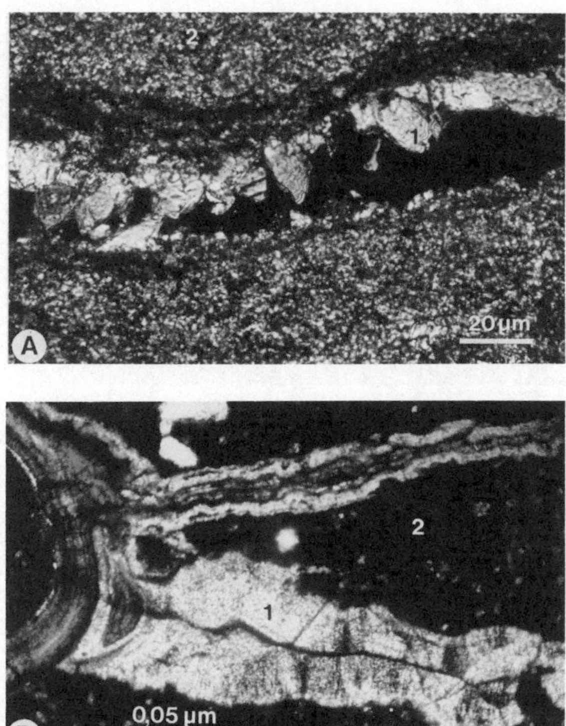

Figure 4.21. Neoformed Plasmic Concentration, Cutans, Close in Mineralogical Nature to the Surrounding S-Matrix. **A**: Cutan of sparitic calcite (1) on the walls of an alveole within an indurated calcitic sheet of crust (2). **B**: Cutan of fibroradiated goethite (1) on the walls of fractures and alveoles within an hematitic s-matrix (2). (Fig. 21-A from Nahon 1976, reprinted by permission of Sciences Géologiques.)

matrix of the calcrete consisted of micritic calcite. The sparite crystals of the coating were better developed on the roof of the cavity than on the floor. This geopetal character is common to this type of coating. Furthermore, the same character occurs also in the neocutans of iron hydroxides, which indicates that the latter are also related to the alveolar cavity. The second example is a coating of well-crystallized goethite with fibroradiated structure perpendicular to the walls of an alveole in an indurated ferruginous horizon of hematite of Senegal.

The above-mentioned examples clearly show that the elements necessary to the formation of cutans are transported by solutions through the pore system. However, these elements indicate a transfer over variable distances.

When the mineralogical nature of cutans is similar or close to that of the walls supporting them, the source of the transferred elements is probably the enclosing s-matrix. This situation is obvious in the example of the sparitic calcite cutans of Thies (Nahon, 1976). The polarity of the stalagtitic sparite indicates that it probably originated from diffusion of a solution through cryptocavities of the enclosing calcitic s-matrix, followed by precipitation along the margins of the cavity. This situation probably also occurred in the goethite cutan. Transfers, therefore, took place over short distances through the enclosing plasma (interplasma microtransfers).

Subcutanic Features: Neocutans and Quasicutans

Subcutanic features can be defined according to Brewer (1964) as individualized plasmic concentrations related to the cavity system, but within the plasma. Thus, neocutans develop against the wall of the cavity, but in the plasma, whereas quasicutans also develop in the plasma, but slightly behind with respect to the wall of the cavity. Excluded from subcutanic features are plasmic separations considered above as simple reorientations of weathering plasmas in relation with cavity discontinuities. Thus defined, subcutanic features correspond well to transfers and localized accumulations of material.

The difficulty generally lies in the interpretation of the direction of transfer and of accumulation and, hence, in their genetic significance. Several situations can occur.

The neocutan is directly related to the coating on the wall of the cavity. In this case, the root of a cutan penetrates the cryptoporosity of the enclosing plasma. Hence, this neocutan is an integrant part of the cutan and represents a deposit or precipitation of material from solution circulating in the macro- or micropore system, this is followed by penetration through diffusion in cryptocavities of the plasma surrounding the cavity. Penetration through diffusion is not necessarily limited to cryptocavities, but can replace a portion of the mineral constituents of the surrounding plasma, or even a fraction of the skeleton. This is, in fact, an epigenetic replacement of the s-matrix comparable to that described for glaebulization. Such a case was reported to occur commonly in the calcretes of Morocco (Fig. 4.22).

If plasmic concentration, which takes place from the cavity toward the enclosing plasma or in the opposite direction, involves different elements that participate in the same transfer process, a real chemical chromatography can appear at the microscopic scale. Spectacular examples of this situation also occur in the calcretes of the Bargny Plateau of western Senegal. Thus margins of fissural or alveolar cavities often display successions of plasmic segregations of oxyhydroxides of iron and manganese and of calcium carbonate (Fig. 4.23). The arrangement of any of these constituents as quasicutan, neocutan, or cutan determines the direction of transfer. In case A of Figure 4.23, the iron oxyhydroxide is in quasicutan position, the manganese oxyhydroxide is in neocutan position, and calcite is in cutan position. The direction of transfer is from the enclosing plasma toward the macrocavity. In other words, segregation between elements occurs in the order of their increasing mobility during their transfer in alkaline environment. In case B of Figure 4.23, the calcite cutan is missing and the iron oxyhydroxide is in

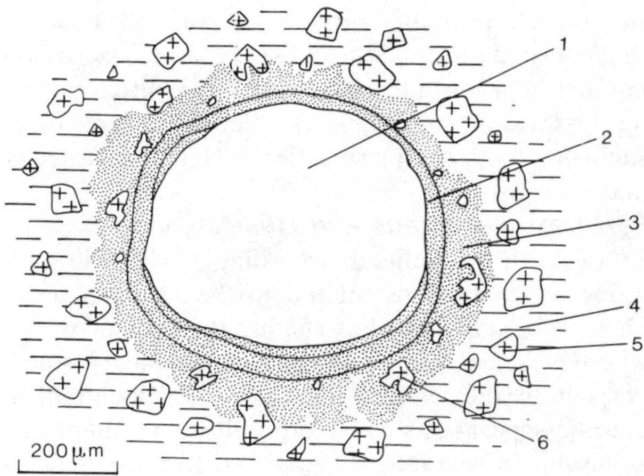

Figure 4.22. Cutan and Calcitic Neocutan. (1) Tubular microvoid; (2) calcitic cutan with micritic crystals lighter-colored than in (3); (3) calcitic neocutan with micritic crystals; (4) surrounding weathering argilliplasma; (5) angular to rounded grains of skeleton (quartz); (6) grains of skeleton with corroded margins displaying embayments invaded by the calcitic plasma of the neocutan.

the position of the neocutan. Transfer is therefore directed toward the enclosing plasma, as shown by the growth arrangement of manganese dendrites in quasicutan position.

Finally, neocutans or quasicutans can result from a plasmic concentration of material by secondary segregation of certain constituents of the cutan. This differentiation is subsequent to the formation of the cutan, and the direction of transfer is always toward the enclosing plasma. This type of example, which belongs, in fact, to late processes of transformation that can affect a cutan, is discussed below in more detail.

Transformations Affecting Cutans

Several types of transformation can affect cutans or crystallarias after their deposition; they either belong to essentially mechanical processes or are of crystallo-chemical nature.

Mechanical Transformations

Mechanical transformations can affect argillaceous cutans. In Chapter 3, the latter were shown to develop within themselves, under processes of wetting and drying, strong mechanical constraints responsible for most of the rearrangements and reorientations of their constitutive argillaceous particles.

Argillaceous cutans (called argillans), when formed by argillaceous particles reacting to phases of wetting and drying, are regulated by the same processes and can develop constraints within themselves that lead to their fragmen-

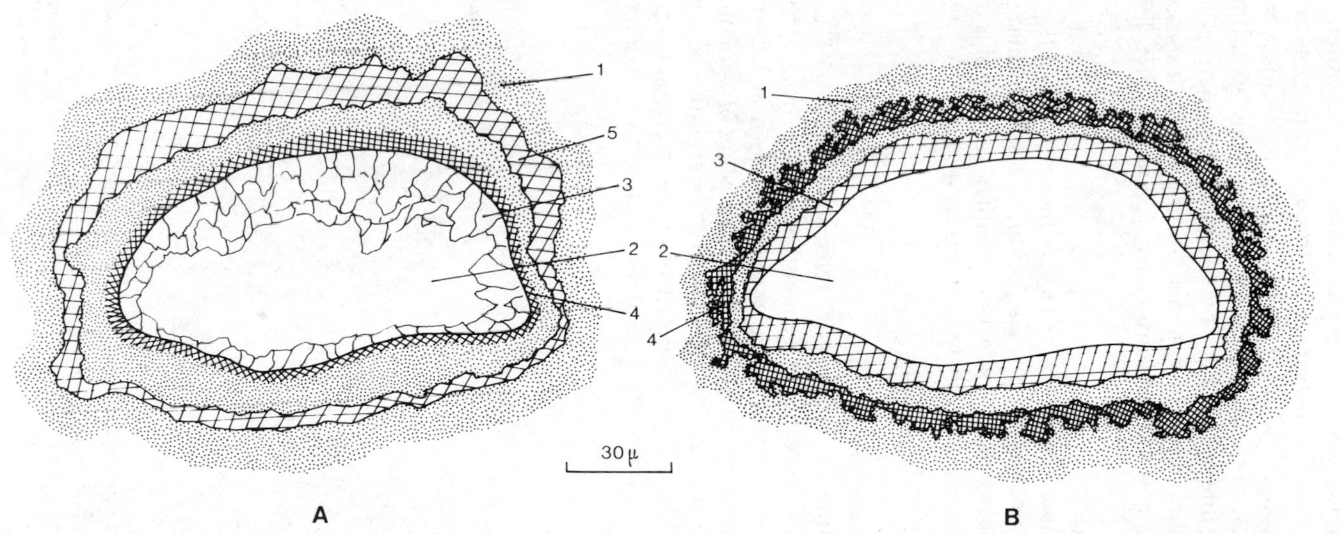

Figure 4.23. Schematic Geometric Relations between Quasicutans and Neocutans of Iron or Manganese Oxyhydroxides and Calcitic Cutan in Calcretes of the Bargny Plateau (Senegal). **A**: (1) Enclosing calcitic plasma with micrite crystals; (2) alveolar microcavity; (3) calcitic cutan with sparite crystals oriented perpendicularly to the walls of the cavity (crystallaria); (4) black neocutan of manganese oxyhydroxides impregnating the micritic plasma; (5) ochre to brown quasicutan of iron oxyhydroxides (goethite) impregnating the micritic plasma. **B**: (1) Enclosing calcitic plasma with micrite crystals; (2) alveolar microcavity; (3) ochre to brown neocutan of iron oxyhydroxides (goethite) impregnating the micritic plasma; (4) black quasicutan of manganese oxyhydroxides (birnessite?) impregnating the micritic plasma.

tation into microaggregates. It is a *fissuration of cutans by internal constraint* (Fig. 4.24).

If the plasma enclosing the cavities in which cutans are deposited is also argillaceous, the plasma itself can be submitted to constraints generated by hydric variations. This leads to a general rearrangement of the s-matrix. Certain pedologic features, particularly cutans, undergo this evolution and hence become gradually integrated into the reorganized s-matrix. It is a *mechanical fragmentation of cutans by external constraint* (the fragments of cutans integrated into the new s-matrix are called "papules" (Fig. 4.25).

Geochemical Transformations

After their deposition, cutans and crystallarias can undergo, under the effect of particles in suspension or of elements in solution, late transformations of a geochemical nature. These are of two types: either transformations by segregation or crystallo-chemical transformations resulting from loss or accumulation.

TRANSFORMATIONS BY SEGREGATION

This type of transformation is frequent in soils and alterites of tropical regions. At least one of the constituent elements separates from a zone of the original cutan and reconcentrates in another. A typical example is represented by the coatings of cavities (Fig. 4.26) consisting of kaolinite and iron oxyhydroxides, amorphous or crystallized (goethite), which are commonly called ferriargillans. These deposits of argillo-ferruginous plasma are either homogeneous in color (red) and in composition (contents of

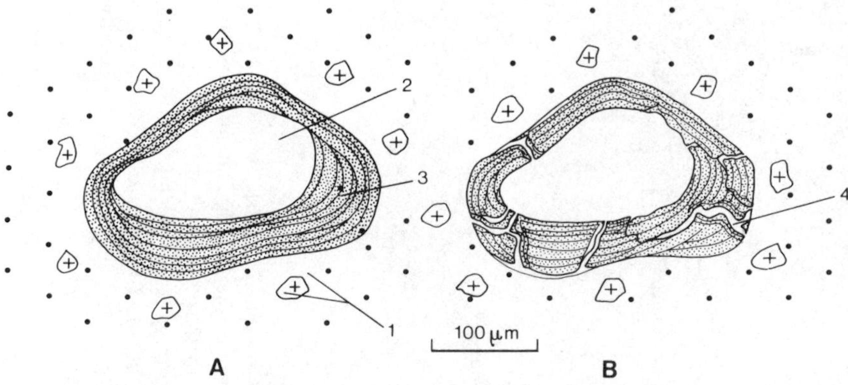

Figure 4.24. Schematic Illustration of the Fissuration of an Argillaceous Cutan under Internal Constraint. **A**: stage 1, zoned argillaceous cutan. **B**: stage 2, fissured zoned argillaceous cutan. (1) plasma and grain of enclosing skeleton; (2) microvoid; (3) argillaceous zoned cutan; (4) secondary fissures with marginal reorientation of the cutan argillaceous particles.

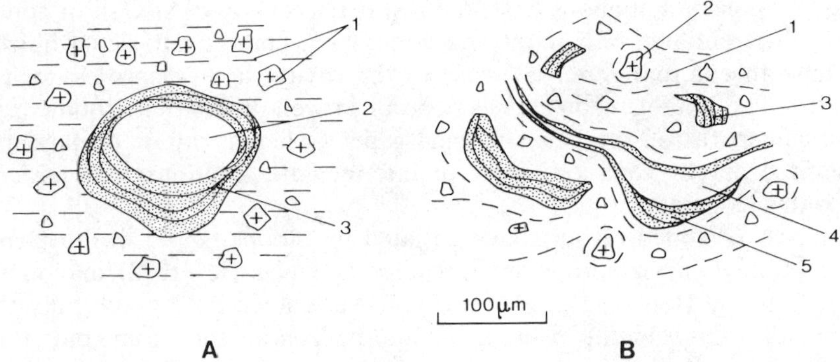

Figure 4.25. Schematic Illustration of the Integration of a Zoned Argillaceous Cutan into a Reorganized S-Matrix under the effect of Mechanical Constraints. **A**: Stage 1. Zoned argillaceous cutan within the unaffected s-matrix. (1) Grain of the skeleton (important coarse-grained fraction); (2) microvoid; (3) zoned argillaceous cutan. **B**: Stage 2. Argillaceous cutan (papules) integrated into the reorganized s-matrix. (1) Reorganized plasma; (2) reorganized grains of the skeleton (dominant fine-grained fraction); (3) fragment of zoned argillaceous cutan integrated into the s-matrix (= papules); (4) fissural void; (5) new generation of argillaceous cutan.

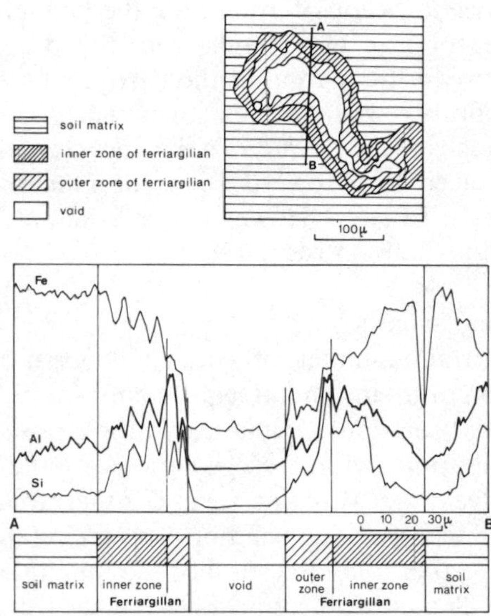

Figure 4.26. Distribution of the Main Chemical Elements Shown by a Microprobe Profile of the Various Zones of a Ferriargillan (modified from Bocquier and Nalovic, 1972; and Nahon and Bocquier, 1983). (Reprinted by permission of Sciences Géologiques.)

Fe_2O_3 are constant at about 12%, the rest being SiO_2 and Al_2O_3), or zoned, that is, differentiated into an internal zone, more colored and richer in iron, and into an external zone, adjacent to the cavity, lighter and less rich in iron. The red homogeneous deposits occur in well-drained environments in the upper parts of profiles and landscapes, whereas those displaying a constant order of zonation occur in intermediate locations that undergo temporary clogging.

Analysis of zoned ferriargillans initiated by Stoops (1967) in the Congo was continued in Cameroon by Bocquier and Nalovic (1972) and in the Ivory Coast by Boulangé et al. (1975). It was shown petrographically that the boundary between the external and internal zones varies from gradational to abrupt and that it can intersect the orientation of the argillaceous particles (microbeds), indicating that in such a case the redistribution of iron corresponding to the zonation is much later than the deposition of the argillo-ferruginous plasma. Microprobe analysis (Fig. 4.26) shows that kaolinite occurring in the two zones does not undergo weathering and that iron content gradually increases from the margins of the cavity toward the internal zone. This increase of iron content, related to that of kaolinite in the external zone, becomes independent of the behavior of kaolinite in the internal zone of the ferriargillans. Therefore, in these zoned ferriargillans, a redistribution of iron takes place after their deposition corresponding to a deferruginization of the external zone adjacent to the cavity, and to a concentration of the iron toward the internal zone bordering the enclosing s-matrix. It is a *centripetal diffusion* of iron inside the kaolinitic plasma. Such an intraplasma transfer of iron may result from the difference of oxido-reduction potential between the margin of the cavity, where more reducing conditions occur temporarily when the cavity is water-saturated, and the contact of the enclosing s-matrix, where oxidizing conditions still occur. Such diffusions of reduced iron toward zones of greater oxidoreduction potential were recognized in soils (Howeler and Bouldin, 1971) and were also reproduced experimentally (Vizier, 1983).

CRYSTALLO-CHEMICAL TRANSFORMATIONS

After deposition, a certain number of cutanic deposits undergo a very appreciable modification of their mineralogical composition, which is a function of the nature of the solutions circulating in their pore system. Cutans are no longer in equilibrium with these solutions. Some of the mineral constituents are dissolved, and one or several other mineral phases, in equilibrium with the new percolating solutions, reprecipitate and take their place. In this situation, either one or several of the minerals that precipitate consist of a portion of elements not removed from the environment of the original cutan (neoformation by relative accumulation), or one or several of the minerals that precipitate do not use any of the elements forming the original cutan (neoformation by absolute accumulation). The transformation is called *subtractive* when, with respect to the concentration of the percolating solutions responsible for the disequilibrium, the formation of the new cutan occurs by relative-accumulation of ions taken essentially in situ and not from the solution. Inversely, the transformation is called *additive* if the formation of the new cutan takes its elements from those brought to equilibrium by the solution. In this case, a phase is replaced by another of very different composition: It is an epigenetic replacement.

Neither the subtractive nor the additive transformation is a simple segregation between two different constituents within a coating as described above. Both are a crystallo-chemical modification of their mineralogical phases.

SUBTRACTIVE TRANSFORMATIONS

The most typical examples of this type of cutanic transformation were described by Boulangé et al. (1975) and by Mpiana (1980) in bauxitic profiles of tropical and equatorial Africa. The first example pertains to the evolution of argillo-ferruginous cutans (argilloferrans) in bauxitic horizons of Lakota, Ivory Coast (Boulangé et al., 1975). This bauxite was formed by in situ weathering of a muscovite–biotite granite of the region of Lakota, Ivory Coast. Weathering preserved the original structure of the granite from the base of the profile (weathering horizon) all the way up to the indurated bauxite horizons. The latter display numerous cutans in which oxyhydroxides of iron and aluminum alternate with kaolinite.

Argilloferrans consist of an essentially kaolinitic zone (yellowish) and one that is essentially goethitic (brown). The ferruginous zone is always located in the internal portion adjacent to the enclosing plasma (the latter consists of gibbsite resulting from weathering of feldspars with preservation of the shapes of the original parent crystals). The boundary between ferruginous and kaolinitic zones is often hazy and sometimes oblique to the bedded structure of the deposit; that is, it intersects the orientation of the particles. This demonstrates, as in the previous example, that the iron-enriched zone represents a concentration by intraplasmic transfer, subsequent to deposition of the cutan.

The kaolinitic zone of the cutan can become discolored *laterally and gradually* and particles of kaolinite are replaced at the same time by minute fibroradiated crystals of gibbsite (Fig. 4.27 A and B). Boulangé et al. (1975) demonstrated that this situation results from a desilicification by dissolution of kaolinite and reprecipitation of alumina in situ as gibbsite. This explains the formation in that particular bauxitic horizon of most of the complex cutans in which a gibbsite zone overlies an iron oxide zone.

In summary, kaolinite cutans that can originate from deposition of particles from a suspension may be subsequent desilicified in undersaturated microenvironments, leading to the neoformation of cutans of aluminum hydroxides.

The second example pertains to the evolution of cutans of amorphous ferruginous and aluminous hydroxides described by Mpiana (1980). His study of three bauxitic profiles developed, respectively, on Eburnean schists

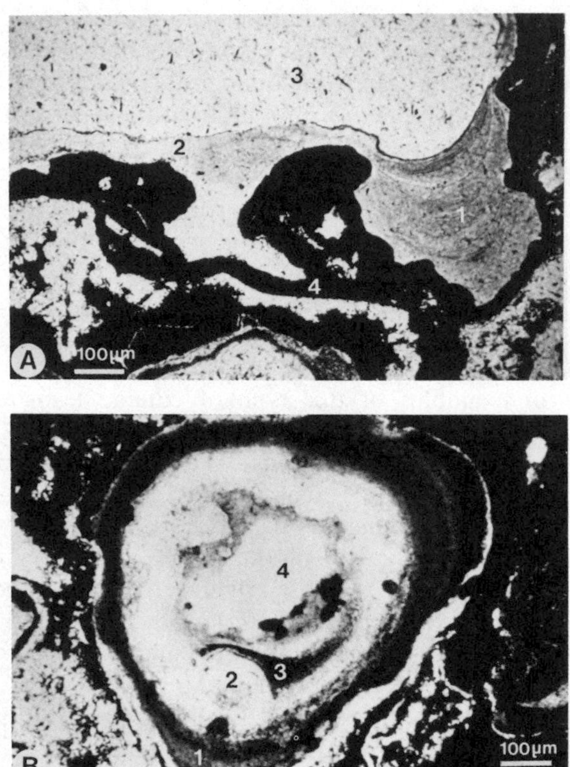

Figure 4.27. Geochemical Transformation of Argilloferranes into Gibbsitic Cutans in Bauxitic Horizons of Lakota, Ivory Coast (after Bocquier et al., 1984). **A**: Zoned argilloferran (1) grading laterally into a cutan with gibbsite crystals (2), (3) is a microvoid. Notice the iron-enriched internal zone of the coating (4) **B**: Internal portion of the iron-enriched cutan (1) External portion of the cutan consisting of gibbsite (2) Notice a relict of original argilloferrane (3) within the gibbsite. (4) is an open void. (Photographs Courtesy B. Boulangé. Fig. 27-A reprinted by permission of IAG, University of Sao Paulo.)

(region of Bénené, Ivory Coast), on Middle Precambrian granite (region of Lakota, Ivory Coast), and on Mesozoic-Cenozoic basalt (region of Ngaoundal, Cameroon) showed that amorphous alumino-ferric complexes formed the essential part of cutanic coatings or fillings of bauxitic horizons at the top of profiles (pseudobrecciated or pisolitic bauxitic crusts).

These cutans of amorphous materials have a light yellow to red color in plane polarized light and are black under crossed nicols. Most commonly, these materials are devoid of any particular orientation. Nevertheless, when an orientation is visible, it is parallel to the walls of the original cavity or displays a finely nodular microstructure. X-ray diffraction analysis confirms the noncrystallized state of these materials. Diagrams show only two broad peaks (between 6 and 13 Å with a maximum around 8 Å, and between 13 and 33 Å with a maximum around 20 or 22 Å).

These cutans show variations of chemical composition not only from one studied horizon to another, but also from one place to another within the same coating. In general, variations of composition are as follows:

Al_2O_3 = 34–49%
Fe_2O_3 = 10–27%
SiO_2 = 0.5–3%

Most commonly, a ferruginous hematitic cutan, sometimes alternating with coarsely crystallized and discontinuous gibbsite cutans, develops against the enclosing plasma. Inversely, cutans of amorphous materials are always located in the most external position and represent the most recent filling (Fig. 4.28).

Under SEM these amorphous materials show a massive finely granular nanostructure grading into filamentous tubular shapes associated with

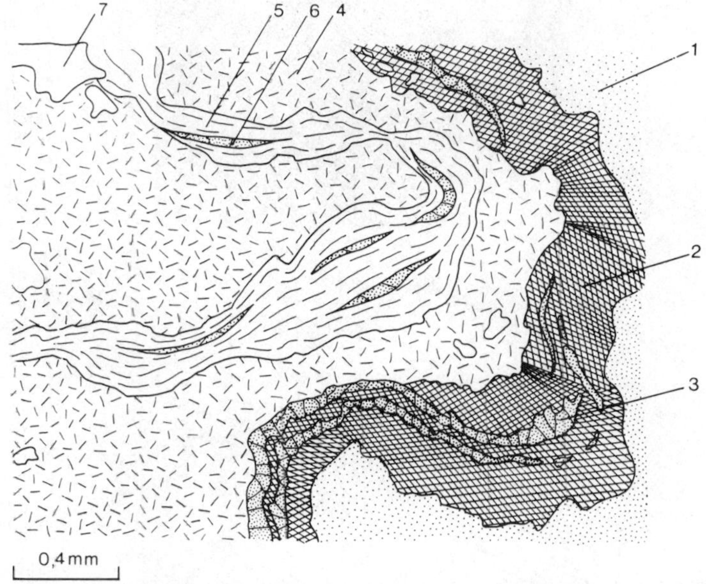

Figure 4.28. Cutans of Amorphous and Crystallized Ferruginous and Aluminous Materials in Bauxitic Horizons Developed over Schists in the Ivory Coast. (Synthetic sketch based on investigations of Mpiana, 1980.) (1) Surrounding gibbsitic plasma; (2) coating of brown crystallized iron oxyhydroxides; (3) coating of well-crystallized gibbsite (crystals perpendicular to walls); (4) coating of red amorphous alumino-ferruginous materials devoid of visible orientation; (5) yellow to gray differentiation within amorphous alumino-ferruginous materials, clearly visible orientation; (6) colorless microbanding of finely crystallized gibbsite; (7) microvoid.

spherules (< 0.5 μm, Fig. 4.29 A). These amorphous materials are locally discolored and in such a case display an orientation. Under TEM, these oriented zones appear more finely porous and display a differentiation of minute gibbsite monocrystals from a background of amorphous clotted of

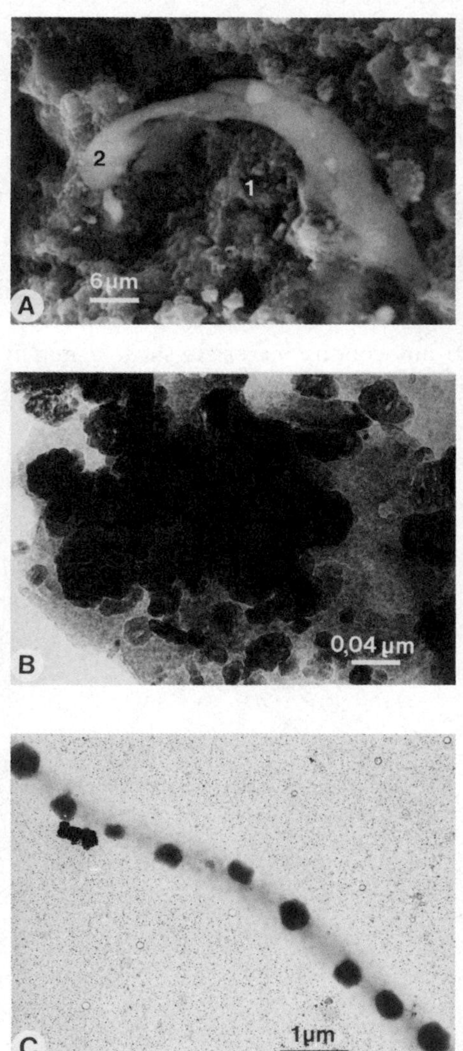

Figure 4.29. Evolution of Cutans of Amorphous Ferruginous and Aluminous Hydroxides into Gibbsite (from Mpiana, 1980). (Photographs courtesy of C. Parron and Mpiana) **A**: Amorphous cutan showing a massive finely granular nanostructure (1) and filamentous shapes (2). Observation under SEM. **B**: Differentiation of minute gibbsite from a background of amorphous clotted masse. **C**: Differentiation of minute gibbsite monocrystals from a background of amorphous filamentous masse. (**B** and **C** are observations under TEM.)

filamentous masses (Fig. 4.29 B and C). This confirms observation under the petrographic microscope and shows a *gradual and lateral transition* from an amorphous yellow zone to a lighter zone devoid of iron and, eventually, to fine bands of well-crystallized gibbsite displaying the same orientation as that of the amorphous materials within which they developed. This evolution traced with the microprobe shows clearly that desilicification followed by deferruginization are gradual and complete.

In short, cutans consisting of amorphous alumino-ferruginous complexes gradually lead, by leaching of silica and iron, to generation of gibbsitic coating. Silica is eliminated from the microenvironment in which these transformations occur, whereas iron is concentrated toward the more internal zones of the cutan where, indeed, colorations (hence iron contents) become stronger (red to brown).

Moreover, alternations of cutans with oxides or oxyhydroxides of iron and hydroxides of alumina located in the more internal position against the wall could represent the first generations of amorphous alumino-ferruginous coatings, subsequently clearly differentiated into bands of hematite or aluminous goethite and of gibbsite.

In summary, examples selected to illustrate crystallo-chemical transformations undergone by cutans after their deposition show that these transformations always begin with a segregation of iron oxyhydroxides, which are transferred toward the more internal parts of the cutan, followed by a reorganization of new crystalline structures.

Thus, for particles of kaolinite, Chauvel (1976) showed that ferric hydrates adsorbed at their surface could make these particles inactive to the aqueous phase. Similarly, Herbillon (1961) and Herbillon and Gastuche (1962) demonstrated the inhibiting role of silica or of ferric ions in the crystallization of amorphous compounds of alumina, and the fact that elimination of these foreign ions (deionization according to Herbillon, 1961) determines the crystallization process toward gibbsite.

Kaolinite particles, therefore freed from protecting iron hydrates become accessible to aqueous solutions and are dissociated; silica is eliminated, aluminum precipitates in situ as gibbsite. Thus, amorphous materials, cleansed of their silica and iron, may reorganize into crystallized aluminum hydroxides.

ADDITIVE TRANSFORMATIONS

Alterites with alumino-calcic phosphates and aluminum phosphates, developed over argillo-phosphatic sedimentary rocks, often display in the micro- and macroporosity of the weathered rock coating of kaolinite, aluminum phosphate, or alumino-calcic phosphates. Several such examples were described by Altschuler et al. (1956, 1958) in the Bone Valley Formation of Florida and by Flicoteaux et al. (1977) and Flicoteaux (1982) in the Lam Lam Formation of Senegal.

In the latter example, certain argillans (consisting of kaolinite) or argilloferrans (consisting of kaolinite plus goethite) display lateral and vertical variations. These coatings can grade into a cryptocrystalline crystalliplasma of millisite. The boundary between kaolinite of the argillan or ferroargillan and millisite is gradational; iron hydroxide diffuses but the morphology and zoned aspect of the original coating are preserved by millisite crystals. Thus, the transition leads to a real millisite coating that could be called "millisan" (Fig. 4.30 A). All stages of vertical and lateral gradation can be observed between argillan and millisan, while the latter encloses patches of argillan. Furthermore, comparison between millisan coatings and those consisting of aluminum phosphates (wavellite) shows that millisans preserve a zoned aspect and, like argillans, are thicker at the bottom of cavities (Fig. 4.30 B). Inversely, wavellite coatings (wavellans) display fibroradiated crystals that are better developed in the upper part of cavities. Assuming that the original argillan is devoid of phosphorus, the zoned coatings of millisite (millisans) can be interpreted as resulting from the epigenetic replacement of preexisting argillans (kaolinite) or argilloferrans (kaolinite plus goethite). The detailed processes of this epigenetic replacement have been shown in the laboratory experiments by many workers on the reactions clay minerals-phosphate. Coleman (1944) showed, for instance, the minor importance of the type of clay mineral on phosphorus fixation. According to Stout (1939) in Segalen (1973), Black (1943), Low and Black (1948, 1950), Cole and Jackson (1951), Wey (1953, 1955), Kittrick and Jackson (1954, 1955), Hemwall (1957), Wada (1959), Tamini et al. (1964), Taylor and Gurney (1965), Muljadi et al. (1966), Webber and Clarke (1969), and Hudcova (1970), reactions phosphates-clay minerals take place in two stages. The first one — fast — belongs to an adsorption process; the second one — slower and more continuous — leads to destruction of the clay mineral lattice, particularly in the case of kaolinite. These reactions depend not only on the concentration of solutions used in the experiments, but also on the size of the particles of clay minerals and on the pH at a given temperature.

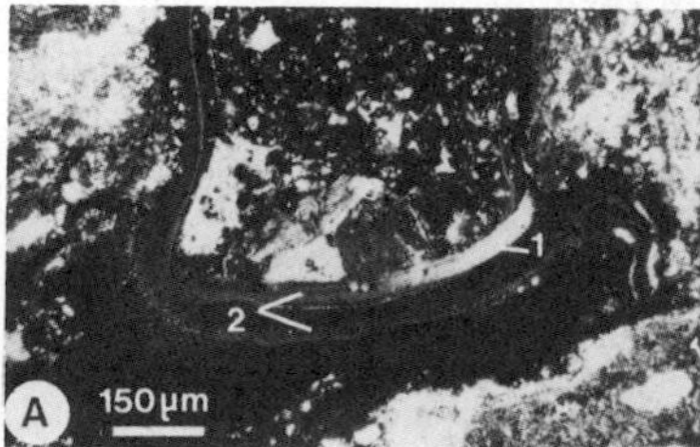

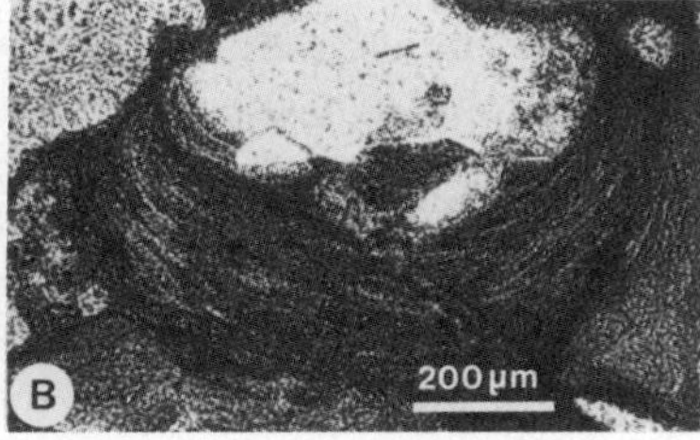

Figure 4.30. Transformation of Argilloferrans (Kaolinite plus Goethite) into Cutans of Millisite of Additive Transfers of Phosphorus (from Flicoteaux et al., 1977). **A**: Lateral transformation of argilloferran (1) into millisan (2). Observation under petrographic microscope (crossed nicols). **B**: Complete transformation of argilloferran into millisan. Note original structure of argilloferrane which is preserved. Observation under petrographic microscope (plane light). (Reprinted by permission of Sciences Géologiques.)

Adsorption of phosphate takes place at the OH level (Kittrick and Jackson, 1956), or at the Al level (Wey, 1953). However, the essential process of this epigenetic replacement is congruent dissolution of kaolinite and precipitation of alumino-calcic phosphate. This was called "phosphatolysis" by Low and Black (1948) and was tested by them in the laboratory in 1950.

Millisite of coatings can in turn evolve by simple hydrolysis into crandallite and/or wavellite.

Numerous other cases of additive transformation are quoted in the literature. In most cases, the original coating or filling consists of an argillaceous plasma. Such is the case of the epigenetic replacement of certain smectites by calcite in calcretes, or of the epigenetic replacement of kaolinite by hematite in ferruginous crusts. In both cases, this epigenetic replacement requires dissociation of the preexisting clay mineral and precipitation of the replacing mineral. Each instance seems to imply the existence, at the surface of the clay bed, of strongly alkaline conditions (calcretes), or strongly acid ones (ferruginous crusts) that would lead to dissolution. In fact, it is the same type of process that regulates glaebular accumulation within an argillaceous s-matrix.

HISTORICAL AND GEOCHEMICAL DIFFERENTIATIONS OF ACCUMULATION

Observation of all the accumulations developed in a horizon, a profile, or a sequence frequently shows that mineralogical composition and structures vary from one place to another in a horizon, from bottom to top in a profile, and from downstream to upstream in a sequence. In other words, *an order of differentiation of structures exists*. This order can be historical and geochemical if these structures develop successively, or it can be only geochemical if they develop simultaneously.

Structures Resulting from Successive Accumulations

Petrographic analysis shows the existence of one or several generations of plasmic concentrations whose differentiation requires different geochemical conditions that cannot possibly be considered as simultaneous.

A typical example is the juxtaposition in weathering horizons of the Pleistocene basalts of Yoff in Senegal (Lappartient, 1970) of ferruginous and calcitic glaebules. Furthermore, calcitic glaebules can, during their growth, enclose one or several ferruginous glaebules. These two generations of glaebules show clearly by their arrangement and texture that they formed in situ in the weathering s-matrix of basalts (Lappartient, 1970) and that they are therefore successive.

In this case, *differentiation* of plasmic concentrations generated in the s-matrix is *historical* and their relative chronology can be clearly established (Fig. 4.31 A).

Structures Resulting from Simultaneous Accumulations

It is not unusual to observe in different horizons of the same profile, or even in a single horizon, plasmic concentrations of different mineralogical composition. Three cases are described below.

The first one is a ferralitic profile developed over a granite–gneiss in the

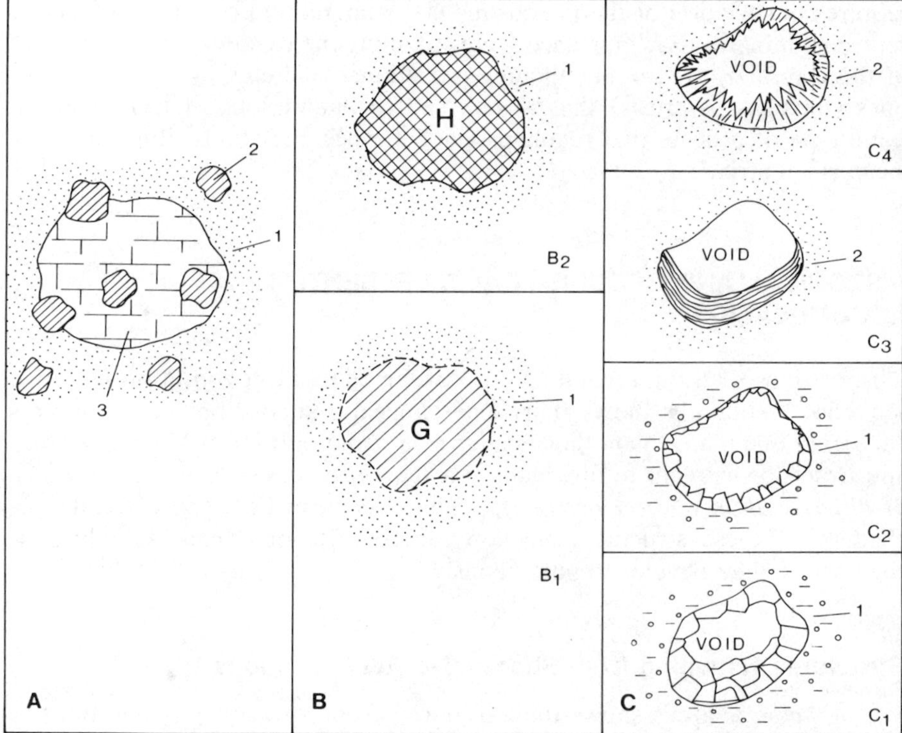

Figure 4.31. Differentiation Order of Structural Units of Concentration. **A**: (1) Argillaceous s-matrix; (2) goethitic glaebule with preserved texture of basaltic rock; (3) calcitic glaebule. **B**: *B1. Weathering horizon*: (1) Argillaceous s-matrix; (G) goethitic glaebule with hazy boundaries. *B2. Base of nodular horizon*: (1) Argillaceous s-matrix; (H) indurated hematitic glaebule. **C**: (C1) Parent rock (1) with fissural voids and tubules with calcitic filling. (C2) Parent rock (1) with fissural voids and tubules with filling of apatite crystals (associated with euhedral quartz crystals). (C3) Weathering s-matrix (2) with tubular voids coated with an association of kaolinite, goethite and alumino-calcic phosphates. (C4) Weathering s-matrix (2) with tubular voids coated with aluminum phosphates.

Chaillu Massif of the Congo (Muller et al., 1980). The lower part of the profile shows two superposed horizons: a weathering horizon overlain by a nodular one. Whereas goethitic glaebular concentrations developed in the lower horizon, the base of the overlying horizon shows hematitic glaebules. The authors show clearly that these nodules are synchronous and originate from the same pedologic differentiation but within two different horizons, the physicochemical conditions of the environment ranging from a lower permanently saturated horizon to an overlying one not saturated permanently. This situation accounts for the simultaneous formation of iron oxyhydroxide at the base of the profile and of iron oxide higher up (Fig. 4.31 B).

The second case is a lateritic profile developed over metabasic rocks in the region of Diouga, northern Burkina Faso (Ambrosi, 1984; Ambrosi et al., 1986). Two types of ferruginous plasmic concentrations developed in the same horizon, called "mottled clay," at the expense of a weathering s-matrix consisting essentially of kaolinite. The first is purplish and hematitic, the second yellow and goethitic. These two types of glaebules are juxtaposed and never intersect, which seems to indicate their simultaneous generation. An accurate petrological and mineralogical investigation completed by the authors clearly shows that the weathering kaolinitic s-matrix, in which glaebular concentration occurs, consists itself of the juxtaposition of two domains: One is characterized by tangled vermicular crystals of kaolinite with a size ranging from 30 to 100 μm, and the other by tangled vermicular kaolinite with a size smaller than 10 μm. Ferruginous hematitic concentrations developed at the expense of the plasma consisting of the smallest size kaolinite; goethitic concentrations, on the other hand, appear where kaolinite vermicular crystals are of larger size and provide a much coarser interstitial porosity than in the domains of smaller kaolinite crystals. Tardy and Nahon (1985) showed that the distribution of hematite and goethite can be controlled by the activity of water: When the solution becomes saturated with respect to iron compounds, it should precipitate goethite in waters of great activity, namely, where porosity is largest, and hematite where porosity is smallest. Thus, in the mottled clay horizon, which is not a permanently saturated horizon, distribution of different porosity values within the kaolinitic s-matrix plays an important role in the simultaneous distribution of hematite and goethite.

The third case is a lateritic weathering profile on phosphatic sedimentary rocks of the Eocene of Lam Lam, western Senegal (Flicoteaux et al., 1977). The entire profile, in which differential weathering at the expense of alternating beds of calcium phosphates and smectites plus quartz has generated a weathering s-matrix of aluminocalcic phosphates and kaolinite plus quartz, displays a functional and anastomosed micro- and macropore system consisting of fissures, tubules, and alveoles. These cavities are the location of plasmic concentrations, which occur as coatings (cutans) on their walls. However, from top to bottom of the profile, structure and composition of plasmic concentrations vary. Thus, in the strongly leached upper horizons of alumino-calcic phosphates, cutans are complex and consist of fan-shaped rodlets of aluminum phosphate (wavellite). In the intermediate horizons, coatings consist of small vermicular crystals of kaolinite and of minute particles of goethite associated with coatings of alumino-calcic phosphates (crandallite and millisite). In the lower horizons, still of calcium phosphate, cutans consist of automorphic quartz crystals and of fibroradiated apatite needles. Finally, in the smectitic shales of the base of the profile, coatings consist of calcite crystals (Fig. 4.31 C).

In short, from the top of the weathering profile down to the parent rock, plasmic concentrations express the different stages of concentration of leaching solutions. The latter transport dissolved products through the entire pore system and generate on their way, according to the nature and concentration of the ions they contain, various mineralogical neoformations, which are organized vertically. The less soluble aluminous compounds crystallize higher in the profile, whereas the more calcic compounds are deposited below. This succession is symmetrical to that observed during the isovolume weathering of phosphatic deposits: The most calcic minerals hydrolyze first and the most aluminous survive longest toward the upper part of the profile.

This vertical zonation of structure and composition of neoformed plasmic concentrations in the micro- and macropore system is therefore *geochemical.* It is initiated by concentrations of solutions that percolate through the profile and are responsible for its pedologic differentiation. This geochemical zonation was reported earlier in other weathering profiles of phosphatic rocks by Altschuler et al. (1956, 1958) and Altschuler (1973).

Thus, within a horizon, a profile, or a sequence, plasmic concentrations developed in the s-matrix or in the pore system may be either well organized in time and space and hence with a historical and geochemical chronology, or well organized in space with chronology that is only a geochemical differentiation. In the latter case, distribution of plasmic concentrations reflects a real chromatography of chemical elements essentially controlled by the concentration of the percolating solution.

Distributions of plasmic concentrations in a profile indicate the direction in which pedologic differentiation proceeds. However, as further discussed in the next chapter, differentiation of most of the thick pedologic profiles can be related to a span of time sufficiently long for variations of environmental conditions, a function of bioclimatic evolution, to be recorded by the structures of soil, particularly at the level of pedologic features. Transfers and accumulations are endlessly superposed, thus generating a sequence of structures where each one results from the evolution of the preceeding one, which, in turn, it tends to intersect and eventually to wipe out. If each of these structures represents a stage of an evolution, it remains always an intermediate stage between a less evolved and a better evolved structure.

Structures resulting from endlessly repeated transfers and accumulations generate variably complex soil or alterite structures. Thus, pedologic features become, by means of their alternation and multiplication, the predominant structures. In simpler cases, when this multiplication is continuous and controlled by the same geochemical process, an indurated, continuous, and monomineralic horizon is generated. But when this multiplication results from alternating geochemical conditions (hence processes) succeeding each other, then the resulting structures are more complex. Such is the differentiation by geochemical evolutions of pseudoconglomeratic and pseudobrecciated structures.

Pseudobrecciated and Pseudoconglomeratic Structures of Accumulation

The example presented here summarized the major results of the works by Flicoteaux et al. (1977) and Flicoteaux (1982) on lateritic weathering profiles of Eocene argillo-phosphatic sedimentary rocks of Lam Lam in Senegal.

Lateritic Profiles on Phosphatic Rocks of Lam Lam

Parent Rock

The parent rock consists of regular alternations of beds of calcium phosphate (calci-fluorapatites), associated or not with grains of detrital quartz, and of argillaceous silty beds of smectitic plasma (with traces of illite) with abundant detrital quartz (size ≈ 3–10 μm), as well as scattered phosphatic grains.

The microtexture of phosphatic beds is essentially characterized by an assemblage of bioclasts (entire or broken tests of foraminifers, fragments of ostracod valves, pellets, and teeth of marine vertebrates) cemented by microcrystalline apatite. Argillaceous silty beds display finely undulated laminae.

Isovolume Weathering Horizon

Overlying the parent rock, and over a thickness of about 6 m, a sequence develops gradually characterized by an abundance of alumino-calcic phosphatic minerals and by the preservation of sedimentary structures.

The essential mineralogical phase of weathered phosphatic beds consists of a mixture of millisite–crandallite at the base of the horizon and of crandillite alone toward the top. The bioclastic microtexture is preserved during these transformations; only an appreciable increase of microporosity is noticeable. Furthermore, quartz grains are strongly corroded. Numerous cutans of kaolinite plus goethite coat micropores. This leads, under the naked eye, to the differentiation of *gray platelets of alumino-calcic phosphates*.

Argillaceous silty beds, gray in the parent rock, here display colors of ochre and brown. The smectitic argillaceous fraction is transformed into kaolinite (90–100%) associated or not with illite (0–10%) at the base of the profile. Streaks and small glaebules of goethite develop upward in increasing number in the kaolinitic plasma together with glaebules of millisite.

All of these transformations result from a weathering process that increases in intensity upward. Differential weathering of superposed lithological facies is thus displayed in which aluminum liberated in part by weathering of the smectitic phase combines with phosphorus and with a portion of the noneliminated alkalis and alkaline earths, leading to the formation of platelets of alumino-calcic phosphate. Such a process is similar to that described in Senegal by Capdecomme (1952), Capdecomme and Kulbicki (1954), and Slansky et al. (1964); in Nigeria by Russ and Andrews (1924); in Florida by Altschuler and Boudreau (1949), Cathcart and Mc Greevy (1953), Cathcart et al. (1953), Altschuler et al. (1956). Altschuler et al. (1958), Owens et al. (1960), Cathcart (1966), and Altschuler (1965, 1973); in Siberia by Zanin (1968); and in Central African Republic by Bigotte and Bonifas (1968).

In summary, the essential structure generated by this type of weathering, which preserves original structures, is *alumino-calcic phosphate in gray platelets*, the weathering s-matrix consisting of a plasma made of a meshwork of small crandallite needles with microporosity and rare corroded quartz grains.

Differentiation of the Pseudobrecciated Structure

Alumino-calcic phosphates in gray platelets with original structure preserved may, in turn, evolve into fine to coarse brecciated structures. This evolution is *lateral and gradual* over 1 or 2 m and occurs at the expense of the weathering s-matrix of 'alumino-calcic phosphate in gray platelets.' The different steps of this lateral transition were observed in the field and under the microscope, and are summarized below.

TRANSITION OF ALUMINO-CALCIC PHOSPHATES IN GRAY PLATELETS TO ALUMINOUS PHOSPHATES IN WHITE PLATELETS

During this lateral transition, micro- and macropores (fissural cavities, tubules, and alveoles) of the gray crandallite s-matrix show an increasing development of coatings of wavellite $[(Al_3(PO_4)_2(OH)_3(H_2O)_9H_2O)]$ rodlets growing perpendicular to the walls of cavities. Furthermore, numerous white glaebules of wavellite appear within the gray crandallite plasma; these glaebules consist of very well-crystallized wavellite in rodlets, 5−10 μm in size and arranged in sheaves. These glaebules gradually increase in number and in size. Crystals of wavellite in sheaves increase in size by autoepitaxic overgrowth at the expense of the crandallite s-matrix. Multiplication and size increase of the glaebules build a real anastomosed meshwork of wavellite sheaves of a size reaching 0.5−1.0 mm. This leads to an almost complete replacement of the crandallite and of the quartz grains of gray s-matrix by white sheaves of wavellite. The original bioclastic microtexture is preserved during this transition, indicating that the process is an epigenetic replacement with preservation of original structures. Furthermore, micro- and macropores are entirely filled by wavellite. This situation demonstrates that most of the precipitated wavellite originates from an absolute accumulation after transfer of Al^{3+}, H_3PO_4, and H^+ ions by the solutions and simultaneous elimination of Ca^{2+} and of silica as H_4SiO_4, respectively, from crandallites and from quartz.

The final result is *formation of massive white platelets of wavellite*, these platelets contain centimetric residual patches of the original crandallite gray s-matrix.

TRANSITION OF MASSIVE WHITE PLATELETS OF WAVELLITE TO OCHRE SPONGY ALUMINO-CALCIC PHOSPHATES

Wavellite platelets may, in turn, contain or be intersected by ochre and very finely porous structures leading to a general spongy aspect. Transitions occur in all directions. First, minute lenticular crystals of crandallite, of the order of 5 μm in size, develop between the sheaves of weavellite, and later between the rodlets forming these sheaves. Gradually, these crystals of crandallite replace the wavellite with an orientation perpendicular to the central axis of the rodlets, thus preserving the original aspect of the latter. It is a real pseudomorphic replacement of rodlets of wavellite by crandallite (retrogradation from the most stable mineral to a less stable one). In the most evolved zones, the axis may remain empty. In general, this transition between wavellite and crandallite is accompanied by a great increase of porosity, particularly between rodlets, which implies that a portion of the wavellite has been dissolved without having been replaced by crandallite.

Moreover, this new crandallite s-matrix can show the development of goethitic glaebules with hazy margins whose elongated shape is controlled by the arrangement of crandallite crystals and their interstitial cryptoporosity. In addition, micro- and macropores show thin coatings of yellow kaolinite over lenticular crandallite. Scattered goethite glaebules and kaolinite coatings are responsible for the ochre color of this structure to the naked eye.

This situation eventually leads to a coarse brecciated structure in which domains of white platelets of wavellite occur as patches of a few microns to 5 cm in size within spongy ochre domains of crandallite.

These structures can, in turn, be intersected and isolated into masses of variable size and shape by a beige to dark brown banded structure that takes on the shape of a rather tight network. Under these conditions, the phosphatic rock displays a finely brecciated structure.

BEIGE TO DARK BROWN BANDED STRUCTURES

A network of fine tubules and fissures, 1 μm to 2 mm in size, develops at the expense of the above-described structures. These fissural cavities and tubules are undulose, subhorizontal, and anastomosed by oblique fissures and minute cracks. Tubules and fissures, probably generated by congruent dissolution of phosphatic plasma, interconnect alveoles vesicles of all sizes, these, in turn, form the cavities characteristic of the different structures previously described. Thus, these alveoles can increase in size and reach an average diameter of several millimeters. Vertical or oblique tubules, 1 μm to 1 mm in diameter, are added to the cavity network. Along the margins of all these cavities, there occur complex cutans that consist of several distinct mineralogical phases. Coatings of kaolinite (argillans) are abundant, associated or not with goethite (argilloferrans), and grading both vertically and horizontally into coatings of millisite. This situation results from an epigenetic replacement of kaolinite by alumino-calcic phosphate. Frequently, in the direction of the cavity and overlying the millisite cutan, a very thin fringe of crandallite may exist, itself overlain by minute fibroradiated crystals of wavellite. All these cutans may, in turn, be unconformably overlaid by a new generation of argilloferrans, which themselves can grade into a millisite cutan. Goethite and hematite are also present. Cryptocrystalline goethite occurs essentially at the scale of cutans and can intersect unconformably preexisting cutans. Hematite occurs as abundant glaebules within crandallite s-matrix or wavellite s-matrix. These glaebules can be locally abundant and become reciprocally anastomosed.

Pseudobrecciated Structures Resulting from an In Situ Pedological Differentiation

Aluminous phosphatic rocks with complex brecciated structure of the Thies Plateau in Senegal were described often under the general term of "phosphatic lateritoids" (Besairie, 1943; Arnaud, 1945; Tessier, 1950, 1952, 1954, 1965; Flicoteaux and Tessier, 1971). The brecciated structure of these lateritoids was explained by allochthonous reworking of phosphatic material, or by collapse and compaction (Capdecomme, 1952).

Observations presented in the work by Flicoteaux et al. (1977), where microscopic data support examination with the naked eye (Fig. 4.32 A, B and C), confirm the role of weathering. These observations demonstrate a pseudobreccia origin devoid of transport or of any displacement of clasts. The arguments are as follows. Weathered beds with preserved original structure slowly grade into brecciated structures. Clasts of the breccia preserve their original orientation and grade laterally into nonbrecciated equivalents (platelets). These clasts are isolated from one another by the endlessly repeated play of transfers and accumulations. Subtractive transfers open new alveolar, tubular, and fissural micro- and macroporosity. Accumulation takes place either as cutans in the pore system, or within the various s-matrixes by epigenetic replacement. These replacements are also able to

Figure 4.32. Brecciated Structures of Aluminous Phosphatic Rocks from the Thies Plateau in Senegal (from Flicoteaux et al., 1977). **A**: Hand sample showing a layer of ocher silt (1) consisting of kaolinite quartz and crandallite between two layers (2) of white alumino-calcic phosphates (crandallite). Original structure of the parent sedimentary rock is preserved. **B**: Hand sample showing the vertical and lateral evolution of the rock described under A toward brecciated facies. The aluminous phosphate with wavellite (1) forming white platelets is locally replaced by a spongy gray to ocher facies with late crandallite (2) or with late millisite (3). A banded facies (4) with a subsequent generation of millisite cuts across the other facies. **C**. Hand sample showing the evolution of brecciated facies toward microbrecciated facies. The zones with wavellite (1) and with spongy crandallite (2) are cut in situ into small relicts by the banded facies (3). (Photographs Courtesy R. Flicoteaux.)

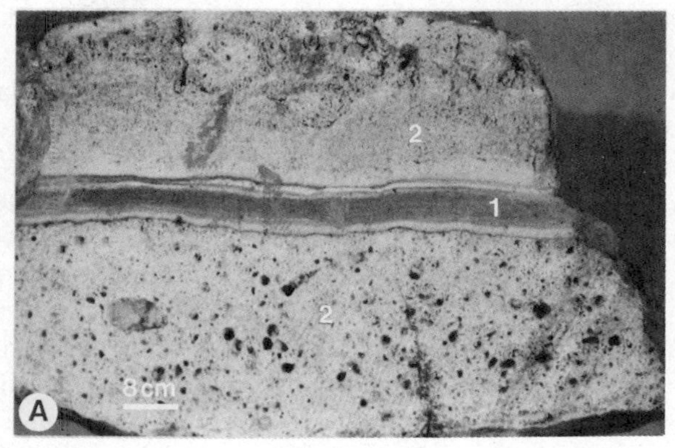
2
1
2
8 cm
A

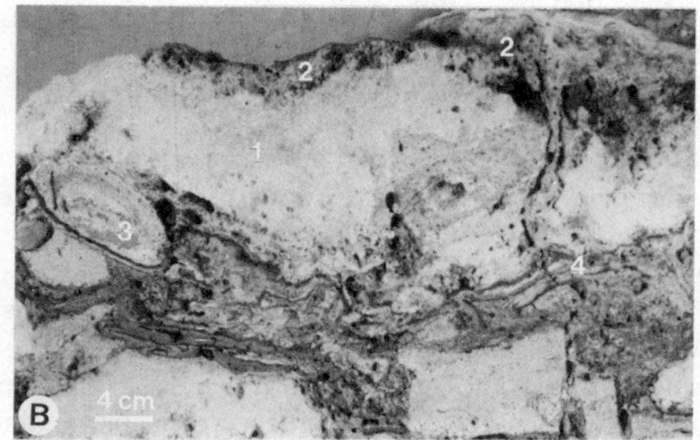
2
2
1
3
4
4 cm
B

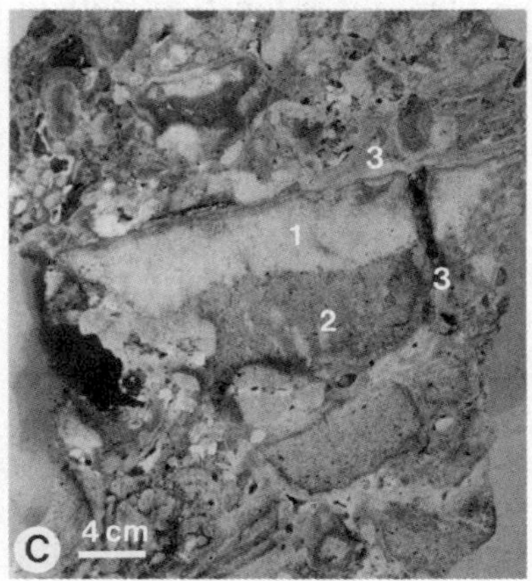
3
1
3
2
4 cm
C

affect argillaceous plasmic concentrations of the pore system, leading from argilloferrans to millisite coatings.

In summary, ceaseless repetition of transfers and accumulations is responsible for alternations in the nature of plasmic concentrations. Indeed, in numerous instances, aluminum phosphate is replaced by alumino-calcic phosphate. It is an inverse evolution compared to that of a gradual leaching weathering, in which alkalis and alkaline earths are eventually completely depleted. The situation here is a real retrogradation by pseudomorphic replacement.

This succession of structures is controlled by concentration of H_3PO_4, H_4SiO_4, Ca^{2+} and H^+ ions in solutions. Thermodynamic stability diagrams that take into account these concentrations allow us to estimate changes of concentration of percolating solutions during pedological differentiation. Differentiation leads first to systems with preserved original structures and subsequently to those with brecciated structure.

A thermodynamic model was proposed earlier by Nriagu and Dell (1974) and Nriagu (1976) for the estimation of enthalpies and free energies of formation of phosphatic minerals. This method is based on the sum of free energies of formation of each simple component. GIBBS free energy for a given hydroxyphosphate is equal to the sum of the three terms ΔG_f° hydroxyphosphate $= \Sigma_i \, \Delta G_f^\circ \, (ri) + \chi \Delta G_f^\circ \, H_2O - Q$ in which $\Sigma_i \, \Delta G_f^\circ \, (ri)$ is the sum of GIBBS free energy of hydroxides or of the simple phosphates forming the hydroxyphosphates; χ is the number of water moles considered in the phosphate complex; and $Q = 2.303 \, RT \, \Sigma_{n_i} \log n_i$, where n_i is the stoichiometric coefficient of the i hydroxide of the simple phosphate. In order to obtain a better relationship with observed natural facts, Viellard (1978), using as a starting point the estimation method of Nriagu, modified the term Q in agreement with the new data obtained by Tardy and Garrels (1976, 1977), Tardy and Gartner (1977), and Tardy and Vieillard (1977). Following Vieillard (1978), Vieillard et al. (1979) reached the following estimates of G_f°:

Fluo carbonate apatite $Ca_{10}(PO_4)_5CO_3F_3$	$\Delta G_f^\circ = -2983.3$kcal/mole
Ca-millisite $Ca_{1.5}Al_6(PO_4)_4(OH)_9 3H_2O$	$\Delta G_f^\circ = -2614.0$
Crandallite $CaAl_3(PO_4)_2(OH)_5H_2O$	$\Delta G_f^\circ = -1336.7$
Wavellite $Al_3(PO_4)_2(OH)_3(H_2O)_4H_2O$	$\Delta G_f^\circ = -1330.0$
Augelite $Al_2PO_4(OH)_3$	$\Delta G_f^\circ = -660.9$

From the obtained G_f° values, Vieillard (1978) was able to prepare a whole series of thermodynamic diagrams that take into account concentrations of $[H_3PO_4]$, $[H_4SiO_4]$, $[Ca^{2+}]$, and $[H^+]$ (Fig. 4.33). These diagrams do not show the stability field of millisite, considered in this thermodynamic approach as a Ca-millisite, because it does not appear, as seen in nature, between the stability fields of apatite and crandallite. Furthermore, it is known that

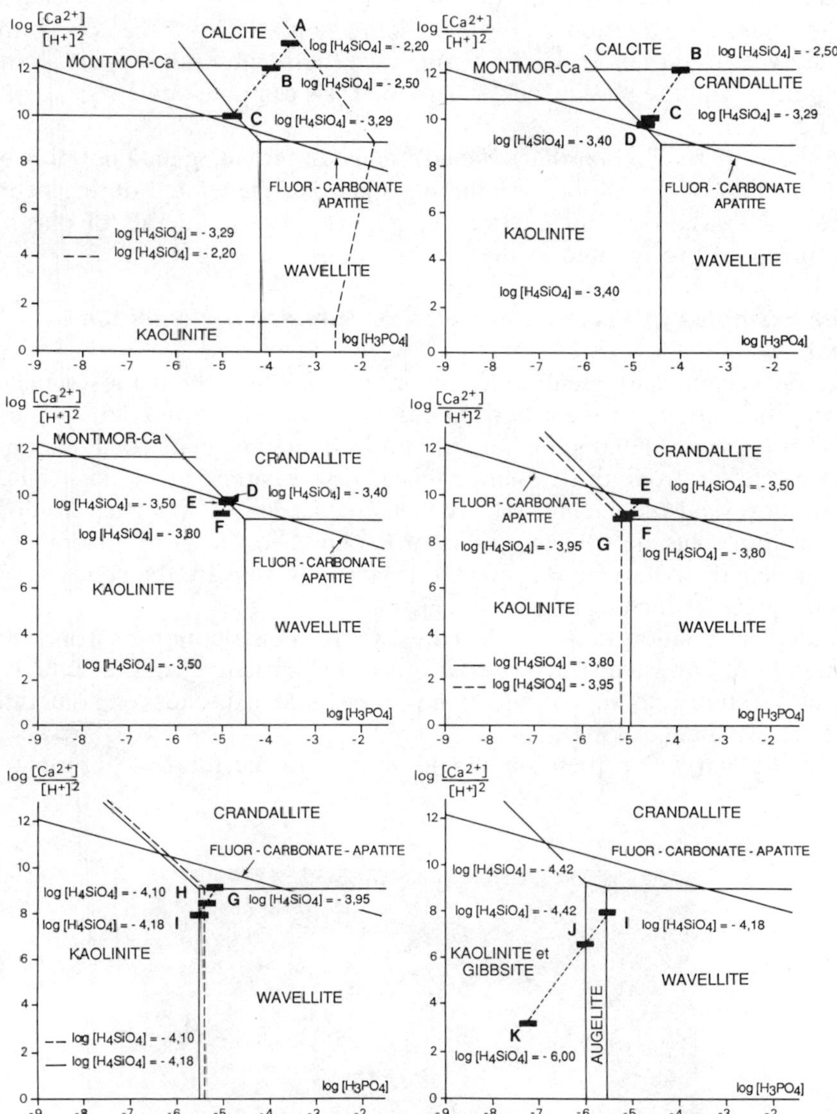

Figure 4.33. Two-Dimensional Activity Diagrams (log $[Ca^{2+}]/[H^+]^2$, log $[H_3PO_4]$) for Different Values of log $[H_4SiO_4]$ (modified after Vieillard et al., 1979). From K to A is the pathway corresponding to the geochemical evolution of the percolating solution from the top to the base of weathering profiles. (Reprinted by permission of the Mineralogical Society of America.)

in nature millisite is not only calcic, but also contains sodium and probably iron.

These diagrams clearly show the pathway from K to A, corresponding to the geochemical evolution of the percolating solution from the top of the profile, where the most leached minerals (here gibbsite) appear, until equilibrium with the parent rock, where minerals can precipitate secondarily (here calcite).

In short, these diagrams have the advantage of providing data not only on the formation of the various weathering s-matrixes generated during pedological differentiation, but also on the nature of the minerals of plasmic concentrations precipitated in the pore system.

Other Examples of Pseudobrecciated or Pseudoconglomeratic Structures

Pseudobrecciated and pseudoconglomeratic structures belonging to pedological differentiation have also been described for other conditions. Nahon (1971, 1976) showed that in the case of iron, certain crusts with conglomeratic structure result from the evolution in situ by weathering of an original ferruginous sandstone. Endlessly repeated episodes of deferruginization, alveolization, plasmic concentrations of kaolinite in the pore system, and their epigenetic replacement by hematitic glaebulization are the geochemical driving forces of such an evolution (Fig. 4.34).

A similar situation occurs in bauxites where redistribution of iron leads from an initial massive structure to a brecciated structure, and, eventually, by exaggeration and repetition of the process, to a pseudoconglomeratic structure (Boulangé, 1984).

In short, particular attention should be paid in the future to brecciated

Figure 4.34. Pseudoconglomeratic Structure of a Ferruginous Crust. (1) Relict of indurated massive ferruginous facies with preserved original structure of the parent sandstone. (2) Pisolitic facies generated by the in situ geochemical evolution of facies 1.

and conglomeratic structures that might originate from a geochemical evolution due to repeated alternation of transfers and accumulations taking place in the s-matrix and the micro- and macropore system. These differentiations are controlled by the concentration of percolating solutions but also, as discussed above, by the size of cavities where accumulation occurs (cryptocavities in the s-matrix, micro- and macrocavities of the pore system).

Thus, pedologic features become, by their multiplication, predominant structures. They intersect and then wipe out preexisting structures. This is also a restructuring of weathering, but contrary to restructuring affecting argillaceous horizons, this one results essentially from an alternation of dissolutions and reprecipitations of mineralogical phases, whether crystallized or not.

Dissolutions and reprecipitations that act simultaneously and/or successively are more frequent in that part of profiles which is not saturated permanently, that is, in the upper part of profiles that is more sensitive to pedobioclimatic variations. Thus, pseudobrecciated and pseudoconglomeratic structures, when they exist, characterize upper horizons. It is the equilibrium between transfers and accumulations that permits differentiation and durability of such structures. However, nearer to the interface with the atmosphere, the equilibrium is broken and subtractive transfers gradually take over. Observed accumulations are only relative accumulations, namely, reorganizations in situ of the least mobile elements. This process is a *degradation* of preexisting structures. Thus, differentiation of pseudobrecciated and pseudoconglomeratic structures, when it occurs, is the geochemical stage that precedes the real degradation of horizons at the top of the profile.

DEGRADATION OF ACCUMULATION STRUCTURES

The example of ferruginous crusts of the Ndias Massif of Senegal (Nahon, 1976) is discussed again here, followed by the example of degradation of manganesiferous crusts of Ziémougoula in the Ivory Coast (Beauvais and Nahon, 1985).

Pisolitic Structures: Reorganization of Degradation?

Banded Structures of Reorganization

Nodules generated by glaebular accumulation in which fragments display pseudobrecciated or pseudoconglomeratic structures can be characterized along all their margins by the development of a finely banded structure called cortex. If well-individualized, this structure leads to pisolites or pisolitic concretions. If nodules generated by glaebular accumulation are very abundant and reciprocally anastomosed, cortexes develop by taking advantage of all their discontinuities (cracks, fissures, and tubules) and separate the nodules into smaller units. This leads to the change of a pseudoconglomeratic structure into a pisolitic one, in which the pisolites are welded to one another. The most frequent example belongs to ferruginous crusts: Nodules resulting from a ferruginous glaebulization within a kaolinitic plasma always consist of aluminous hematite (Tardy and Nahon, 1985; Ambrosi et al., 1986), and the banded cortexes, developed secondarily and concentrically at the expense of the nodules, always consist of aluminous goethite. Hence, the latter results from dissolution of aluminous hematite and from reprecipitation with reorientation. Furthermore, the amount of aluminum substituted to the oxyhydroxide is always much greater than that of the hematite of nodules. A precise analysis completed on a sample from Diouga (Burkina Faso) allowed Ambrosi and Nahon (1986) to calculate the geochemical balances corresponding to the change of hematite with 7 mole fraction percent of Al_2O_3 of the nodules to goethite with 21 mole fraction percent ALOOH of the cortexes. This evolution can be expressed by the following equation:

$$[23.02Fe_2O_3 + 1.74\ Al_2O_3]\ \text{hematite} + 8.09Al^{3+} + 40.94H_2O \rightarrow [46.04FeOOH + 11.57AlOOH]\ \text{goethite} + 24.27H^+.$$

This equation shows that for a constant content of iron, a gain of 70% aluminum is necessary. Clearly, formation of the cortex requires an important absolute accumulation of aluminum brought in by solutions. These conditions indicate that generation of such a cortex does not correspond to an incipient degradation but to the last stage of glaebulization.

Banded Structures of Aggradation

Individualization of pisolites by centrifugal restruction of a nodule was also proposed by Du Preez (1954), Sherman and Kanehiro (1954), Bruckner (1957), Alexander and Cady (1962), Frankel and Bayliss (1966), Mitsuchi (1976), Furakana et al. (1976), Eswaran et al. (1977), and Boulangé (1984). Recently, Bocquier et al. (1984, p. 130) were able, on the basis of the above-mentioned authors, to summarize the differentiation of such nodules as follows: "The nodule increases its volume by adding to its periphery one or several cortical layers at the expense of the surrounding internodular plasma." Indeed, in this differentiation, reorganization of ferruginous plasma plays an important role (Boulangé and Bocquier, 1983).

Obviously, in this case also, growth of these nodules cannot be attributed to a degradation but rather to an aggradation.

Degradation by Geochemical Fragmentation

In the Ndias Massif in Senegal (Nahon, 1976), iron crusts with pisolitic structures that crop out or are covered by a thin sandy horizon display several structures. The most evolved and the most abundant corresponds to pisolites, whereas relict nodules of hematite occur also in which cortexes develop. Finally, lithorelicts characterized by ferruginous sandstones, which formed the major part of the massive structure from which pseudo-conglomeratic and subsequently pisolitic structures developed, also exist.

Near the surface, this crust displays numerous fissures, cracks, and tubules, which develop laterally and vertically to the extent of forming an anastomosed network that breaks up the crust into decimetric to centimetric fragments. Simultaneously with the development of this network of cavities, argillo-ferruginous coatings appear on the walls of cavities. With the increase in size of these cavities, coatings become thicker until they generate a reddish argillo-silty matrix in which fragments of the crust float. Thus, a pebbly ferruginous horizon considered to be the degradation product of the indurated ferruginous crust is formed. The main stages of this degradation can be established by petrographic analysis.

The unconsolidated pebbly ferruginous horizon shows under the petrographic microscope an important development of porosity (Fig. 4.35). Relationships between fragments of the crust and reddish argillosilty matrix

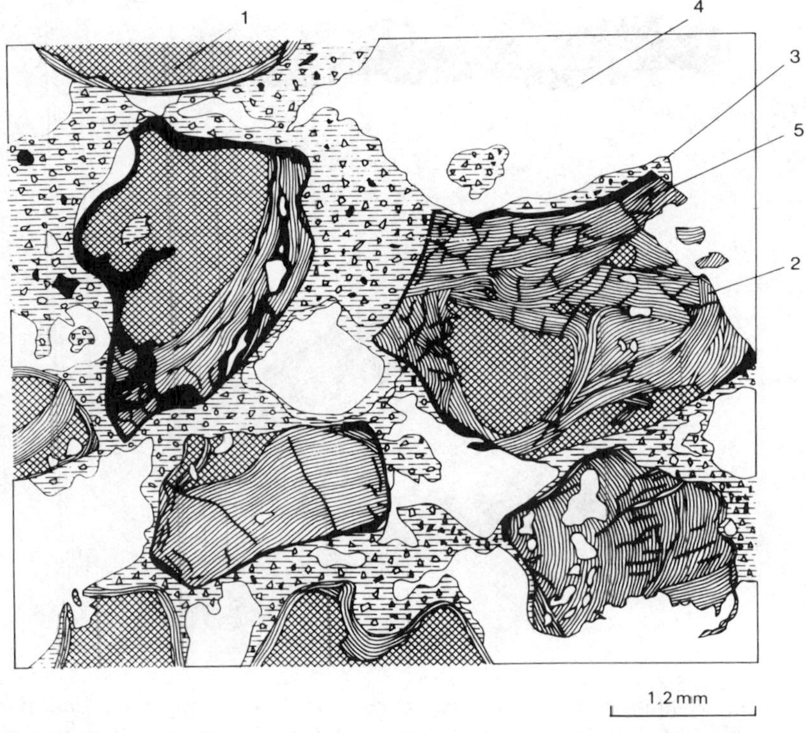

Figure 4.35. Microstructure of Unconsolidated Pebbly Ferruginous Horizon (modified from Nahon, 1976). (1) Hematitic core of loose pisolites or of ferruginous fragments; (2) goethitic banded cortexes; (3) reddish argillo-silty matrix; (4) voids; (5) network of cracks and fissures affecting the cortexes. (Reprinted by permission of Sciences Géologiques.)

can be designated as intertexic (Nahon, 1976), whereas a fine alveolization or fissuration affects the fragments themselves.

Examination of the network of cracks and fissures, while transition proceeds between ferruginous crust and unconsolidated degradation horizon, emphasizes the following facts. Walls of fissures and cracks do not correspond to one another from the initial stages, this situation leads to assume that they result from a dissolution.

When fissures, cracks, or tubules affect structures still rich in corroded quartz grains, the following features appear during the transition and, hence, during the opening of fissures (Fig. 4.36 A): An argillaceous coating consisting

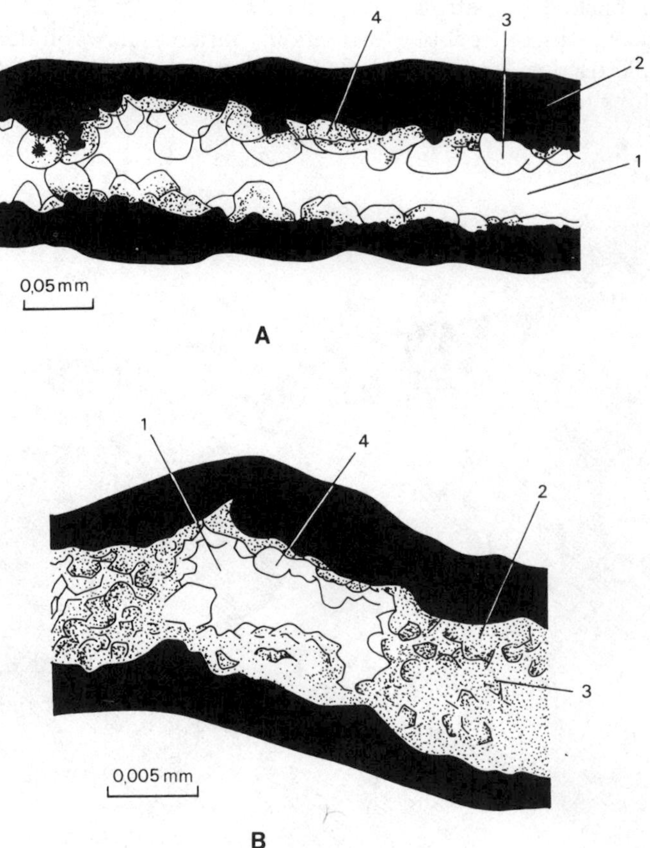

Figure 4.36. Evolution of Fissures Affecting Ferruginous Fragments and Banded Cortexes of Pisolites Still Rich in Relicts of Quartz Grains in an Unconsolidated Pebbly Ferruginous Horizon. **A**: (1) Fissure; (2) ferruginous walls of the fissure; (3) colorless platelets of kaolinite; (4) yellow to orange colored platelets of kaolinite. **B**: (1) Residual void; (2) ferruginous walls; (3) abundant platelets of kaolinite colored by goethite (argillo ferruginous matrix); (4) colorless platelets of kaolinite. (From Nahon 1976. Reprinted by permission of Sciences Géologiques.)

of colorless platelets of kaolinite is arranged perpendicularly to the walls; when the argillaceous platelets are more numerous, those located toward the walls become colored in yellow or orange red by goethite; when argillaceous coatings become generalized (Fig. 4.36 B), platelets are abundant and display only a residual cavity in which a few colorless platelets remain. Thus, while transition proceeds, an argillo-silty plasma develops. The silty fraction consisting of strongly corroded quartz grains and of minute fragments of ferruginous plasma is essentially residual. When fissures, cracks, or tubules affect the cortexes of pisolites (and their center) devoid of quartz grains, their walls are coated by minute orange and fibroradiated crystals of *goethite*. In such a case (Fig. 4.37), microprobe profiles for the major elements Al,

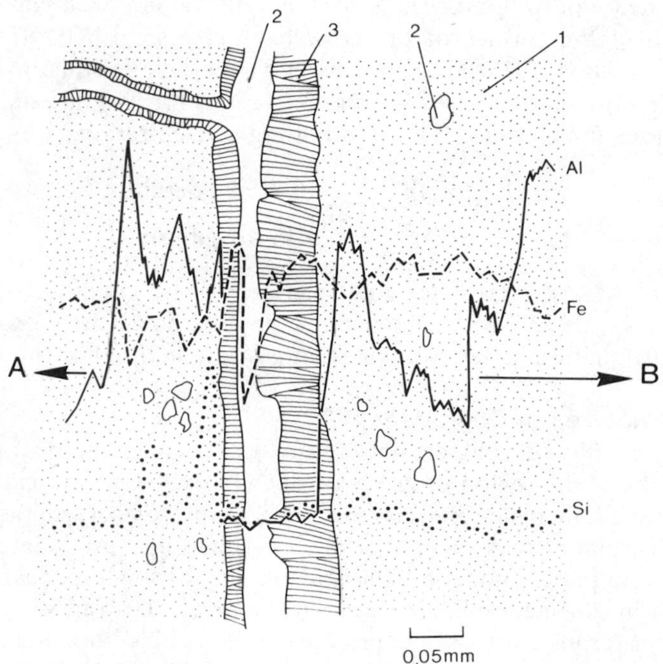

Figure 4.37. Dissolution Fissures Involving the Banded Cortex of Ferruginous Pisolites. Dissolution of Al, Fe, and Si Shown by a Microprobe Profile along A-B. (1) Banded cortex of aluminous goethite; (2) fissural voids or dissolution tubules; (3) crystals of goethite coating perpendicularly the walls of voids. The distribution of Al and Fe in the banded cortex varies according to the arrangement of the concentric banding whose goethite content in moles fraction of A100H ranges between 16 and 22%. The distribution of Si shows very localized peaks corresponding to small concentrations of highly corroded residual kaolinite or to small fan-shaped crystals of a 10.2 Å phyllosilicate rich in Fe, devoid of K, and at equilibrium with aluminous goethite. These conditions were revealed by Amouric et al. (1986) by means of high resolution electron microscopy of the cortex of HREM pisolites.

Fe, and Si show that iron oxyhydroxides that precipitate in the cavities are not or little substituted: Therefore, it is goethite.

Fragmentation of the ferruginous crust thus results from a dissolution of the major ferruginous constituents of the crust, namely, aluminous hematite and aluminous goethite, and of the reprecipitation at a very short distance of pure goethite, or of kaolinite plus pure goethite. A comparison between the amount of precipitated kaolinite and aluminum contained in the dissolved constituents of the crust, taking into account porosity, shows an important loss of aluminum, whereas most of the iron remains in situ.

Degradation of the ferruginous crust with pisolitic structure is, therefore, characterized by three features: increase of porosity, precipitation of pure goethite, and depletion of most of the aluminum. In this pore system opened toward the surface, leaching by undersaturated water is important and the activity of water is close to the unit value. Such an oxidizing environment at the surface of the crust has pH 5.65, Eh 0.887, and log pCO_2 −0.68 (Nahon, 1976). If a comparison is undertaken in this environment between the solubility balances in total ferric ions for goethite and in total aluminous ions for gibbsite, the following log concentrations are obtained:

$$\log\,[\text{Fe total}]^{3+} = -10.4 \text{ for goethite}$$

$$\log\,[\text{Al total}]^{3+} = -6.46 \text{ for gibbsite.}$$

For values of ΔG_f° of −116.7 for goethite (Berner, 1971) and of −275.3 for gibbsite (Parks, 1972), it is understandable that in such a geochemical environment, goethite is less soluble than gibbsite and hence precipitates.

Banded Structures of Degradation

Weathering profiles developed on Precambrian manganese-bearing metamorphic rocks of the region of Ziémougoula (Ivory Coast) are capped by a highly indurated manganesiferous crust. This crust preserved the original texture of the parent rock (Nahon et al., 1984) and may grade laterally and vertically into a pebbly horizon, consisting mainly of small manganesiferous pisolites set in an unconsolidated red silty matrix. The transition between the manganesiferous crust and the pisolites of the pebbly horizon takes place over a thickness of 10 cm. Fissures and smaller cracks develop along the margin of the crust and gradually isolate first decimetric- then centimetric-sized fragments. These fragments become less angular with decreasing size, and a banded peripheral cortex develops in a centripetal fashion at the expense of the central part of the fragments where the original texture of the parent rock is still recognizable. Whereas the crust and the central part of the fragments consisted of lithiophorite (from weathering of parent garnets) and of birnessite plus lithiophorite (from weathering of parent chlorites), the banded cortexes are made of alternating bands of lithiophorite and cryptomelane (Beauvais and Nahon, 1985). The lithiophorite bands of cortexes increase in thickness and in frequency toward the margins of the pisolites.

Eventually, well-developed crystals of gibbsite appear in the voids, which, in places, occur between lithiophorite bands toward the margins of the cortexes. Furthermore, the margins of the pisolites are themselves coated by a thin film of gibbsite. This mineral is also the main constituent of the unconsolidated red silty matrix that binds the pisolites together; it is associated with a small amount of aluminous goethite responsible for the color of the matrix.

The degradation of the manganesiferous crust results from a geochemical fragmentation that leads to the generation of pisolites and of an unconsolidated matrix. During this degradation, original parent structures are replaced by banded structures that become increasingly rich in alumina. This enrichment is relative and results from a greater mobility of the manganese in the solutions at the top of weathering profiles.

CONCLUSIONS: EVOLUTION OF MICROSTRUCTURAL FEATURES OF ACCUMULATION AND TRANSFER

Intrinsic s-matrixes of weathering and of pedoplasmation, once differentiated, represent, in fact, only an evolutionary stage of the weathering profile or the soil at a given moment. With continuous development of external factors in time and space (Kovda, 1933), these natural and dynamic structures corresponding to intrinsic s-matrixes display a gradual change through the endlessly repeated interaction of transfers and accumulations. Thus, new structures are superposed by autodevelopment to preexisting ones. This succession can be organized both in its geometry and history by means of petrographic studies. Only when chronology and dynamics of differentiation of these structures are understood can geochemical processes responsible for their generation be grasped.

This chapter showed an entire evolutionary series of structures. Each structure is characterized by its mineralogical composition and its specific features. It was shown that if unconformities or discontinuities sometimes appear between each structure, they result from a specific geochemical evolution. Thus, it is possible to distinguish inherited structures, accumulation structures, transformation structures, and structures of leaching type.

Inherited Microstructures

Percolation of solutions and related transfers and accumulation, which are responsible for microstructures, take place in all cryptocavities of intrinsic s-matrixes and in macrocavities, regardless of their size and degree of interconnection. Plasmic concentrations thus developed can either preserve a portion of original structures or congeal them in a more or less perennial way. In such a case, they represent petrographic proofs of past structures and are, therefore, inherited microstructures.

Accumulation Microstructures

Plasmic concentrations define new structures called accumulation structures. According to the intensity of plasmic concentration, several successive stages of their differentiation can be distinguished. These structures either replace preexisting ones and represent epigenetic glaebulization of s-matrixes, or they overlap preexisting ones and represent cutanic illuviation of the pore system.

Transformation Microstructures

Accumulation microstructures undergo reorganizations by autodevelopment in contact with solutions. It is a process of rearrangement of material in which solution of one or all of the constituents occurs, followed by recrystallization in situ or at a short distance with reorientation. In general, these new structures destroy preexisting ones.

Leaching Microstructures

Migrations sometimes lead to important losses of material, either by dissolution or in suspension, with a resulting large increase of porosity. If losses of material are localized and discontinuous, massive structures become separated into nodules or sheets. However, if losses of material are continuous, the plasma may be entirely eliminated, leading to a complete redistribution of skeleton grains. All these new microstructures are typical of leaching processes.

Microstructures Are Progressive

If leaching structures develop at the expense of any kind of initial structure of accumulation or of transformation, it is because they express, in the most drastic manner, desequilibria to which all these structures are submitted under the effect either of variations of external factors or of self-development.

Transformation structures always follow in time accumulation structures, since they are essentially induced by self-development. Finally, if inherited structures are destroyed, the process is irreversible. But accumulation, transformation, and leaching structures can be repeated, successively or simultaneously, during the pedologic history; these structures are progressive. The final product is necessarily at a higher structural level, namely, the horizon. Thus, the complex hierarchy of transfers and accumulations of matter eventually generates horizons with nodular, pseudoconglomeratic, or pseudobrecciated structures. Pedologic features detected at the scale of the thin section are followed by morphological features at the scale of the profile, the sequence, and the landscape. The repeated and organized succession of several microstructures eventually leads to differentiation of horizon structures, that is, *pedologic differentiation*.

CHAPTER 5

DIFFERENTIATION AND EVOLUTION OF PEDOLOGIC MANTLES OF TROPICAL AND SUBTROPICAL ZONES

Variations of the nature of constituents and their distribution at all scales of observation lead us to consider weathering profiles and soils as a differentiated superficial mantle displaying several units of organization (or structural units).

All structural units do not develop simultaneously, but follow each other in time and space. They interfere with each other, add or cancel their effects, and replace each other according to a genetic and historical order (Bocquier, 1973). Thus *original structural units* and *derived structural units* can be distinguished. Whenever the former predominate, they define, in three dimensions, *original pedologic sequences* (association of horizons) in which units are generally superposed vertically. The pedologic differentiation is said to be vertical. Original structural units can be reorganized into derived structural units, which, whenever predominant, define *derived pedologic sequences*. Such sequences could result from a vertical differentiation. But, when derived structural units are uncomformable with respect to original structural units, they are distributed or propagated most frequently in a lateral fashion. In such a case the, pedologic differentiation is said to be lateral.

The pedologic differentiation is, in general, visible in the field, only if the related geochemical processes have acted over a sufficiently long period of time (at least a few thousand years). During such a span of time, external bioclimatic conditions can evolve and act over geochemical processes leading to obvious changes of the structures of soils and alterites.

In this chapter an attempt is made to show how these changes are expressed at the scale of horizons and of pedologic sequences. For certains cases, it is necessary to return to the microscopic scale in order to understand generation and evolution of sequences. Finally, general causes of pedologic differentiation are discussed without overemphasizing the theorical approach.

PEDOLOGIC MANTLES OF TROPICAL HUMID LANDSCAPES

Distribution of Major Horizons

Pedologic mantles of tropical humid (equatorial) regions can reach thicknesses of more than 100 m. From the parent rock and over a major portion of its thickness, the pedologic mantle displays the preserved structure of the parent-rock. The differentiation of this lower portion of the alterite is mainly vertical and lithodependent. These means that most of the transfers of solutions responsible for geochemical reactions operate from top to bottom of profiles, thus allowing vertical parentages of weathering products (argilliplasmas and crystalliplasmas) which directly express the nature of the parent rock.

Alteroplasmation (as discussed in Chapter 2) is pseudomorphic and isovolumetric. Volumes and structures are preserved at both the microscopic and megascopic scales. This portion of profiles where relicts of parent materials are still abundant is called *coarse saprolite*. When volumes and structures are visible only at the megascopic scale and relicts of parent minerals are in minor amount with respect to the alteroplasma, the designation of *fine saprolite* is used.

Weathering products forming saprolites in tropical humid regions are characterized by type 1/1 phyllosilicates (essentially kaolinite) and by oxyhydroxides, hydroxides, and oxides of iron and/or alumina. Coexistence and parentages of these phases were shown at different levels of organization and recently, by Muller, (1987), in examples from Cameron.

Saprolite is an unconsolidated and porous material. In its upper portion, fine saprolite can be the location of secondary concentrations of kaolinite or of oxyhydroxides of iron and alumina (Ambrosi and Nahon, 1986; Lucas and Chauvel, 1990), which fill visible porosity or, in the case of oxides, replace a portion of the argilliplasma or of the pedoplasma. These concentrations result from leachings that involve the upper portion of the underlying horizon.

Eventually, an unconsolidated red or yellow material develops in the upper portion of the pedologic mantle and displays a fine structure consisting of microaggregates (or micropeds) of kaolinite and of oxyhydroxides of iron and/or alumina. This microaggregated material may undergo degradation at the surface.

The surficial mantle of tropical humid (equatorial) regions thus consists of

4 to 5 horizons where shape or envelope (their spatial boundary) are more or less unconformable with respect to the surfaces over which they are molded. These horizons are superposed vertically and correspond to a pedologic sequence whose organizations derive from each other during lowering of the surficial mantle in the landscape; namely, each horizon develops at the expenses of the underlying one, but at the same time also feeds the underlying horizon at various levels with the products it liberates.

Among the two saprolite horizons, only coarse saprolite corresponds to a structural unit in which original structures and volumes are preserved at the microscopic and megascopic scales. Pseudomorphs of parent minerals are generated by crystalliplasmas and argilliplasmas, the latter consisting essentially of kaolinite.

In the fine saprolite, the progress of weathering of parent relicts (lithorelicts) provides more and more kaolinite, but simultaneously secondary kaolinite can concentrate in the voids of the saprolite (particularly at the top of the fine saprolite) through illuviation or precipitation from percolating solutions. The amount of clay increases rapidly with respect to the skeleton (parent relicts) and becomes increasingly sensitive to alternations of wetting and drying. However, kaolinite is not a very reactive clay with respect to the aqueous phase; hence, plasma restructuring is limited to the microscopic-scale and large-scale structures and volumes remain unaffected. Through the action of endlessly repeated wetting and drying, argilliplasma is gradually reorganized into pedoplasma with a finely polyhedric structure. This pedoplasmated horizon reaches a variable thickness and can display pedologic features such as nodules of oxyhydroxides of iron and/or alumina. A microaggregated horizon overlies a horizon with polyhedric structure.

These two upper horizons represent structural units derived from saprolite horizons. *Original structural units* and *derived structural units* belong to the same *vertically differentiated sequence* of a pedologic mantle, which develops under bioclimatic conditions existing in tropical humid landscapes. Several transformations can involve such sequences after an important change of drainage conditions (possible tectonic cause), or after a change of bioclimatic conditions, or simply as a consequence of self-evolution of the sequence itself through time. Transformations occur mainly at the expenses of the horizons which cover the sequence: horizon with polyhedric structure and microaggregated horizon. Before considering the transformations of structural units of the sequence, it is necessary to describe these units in detail.

Horizons with Polyhedric Structure and with Microaggregates of Tropical Humid Pedologic Mantles

Numerous examples have been described in Africa and Brazil. A particularly well-studied area is the Paraná basin (States of Saõ Paulo and Paraná) in

Brazil, where soils developed on top of saprolites over basaltic rocks are called *terra roxa*. These soils are actually divided into two unconsolidated horizons: a horizon with fine polyhedric structure called *terra roxa estruturada* and a horizon with microaggregates called *terra roxa legitima*. These horizons were studied by Moniz and Jackson (1970), and Bennema et al. (1970). However, the most recent study by Pédro et al. (1976) shows better the relationships between the organizations of these horizons and is submarized as follows.

Data from the Pedologic Analysis

The *terra roxa estruturada* is of a dark red color and displays a polyhedric structure in corn kernels (Ø ~ 5 mm) consisting mainly of kaolinite plasma (~ 65% of the rock) and of ferric oxyhydroxides of which 50% are amorphous. The grains of the skeleton (quartz and titano-ferriferous opaque oxides) represent only about 15% of the total volume of the rock. Consequently, plasma is largely predominant, is relatively homogeneous with a general dark red color, and displays yellowish zones representing only 25%. Under the petrographic microscope, orientations are visible consisting of lines of mottles or short striae in the dark red zones which are better defined and follow preferred orientations (masepic orientations) in the yellowish zones. Microprobe analyses show little difference in iron content between red zone (13.5%) and yellow zone (11.8%). Voids are mainly fissures and cracks, the largest of which form the boundaries of polyhedric aggregates in corn kernels (Fig. 5.1 A and B). The basic exchange capacity is 10.70 mEq/100 g, a high value for a kaolinitic soil, exchangeable cations consist mainly of Ca and Mg. Saturation index is high: 88.2%.

The *terra roxa legitima* is also of a dark red color and shows a very fine

Figure 5.1. Associations of Iron Oxyhydroxides and Kaolinite: Microaggregation of Ferralitic Soils of the Paraná Basin (Brazil). **A, B, C** and **E**: photomicrographs, crossed nicols; **D**: SEM photograph. (Photographs courtesy of A. Chauvel.) **A**: '*Terra roxa estruturada*.' Quartz grains of the skeleton (Q) or opaques grains (OX) are enclosed within a red to dark red ferruginous argillo plasma (P) with a coarse polyhedric structure. Along the fissures (F) limiting polyhedric aggregates, plasma is largely deferritized, argillaceous, anisotropic, and oriented (OP). **B**: '*Terra roxa estruturada*' with incipient individualization of a finer microaggregated structure (M). The skeleton consists of quartz grains (Q) and of grains of opaque oxides (OX). **C**: Transition between '*terra roxa estruturada*' and '*terra roxa legitima*.' Microaggregation is visible inside the '*terra roxa estruturada*' and emphasized by a kaolinitic, anisotropic plasma. **D**: '*Terra roxa legitima*' or latosol, or microaggregated ferralitic soil. Aggregates have sizes ranging from 100–200 μm as in most ferralitic soils. Microaggregates are interconnected by bridges and their variably reciprocally molded shapes suggest an individualization in situ. **E**: '*Terra roxa legitima*' or latosol, same material as in preceding figure. Microaggregates (M) interconnected by bridges consist of a rare skeleton of quartz grains (Q) and opaque grains (OX), but mainly of plasma (P). Notice the irregular external shapes of the microaggregates.

structure called powdered coffee. The mineralogical composition is the same as in the previous type, except that all iron oxyhydroxides are crystallized. Under the petrographic microscope, the plasma is organized into microaggregates (Fig. 5.1 C, D and E) interconnected by bridges. These microaggregates, of an average size of the order of 100 μm, show irregular external boundaries with protuberances. Plasma is the main constituent of microaggregates, which have a dark red center and a yellow to yellowish-red margin. Plasmic orientations are little or not at all visible in the cloudy dark areas, sometimes designated as agglutinic (Buol and Eswaran, 1978); the yellow parts are oriented. Alveolar voids occupy interaggregate spaces

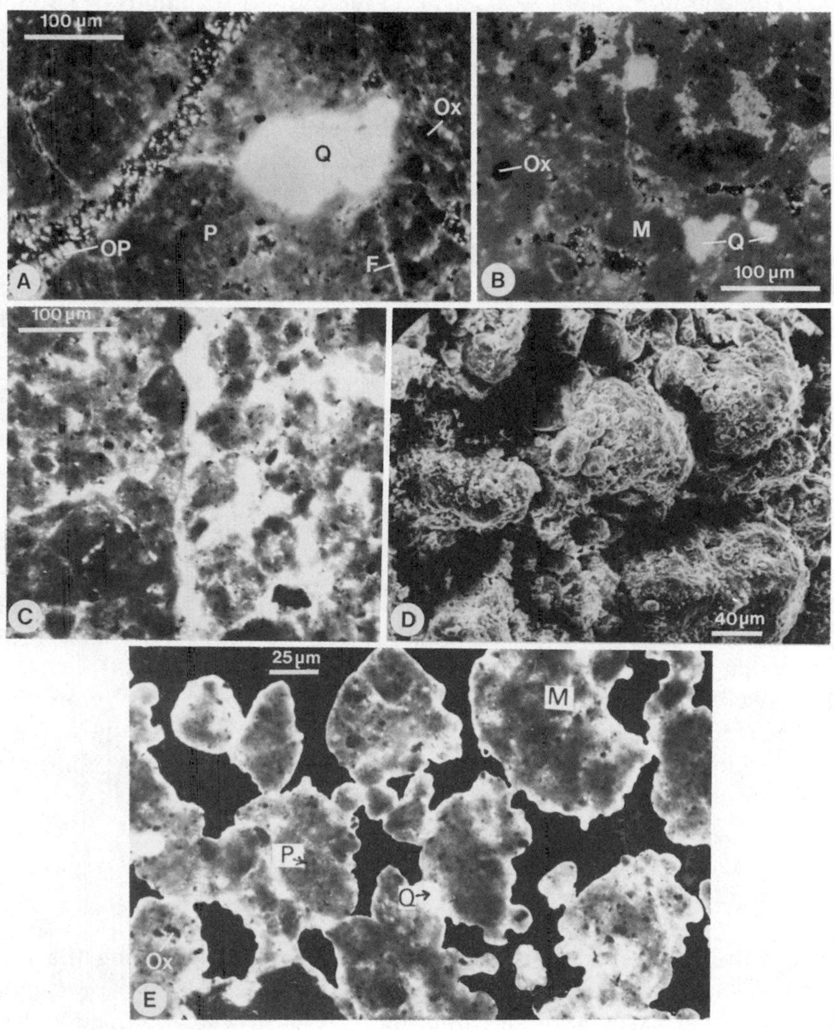

and range in diameter between 5 and 200 μm. The basic exchange capacity is 6.1 mEq/100 g. The rock is practically devoid of exchangeable cations. Saturation index is low: 3.2%

Results of the Analytical and Experimental Approach

The transition between *terra roxa estruturada* and *terra roxa legitima* therefore takes place for the same amount of kaolinite and of iron oxyhydroxides of the plasma by the following three processes: a change of polyhedric structures of the plasma into microaggregates; a crystallization of all iron oxyhydroxides; an important decrease of the exchange capacity (−75%), and a loss of exchangeable cations.

The rare grains of the skeleton play a small role both in the *terra roxa estruturada* and in the *terra roxa legitima*, therefore associations of kaolinite and iron compounds are responsible for the generation of microaggregates. For a better understanding of this process, Pédro et al. (1976) submitted samples of the two types of *terra roxa* to various experiments in order to evaluate the conditions of association of iron and clay previously recognized microscopically and to understand their reactivity to wetting.

The *terra roxa legitima* was submitted to two experiments. The first one is an exchange with KCl N/10 or with hydrogen peroxide, which allows a characterization of a free clay corresponding to the oriented clay of the yellow zones. The second one, a Tamm (1934) deferrizing experiment, allows a separation of oxyhydroxides and clays of the dark red zones, thus making a large portion of the kaolinite free (Fig. 5.2).

The first experiment with hydrogen peroxide or KCl N/10 gives a quantitative value to the results of microscopic observation; namely, the yellow zones of the plasma correspond to free clay, the dark red zones to immobilized clay containing the skeleton.

The Tamm experiment shows that an extraction of Fe_2O_3 close to 0.2% or a complete extraction (13.2%) leads to the same result. Therefore, a small fraction of iron (0.2% Fe_2O_3) appears responsible for the immobilization and masking of most of the kaolinite belonging to the red zones. Besides, microprobe data on the yellow (free) and red (immobilized) clays indicate that variations in the content of Fe_2O_3 are low.

The swelling capacity of these clays was evaluated. This property or potential swelling Ps is quantitatively evaluated as a function of the apparent specific volume in the wet state (V_{ws}) and of the apparent volume in the dry state (V_{ds}) according to Monnier et al. (1973):

$$Ps = \frac{Vws - Vds}{Vds} \times 100.$$

Experiments have been completed on three samples containing the same amount of kaolinite: one of *terra roxa estruturada*, one of *terra roxa legitima*, and one reference sample of kaolinite from Provins (France) doped with Ca (Table 5.1).

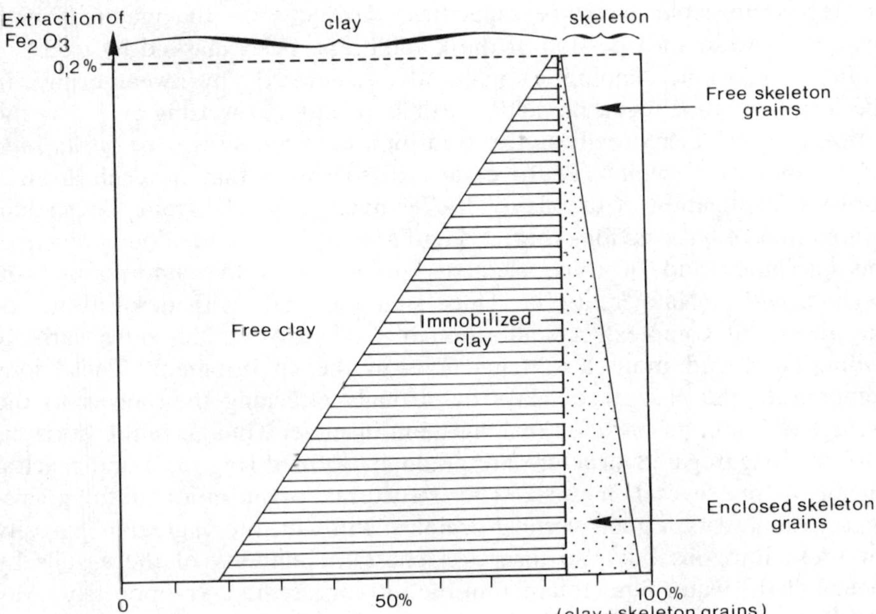

Figure 5.2. Effect of the Tamm Deferritization of Clays and Its Result on Destruction of Microaggregates (modified from Pédro et al., 1976). (Reproduced by permission of INRA Editions.)

Table 5.1. Swelling Capacity of *Terra Roxa* from Paraná Basin (Brazil). (Modified from Pédro et al., 1976. Reproduced by permission of INRA Editions.)

Sample	Potential Swelling Ps
Reference kaolinite	16.0
"Terra roxa estruturada" (Corn kernels)	11.3
"Terra roxa legitima" (Powdered coffee)	5.2

The progressive reactivity to wetting is noticeable. The reactions decrease with the amount of immobilized kaolinite made inactive by iron. The *terra roxa estruturada* and *terra roxa legitima* submitted to a Tamm experiment (which extracts about 0.2% iron) react to swelling in the same manner and display comparable exchange capacities. This confirms the inactivity with respect to water of a portion of the kaolinite particles masked by iron.

In conclusion, kaolinitic materials generated by weathering in the ferralitic domain can be made partially inactive to wetting cycles by the action of iron oxyhydroxides. The transition between *terra roxa estruturada* and *terra roxa legitima* (Pédro et al., 1976) shows that it result from a surficial ferritization of the clay. This ferritization results from the gradual elimination of basic cations (here Ca and Mg) present in the double layer of the kaolinites and of trace elements occurring in the amorphous iron oxyhydroxides (Nalovic, 1971). Thus, simultaneously with desaturation of the absorbant complex, crystallization of oxyhydroxides liberates variably hydroxilated and more acid ferric ions in the environment. These ions compensate the charges of clays by strongly attaching themselves to the surface of kaolinite particles making them inactive. Thus ferralitic horizons with microaggregated structure are gradually formed (Fig. 5.3) and reach a thickness of several meters. The structural organization into micro-aggregates favors a good vertical drainage through interaggregate porosity for these horizons but also insures a constant humidity of these soils by means of the water films retained in the intraaggregate cryptoporosity. Any change of external conditions, tectonic or bioclimatic, leads to a profound

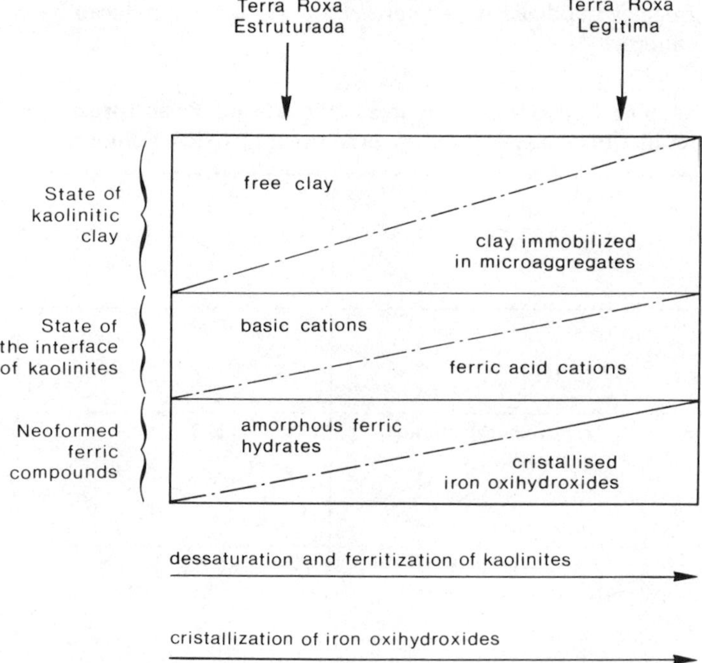

Figure 5.3. Variation of the Degree of Immobilization of Clay Through Micro-aggregation (modified from Pédro et al., 1976). (Reproduced by permission of INRA Editions.)

modification of the structural units of the upper part of these sequences as shown below.

Sequences with Lateral Differentiation Resulting from a Change in Drainage Conditions

Numerous examples recently studied in French Guiana and in Brazil by Turenne (1977), Boulet et al. (1982), Boulet et al. (1984), Lucas et al. (1984, 1986), Chauvel et al. (1987), and Fritsch (1984) show the general occurrence of systems with lateral differentiation of the pedologic mantle of the humid regions of South America, regardless of the type of parent-rock over which the initial weathering sequence developed.

The example selected here shows the transformation of latosols into podzols developed over the sandy argillaceous Barreiras Formation of the region of Manaus, Brazil (Lucas et al., 1984, 1988; Chauvel et al., 1987). Soils sequences were studied in three dimensions and show the relationship between latosol developed over plateaus and podzols, which occupy intermediate surface downslope from plateaus. The geomorphological study of this region shows that between flat plateaus and valley floors there occurs an intermediate surface with a gentle slope; this terminates at the drainage axis by a steep short slope. A close relationship exists between the development of intermediate surfaces and their distance to the head of the valley (Fig. 5.4 A, B, and C): In the upstream part of the valley in the most recent cuts, intermediate surfaces do not exist, whereas they are well-developed in the downstream part of the valley (Fig. 5.5). Upstream, intermediate surfaces are poorly developed and display only the first stages of differentiation of latosols into podzols; whereas, downstream, the complete differentiation can be observed. In extreme cases, the sandy argillaceous sediments of the Barreiras Formation may themselves be invaded by the front of podzolization (Lucas et al. 1984). The most differentiated sequences are described as follows. Over the plateau, the lower part of the pedologic mantle, developed at the expenses of the sediments of the Barreiras Formation, consists of a nodular horizon, reaching about 5 m thickness, made of ferruginous and gibbsitic nodules, 1–5 cm in diameter, scattered in a quartzo-kaolinitic matrix. This horizon of secondary accumulation can be compared with horizon showing fine polyhedric structure described in the Paraná Basin. However, in the case described here, an accumulation of oxyhydroxides of iron and alumina differentiated in the shape of nodules is superposed on the secondary accumulation of kaolinite. A microaggregated horizon (latosol) develops with a thickness of 4–5 m over the nodular areas. This microaggregated horizon is similar to those described in the Paraná Basin. However, in the region of Manaus, this horizon has a yellow-ocher and not a red color. Mineralogical composition explains this difference: less rich in iron oxyhydroxides and on the contrary enriched in gibbsite. The microaggregates of the latosols of the plateaus

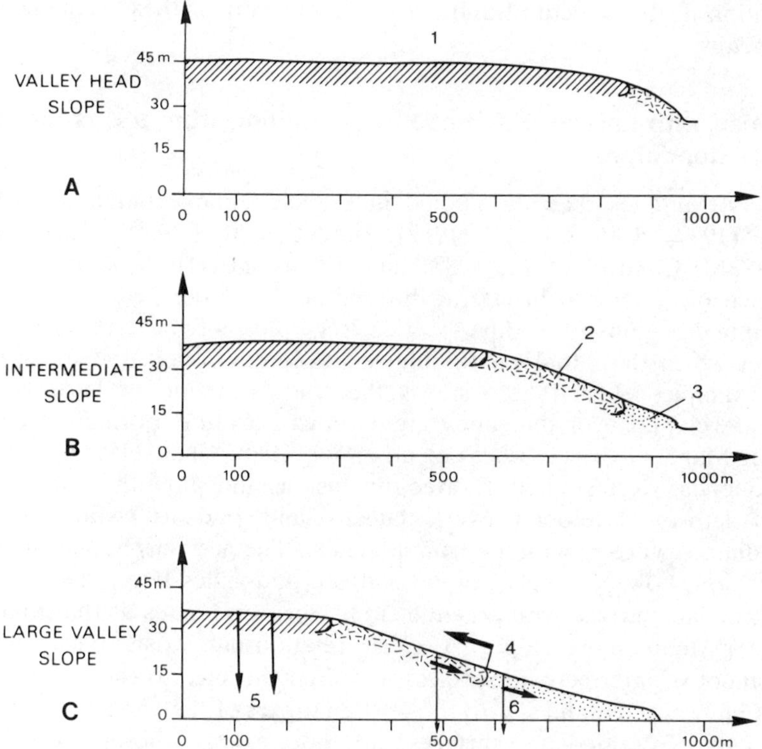

Figure 5.4. Sketch of the Evolution of Slopes (Intermediate Surfaces with Podzols) in the Valleys Which Incise the Plateaus (Supporting Latosols) in the Region of Manaus, Amazonia (modified from Lucas et al., 1984). **A**: Valley Head Slope. (1) Clay soils of the plateau. **B**: Intermediate slope. (2) Intermediate soils; (3) podzol soils. **C**: Large valley slope. (4) Development of the podzolization front; (5) vertical deep drainage with relative leaching of Si; Al remaining as kaolinite or gibbsite; (6) drainage with lateral component; leaching of Al and Si, remains of relict quartz.

contain 80% kaolinite, 2% iron oxyhydroxides, 8% gibbsite, and 10% of relict quartz grains. The uppermost 30 cm of the microaggregated horizon correspond to an A horizon in which quartz reaches 20% of the material and gibbsite tends to disappear. Therefore, the pedologic mantle of plateaus represents a typical vertical differentiation as encountered in tropical humid zones. Of course, this mantle displays a few specific features (nodular accumulations of oxyhydroxides and high gibbsite content of the more fragile microaggregates), which result essentially from the nature of the sedimentary parent-rock.

On the slopes or well-developed intermediate surface (section C of Fig. 5.5) from the upslope latosol to the downslope podzol (Fig. 5.6 A and B), the following lateral variations can be observed: The clay content (fraction

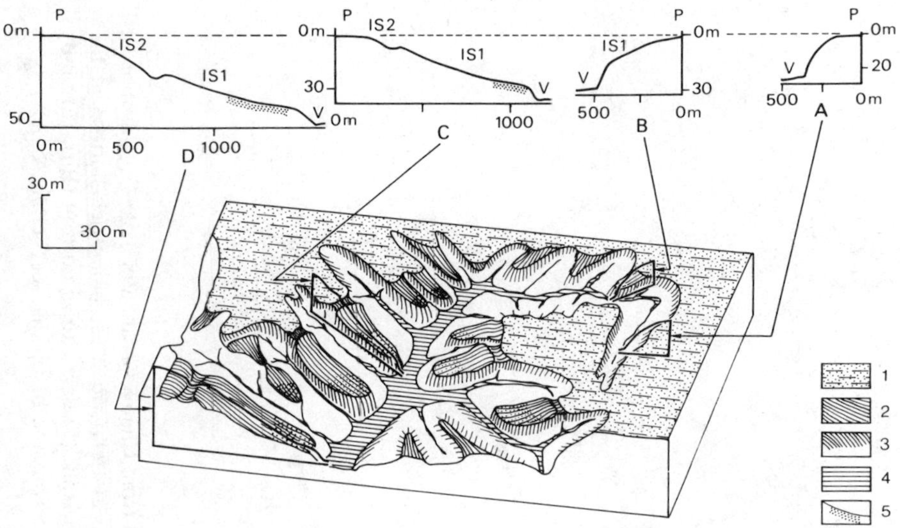

Figure 5.5. Appearance (A) and Development (B and C) of Slopes at the Expense of a Plateau in the Landscape of the Region of Manaus Amazonia (modified from Chauvel et al., 1987). (1) Plateau; (2) intermediate surface; (3) slope; (4) flat valley floor; (5) podzols. (Reproduced by permission of Birkhäuser Verlag A.G.)

< 2 μm) decreases gradually toward the bottom and the top of the surficial cover, as shown by the shape of isoclay content curves (Fig. 5.6B); downslope clay content decreases very rapidly with a lightening of soil color; oxyhydroxides nodules, which were indurated in the latosol, become soft and tend to disappear gradually; organic matter not directly observable with the naked eye upslope because it is combined with clay (type Mull organic matter in the sense of Duchaufour, 1970) concentrates in increasing amount in surficial horizons in the shape of darkish debris not combined with clay (type Mor organic matter in the sense of Duchaufour, 1970); the carbon content (Fig. 5.6 A) shows a decrease at the surface (sandy surficial horizon), but an increase in the immediately underlying horizons to the extent of generating dark-brown humic horizons (Bh) at shallow depth, which can superpose each other downslope (Fig. 5.6 A); the podzol of white sand (2% clay) appears and develops suddenly.

Observations made on this sequence are complemented by studies of other neighbouring ones. Lucas et al. (1984) showed, in particular, that the variations just described on a relatively short intermediate surface are magnified when this surface becomes longer. Indeed, the more extensive the intermediate surface, the greater the development of sandy soils and podzols. In fact, the lateral differentiation of podzols, which begins in the downslope portion of the intermediate surfaces, propagates upslope while the surface extends in that direction.

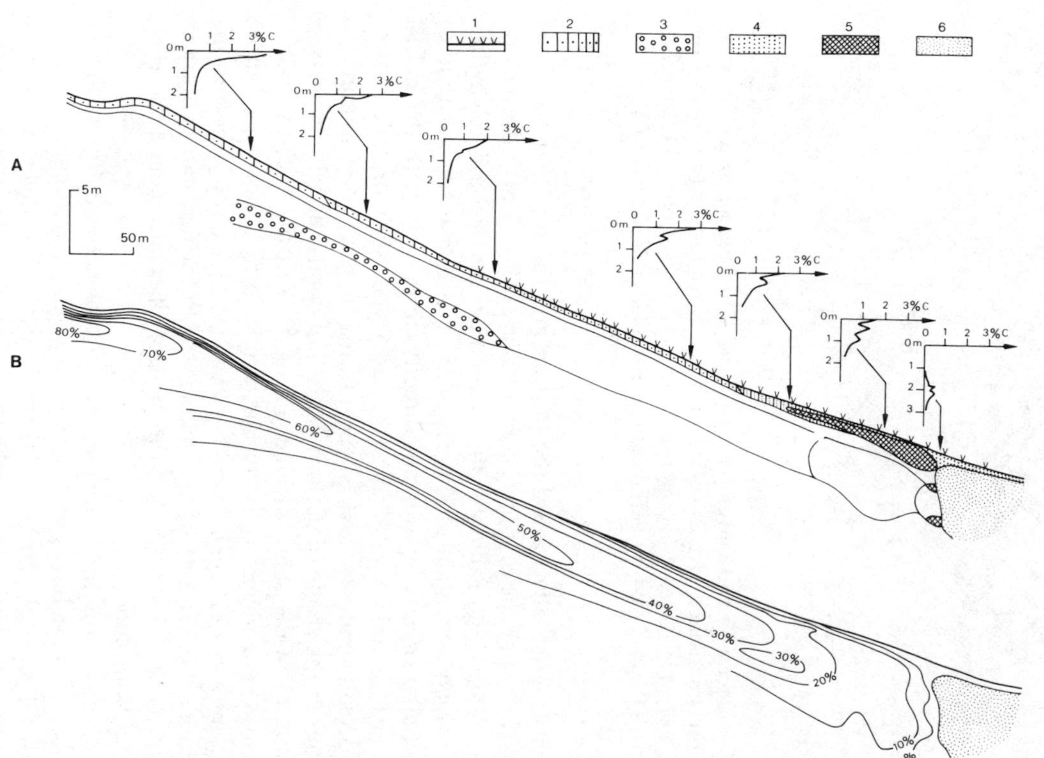

Figure 5.6. Pedological Mantle of Slopes in the Region of Manaus, Amazonia (modified from Chauvel et al., 1987). **A**: Organization of main horizons and vertical variation of Carbon. (1) Surficial accumulation of organic matter (Mor); (2) humusrich horizon at the surface (yellowish brown to dark brown); (3) ferruginous nodules; (4) sandy surficial horizon; (5) dark brown horizon (Bh); (6) sandy podzol horizon. **B**: Pattern of isocontent curves of the clay fraction. (Reproduced by permission of Birkhäuser Verlag A.G.)

Therefore, podzolization can develop at the expense of the original pedologic mantle until the latter disappears, and then can continue its development at the expense even of the sandy argillaceous sediments of the Barreiras Formation. The transition with the podzol was observed in detail (Fig. 5.7) with residual patches and tongues surrounded by a front of accumulation of organic matter (Bh) and even of iron oxyhydroxides (BFe). The differentiation of podzolic mantles parallels to topographic evolution.

In the humid region of Amazonia, two types of sequences develop simultaneously: a sequence with vertical differentiation consisting of microaggregated horizons; and a derived sequence with lateral differentiation consisting of podzolic horizons gradually generated at the expense of the previous sequence. Biogeochemical processes responsible for the transformation of the original sequences into podzols are not very well known. Laboratory experiments have been undertaken (Chauvel et al., 1987) by

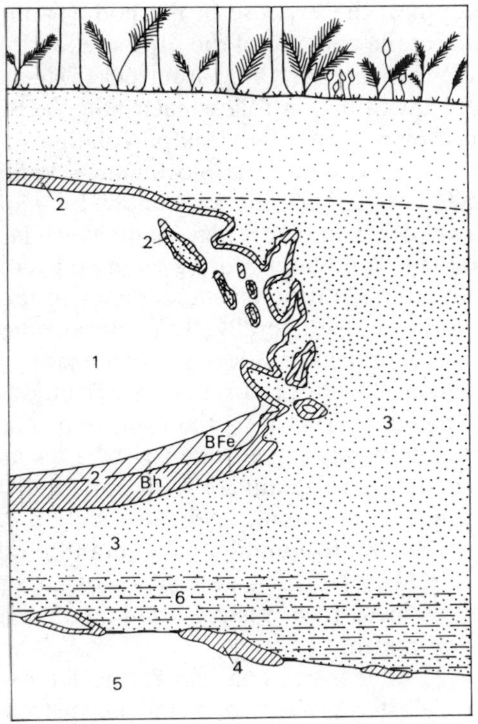

Figure 5.7. Detailed Sketch of the Transition Podzol-Latosol in the Region of Manaus Amazonia (modified from Chauvel et al., 1987). (1) Yellow-red, sandy clayey horizon (latosol); (2) accumulations of organic matter (Bh) and locally of iron (BFe), black to ocher brown; (3) horizon of white sand (podzol); (4) discontinuous horizon of organic accumulation; (5) white clayey sandy material without visible porosity; (6) seasonal perched watertable. (Reproduced with permission of Birkhäuser Verlag A.G.)

letting distilled water percolate through samples of soils collected from different horizons of an original sequence (latosols) of the region of Manaus. The obtained solutions show an export of aluminum (7–30 ppm) only in the first 30 cm of the profile, whereas an export of silica (1–5 ppm) takes place in all horizons. Indeed, organic compounds present at the surface allow solutions through acidocomplexolysis or chelation (Swindale and Jackson, 1956) to destroy a portion of kaolinite, of gibbsite, and of iron oxyhydroxides. This dissolution involves only the few tens of centimeters below the surface of latosols, as can be observed on plateaus. Vertical drainage in latosols is favored by the microaggregated structure and leads to precipitation in the underlying horizon of secondary oxyhydroxides and oxides of iron and alumina in the shape of nodules.

In these original sequences with vertical differentiation an obvious parallelism exists between destruction of microaggregates at the surface and the following simultaneous processes: lowering of microaggregated horizons to a few meters' depth at the expense of the nodular horizons; lowering of the nodular horizons at the expense of the saprolite; and, finally, lowering of the saprolite at the expense of the parentrock. Thus, several fronts of weathering and of transformation follow each other vertically through *self-development* induced by reactions between percolating waters and minerals.

All these processes occur under the bioclimatic conditions of this region of Amazonia. Each cut, by erosion or simply by more rapid chemical weathering along weakness zones, in a plateau on which latosols are evolving, modifies locally drainage conditions. Along the margins of the cut, solutions percolating vertically in the latosol have a tendency to take a new direction, toward the cut, hence to move laterally. This lateral transfer seems to have two major effects: The surficial mantle stops its lowering at the expenses of the parent-rock, and a change occurs in the dynamics of organic matter migration. This change is expressed by the transition of an accumulation of organic matter intimately mixed with clay (Mull) to an accumulation of organic matter consisting of incompletely decomposed debris (Mor). The results are lateral migration of humic acids, destruction of kaolinites and of oxyhydroxides, and their reprecipitation at different depths in the shape of Bh or BFe (Chauvel et al., 1987). The destruction of clays and oxyhydroxides by acidocomplexolysis leads to a relative accumulation of quartz grains as sandy horizons in the podzols.

An increase in size of the erosional cut favors lateral drainage, thus in time slopes are the most transformed in the downstream position of the valley. Intermediate surfaces with podzols therefore evolve simultaneously with the evolution of the hydrographic network. Lateral drainage favors a progressive decrease in the angle of the slopes of the valley and hence a gradual rising of podzols at the expense of latosols. As shown by Lucas et al., (1984), this transformation can extend as far as the disappearance of the plateau and the general occurrence of tropical podzols.

This progressive evolution through time of latosols into podzols, that is, from plateaus into slopes, has been recently studied by Lucas and Chauvel (1990). The transformation of an original sequence of latosols into a derived sequence of podzols is very common as soon as a change of drainage conditions begins. The example of Amazonia shows how this transformation develops starting from the margins of plateaus. However, such a transformation can take place inside a plateau, as commonly observed in the evolution of the pedologic mantle of tropical humid countries.

In another example, in French Guiana, Boulet et al.(1984) described changes of latosols into podzols in lateritic plateaus developed over Archean migmatites. Here the original sequence consists of a thick saprolite overlain by a sandy argillaceous to argillaceous horizon with microaggregated structure, which reaches a few meters thickness. Leached sandy argillaceous horizons develop at the boundary between saprolite and microaggregated horizons, most probably because of secondary accumulation of kaolinite, which clogs porosity in the upper part of the saprolite, thus changing the vertical drainage in favor of a lateral drainage perched above the saprolite. Progressively leached horizons are replaced by podzols at the center of the plateau, and the process, through self-development, leads to expansion of podzolization from beneath and from the center toward the periphery of the geomorphic unit, at the expense of microaggregated horizons. All stages of development between lateritic plateaus with microaggregated horizons, lateritic plateaus with central depression occupied by podzols, and entirely podzolized plateaus were observed by Boulet et al. (1984). The different stages of transformation of microaggregated horizon to a podzol were recently illustrated (Fig. 5.8 A to E) by Lucas and Chauvel (1990).

In tropical humid regions, the original pedologic mantle with vertical differentiation can display upper horizons undergoing transformations that lead to new organizations of podzol type. These modifications result from a change in drainage conditions without change of bioclimatic parameters. Vertical drainage favored by microaggregated structure of the upper part of the original mantle is replaced by lateral drainage. In this case, the role of organic acids seems predominant in the destruction of microaggregates made up of kaolinite plus oxyhydroxides of iron and alumina. The change of drainage in cases of Amazonia and French Guiana may result either from the evolution of the hydrographic network during time (erosional cut or geochemical lowering of plateaus) or from the evolution at the depth of the saprolite porosity (such as clogging through secondary accumulation of kaolinite). In the first case, the transformation is centripetal and extends from the margins toward the center of the plateau. In the second case, the transformation is centrifugal and develops from the center toward the margin of the plateau.

Climatic conditions do not seem to be directly responsible for these transformations. The Amazonia example shows that the gradual size increase of the hydrographic network can itself generate the conditions of its self-development through internal biochemical modification of its slopes. The

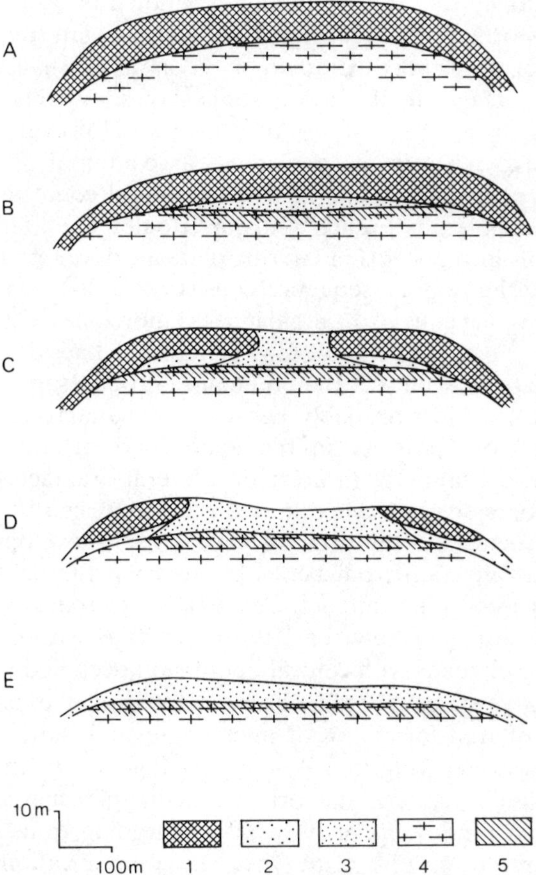

Figure 5.8. Sketch of the Stages (**A** to **E**) of Transformation of Microaggregated Horizons (Latosol) to Podzol in French Guiana (modified from Lucas et al., 1987). (1) Microaggregated horizon (latosol); (2) leached sandy horizons; (3) white sandy podzol; (4) saprolite horizon; (5) newly generated kaolinite that clogs porosity of upper part of saprolite.

French Guiana example shows that the progressive destruction of surficial clays and their precipitation as secondary kaolinite at the top of the saprolite generates, through self-development a horizon sufficiently impervious to modify drainage; this situation induces the internal biochemical transformation of upper horizons.

Therefore, the pedologic systems described above correspond only to the last stages of the same pedologic history entirely explainable through present-day processes, although climatic changes may have certainly affected Amazonia.

The geomorphologic evolution of these regions is therefore controlled by the internal geochemical evolution of the pedologic mantle. Obviously, differentiation of a podzol consisting of white sand can only favor enlargement of interfluves, hence modify the hydrographic network, which itself, through upstream development, allows transformation of lateritic plateaus into podzolized slopes, and so on.

The well-established hierarchy of the units of the pedologic mantle finally generates its own desorganization through self-development, which eventually affects the reorganization of sequences and of geomorphic units. Actually, it is a real feedback between evolutions of organizations at different scales: The evolution of the soil organizations has an effect on the evolution of the pedologic system, which affects the geomorphologic evolution, which, in turn, has an influence on the evolution of the soil organizations. This represents the self-evolution of pedologic landscapes of tropical humid regions.

Derived Sequences Resulting from a Change in Climatic Conditions

Bioclimatic changes, whenever sufficiently long or intense, eventually lead to profound transformations of pedologic systems. Transformation of original sequences is expressed by disorganization of upper horizons, which are the most exposed to such bioclimatic changes.

The example chosen here to illustrate this type of derived system was studied by Chauvel (1976) and Chauvel et al. (1977, 1978) on the pedologic mantle of low plateaus of Casamance in Senegal.

Low plateaus consist of a saprolite developed at the expense of marine sandstones (glauconitic arenites) and shales of Mesozoic-Cenozoic age (Tessier et al., 1975). The saprolite preserves the original sedimentary structures and consists of quartz, kaolinite, and ferric hydrates. It is overlaid by an unconsolidated red microaggregated latosol, which grades laterally into beige sandy ferruginous soils representing horizons resulting from the transformation of latosols.

In the most humid zones of the southwest part of Casamance, only red microaggregated horizons occur. Microscopic study of these red latosols shows that they consist of quartz, kaolinite, and iron oxyhydroxides associated in small globular masses of about 100 μm in diameter, which are the microaggregates. Here also microaggregates result from plasmic concentrations and are similar to those previously described in the Paraná Basin. Quartz grains are enclosed within the red plasma and are more abundant than in the Paraná examples.

When climatic conditions become drier, red latosols, which cover low plateaus, are combined with beige soils, which appear and develop in the centre of plateaus at the expense of red latosols. The most frequent transformation occurs simultaneously in the upper part of red soils and laterally by means of an intermediate product called transitional soil (Fig. 5.9). Megascopically, this transition is shown by a gradation of predominant red

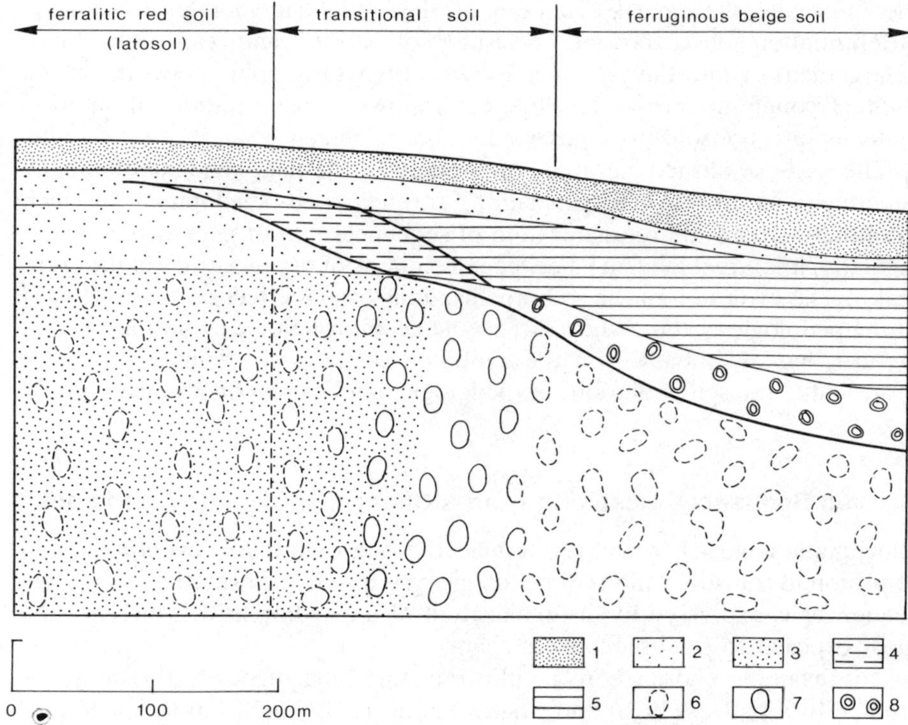

Figure 5.9. Relationship between Red Ferralitic Soils (Latosols) and Beige Ferruginous Soils of Casamance (Senegal) modified from Chauvel (1976). (1) Sandy horizons; (2) sandy argillaceous horizons; (3) horizons with microaggregates; (4) compact clayey horizon; (5) coarse structured brown horizon; (6) ferruginous mottles; (7) ferruginous nodules; (8) indurated ferruginous concretions.

colors to more ochre, yellow, and beige colors, and by surficial horizons becoming more sandy. Ferruginous nodules that might have occurred in the red soil are inherited in the transitional soil and may evolve into pisolites in the upper part of the beige ferruginous soil.

This vertical and lateral transition can be observed under the petrographic microscope. One notices the dislocation followed by the collapse of microaggregates. Fragments are then dispersed and transformed into a new gray and deferretized s-matrix. In it, argillaceous plasma appears mobile and leached from upper horizons, whereas the quartzose skeleton separates from the plasma, breaks up, and reconcentrates to generate predominantly sandy or silty surficial horizons.

These transformations leaching from red soil to beige soil therefore require segregation of plasma constituents (kaolinite and iron oxyhydroxides) and lead to a dissociation of the plasma and of the quartzose skeleton. Hence a structural desorganization takes place without major mineralogical changes.

Furthermore, clay is redistributed between surficial and deeper horizons of the profile and hence a secondary hydromorphosis appears. These transformations are also accompanied by an important compaction of the order of a third of the thickness when the transition of the microaggregated structure to the s-matrix of the beige soil occurs. This compaction is noticeable in the center of plateaus when a small depression indicates the location of beige soils.

This transition between red soils and beige soils corresponds, therefore, to segregation of the three major constituents: kaolinite, iron oxyhydroxides, and quartz. Laboratory experiments Chauvel (1976) and Chauvel et al (1977) have shown the physicochemical processes of such a segregation. Experiments have been conducted on samples of red microaggregated soils. The simple act of submitting such a sample to an extended desiccation followed by a wetting until saturation causes the development of ferriargilanes generated in situ by reorientation of argillaceous particles of microaggregates under the action of the induced hydric constraints. If a Tamm deferritizing experiment is added, the major portion of the microaggregated material is dissociated, and a s-matrix generated in which the clay becomes partly anisotropic, and only a few relicts of microaggregates subsist. The material obtained from all these experiments shows completely changed physical and mechanical properties: swelling potential originally insignificant (1.5%), increased considerably (20.3%), exchange capacity increased by 86% (changing from 4.2 to 7.8 mEq/100g). Apparently, alternation of extreme hydric conditions and of surficial deferritization of kaolinites leads from destruction of microaggregated structure to generation of a dense and oriented argillaceous plasma. Chauvel (1976) and Chauvel et al. (1978) attribute to an *ultradesiccation* the start of this process of transformation of red soils into beige soils. Indeed, red soils display two types of porosity capable of retaining water (Fig. 5.10). First an intermicroaggregates microporosity (diameter of voids 2–100 μm), which insures vertical circulation of solutions within the soil and their retention under weak constraints. This microporosity is maintained by the framework of the microaggregates, which maintains its shape. Second is an intramicroaggregate cryptoporosity located inside the plasmic assemblage of the microaggregates. These cryptovoids (diameter 0.1 μm) retain water with a very high energy. The three-dimensional assemblage of the crystallites of kaolinite maintains this cryptoporosity. However, if conditions of extended aridity occur, considerable water deficits are produced over a great thickness of the microaggregated horizons (Chauvel and Pedro, 1978). These conditions (Fig. 5.11) generate physical constraints responsible for evacuation of the water contained in the cryptovoids and for breaking of interparticles bonds. Therefore, a rewetting generates reorientation of clay particles, hence the observed development of ferriargilanes.

Subsequently this process self-develops and leads to a progressive destruction of microaggregates and to a disassociation of plasma and skeleton.

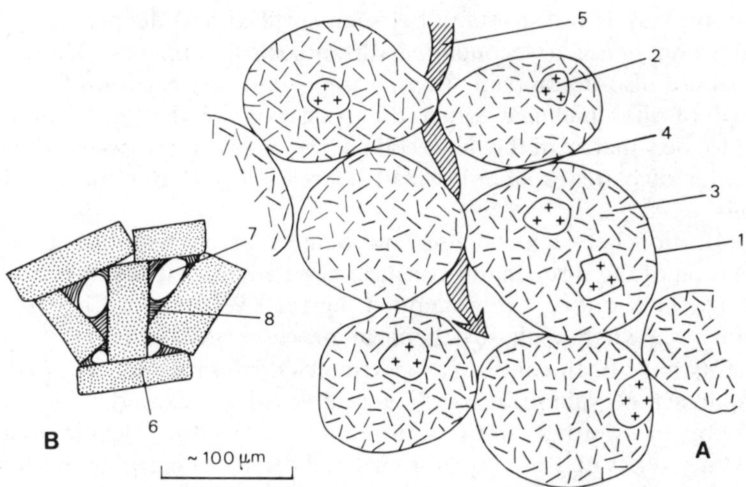

Figure 5.10. Sketch of Microaggregated Structures of Ferralitic Soils (Latosols). **A**: General aspect. (1) Microaggregate of ferritized kaolinite and quartz; (2) enclosed quartz grain; (3) kaolinite agglomerates intraaggregate cryptovoids ($\leqq$ 0.2 μm); (4) interaggregate microvoids; (5) direction of flow of percolating water inside inter-aggregate. **B**: Detailed structure of microaggregates. (6) Crystallites of kaolinite in the ferritized surface; (7) cryptovoids between kaolinites; (8) water meniscus in cryptovoids.

During the ultradesiccation that affects microaggregates, water films that coat ferritized kaolinites become increasingly thin. They can even become dissociated and a very low pH of the order of 2 can develop at the surface of particles, sufficient to mobilize the variably hydroxylated iron attached to the exchange position of kaolinites (Chauvel and Pedro, 1978). Therefore, a separation of clay and iron oxyhydroxides is combined to a disassociation of plasma and skeleton. Kaolinite thus deferritized may subsequently react with the aqueous phase. Iron oxyhydroxides can be mobilized over short distances and reconcentrate in the shape of weakly indurated to indurated glaebules or nodules.

The collapse of microaggregates, the reactivation of kaolinites, and its reorganization into more compact and more impervious new structures lead to a modification of drainage. Vertical drainage characteristic of micro-aggregated horizons changes to a lateral drainage with secondary hydro-morphosis and individualization of diffused mottles of iron oxyhydroxides.

The collapse of structures is also visible in the field in beige ferruginous soils where small depressions develop at the center of plateaus, rain water concentrates in these spots, accentuating contrasts between phases of wetting and drying. Thus, a lateral centrifugal evolution that develops at the expense of the red latosol with microaggregates is added to the initial vertical evolution.

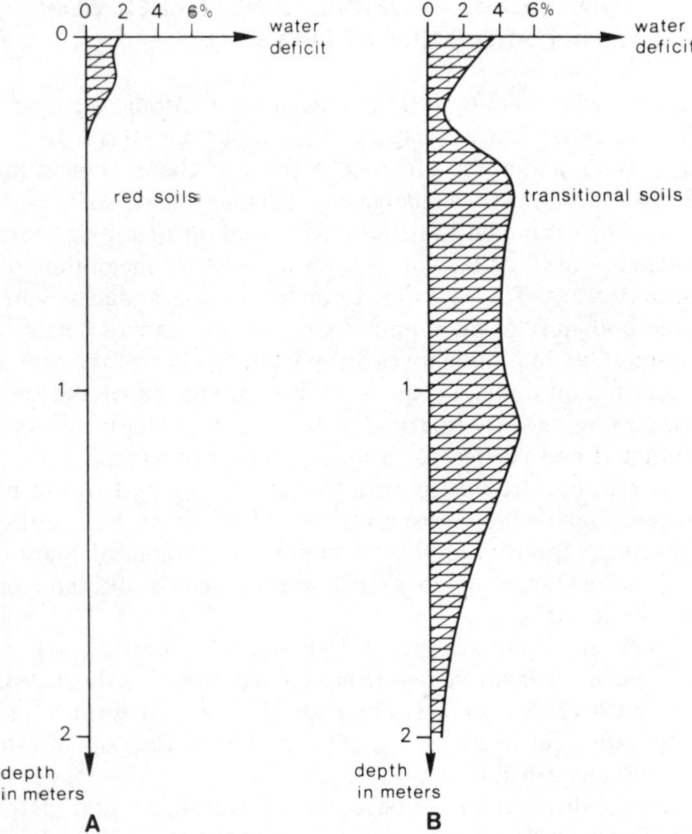

Figure 5.11. Water Deficits Measured in Red Microaggregated Soils (**A**) and in Transitional Soils (**B**) of Casamance, Senegal. The soils considered here are those of Figure 5.9 (modified from Chauvel and Pédro, 1978).

This process can thus, through self-development, extend itself as far as the periphery of plateaus. Furthermore, microdepressions occupied by ferruginous beige soils can join and anastomose each other. The lateral drainage can lead to surficial flow which "is at the origin of the establishment of a hydrographic network in the center of plateaus. Headward erosion and deepening of this network lead to the dissection of plateaus" (Lucas et al., 1984, p. 170).

In this case also, the upper part of the original pedologic mantle evolves from a microaggregated horizon to a ferruginous beige one. The origin of this evolution is a change of climatic conditions from humid toward arid. This change induces new hydric conditions whose effect is desorganization of microaggregates and appearance of new organizations. The process, once started, autodevelops and leads progressively by internal dynamic evolution of the pedologic system to a geomorphological evolution.

PEDOLOGIC MANTLES OF TROPICAL LANDSCAPES WITH ALTERNATING DRY AND RAINY SEASONS

The original surficial mantle with vertical differentiation of tropical humid countries can undergo transformation of its upper part through destruction of microaggregates under the effect of a climatic change consisting of the appearance of a dry season, namely, of a decrease in rainfall.

Chauvel (1976) estimated that the differentiation of a beige ferruginous soil at the expense of a red latosol requires an order of magnitude of time of a few thousand years. The previous examples of Casamance (Senegal) are located at the boundary of the humid domain. If one moves northward from Casamance into Gambia, one moves away from this boundary zone between the permanent humid domain (equatorial in a broad sense) and the domain of alternating rainy and dry seasons, to reach zones where dry seasons are more accentuated and existed for a longer period of time.

In such a case, one frequently sees that the upper part of the pedologic mantle consists of ferruginous crust horizons. Whenever such horizons become generalized, they represent rapidly a major morphological feature of the landscape, showing that ferruginous crusts are characteristic of these particular bioclimatic conditions.

The destruction of microaggregates of latosols allows a separation of kaolinite from iron oxyhydroxides. Iron reconcentrates at short distances in the shape of mottles and glaebules. This process, repeated during a sufficiently long span of time, can generate a mottled clay at the top of which real ferruginous nodules can differentiate.

Consequently, through progressive lateral transformation, latosols are replaced by argillo-ferruginous horizons with mottles and nodules whose colors range from ochre yellow to purple red. In these horizons with lateral and vertical dynamic conditions solutions processes occur. Iron oxyhydroxides remain in these weakly soluble environments of the upper parts of profiles and undergo only very short transfers in the state of ferrous ions. This is the reason an abundance of concentrated iron oxyhydroxides can be directly related to richness in iron of the original parent-rock underlying the saprolite (Leprun, 1979).

Mottled clays, once differentiated, undergo their own evolution. At the surface, this horizon is continuously destroyed by the action of unsatured rain water and of organic matter still present in this landscape of savanna with trees. Kaolinite is destroyed more rapidly than iron oxyhydroxides and oxides. The latter therefore concentrate relatively closer in the shape of purple red, indurated ferruginous nodules consisting essentially of hematite. Dissolution of kaolinite and accumulation of iron oxides as nodules are repeated in time, and an increasing concentration of iron oxides allows a greater dissolution of kaolinite at the places where nodules are formed. Gradually horizons with ferruginous nodules develop at the top of the pedologic mantle. With lowering of the topography, these ferruginous nodules increase in number and become anastomosed into a nodular ferruginous crust called conglomeratic. Kaolinite destroyed by solutions at the top of the profile reprecipitates as a new generation of kaolinite in the lower part of the mottled clay, and even in the upper part of the underlying saprolite. In conclusion, a mottled clay, a nodular ferruginous horizon, and an indurated ferruginous crust become differentiated above the saprolite itself a relict of an evolution in a more humid climate.

Such a derived sequence is identical to original sequences differentiated directly at the expense of the parent-rock under similar bioclimatic conditions. However, there is a difference: These original sequences displays a saprolite horizon thinner than that generated by the transformation of a pedologic mantle previously differentiated under a more humid climate (Fig. 5.12). In summary, under tropical climates with alternating rainy and dry seasons, a lateritic pedologic mantle, consisting of mottled clay horizons and ferruginous crusts, is differentiated.

Wetting predominates in horizons at the top of the pedologic mantle and controls the flat morphology characteristics of these regions. Wetting at depth is not sufficient to allow saprolite horizons to strongly develop at the expense of the parent-rock, consequently, weathering of parent minerals can allow smectites to appear as weathering plasma (Tardy, 1969). Therefore, solutions preferentially affect the upper part of profiles where iron oxyhydroxides undergo dissolutions, transfers and reprecipitations over short distances (a few millimeters to a few centimeters), whereas kaolinite is submitted to the same processes but over greater distances (a few decimeters to a few meters).

All over a geomorphic unit, solutions percolate and generate within iron crusts and mottled clay horizons, dissolutions, recrystallizations, transfers of materials, and neoformations, redistributing iron, alumina, and silica as new generations of iron oxyhydroxides, iron oxides, and kaolinite. New organizations replace earlier ones in crusts as well as in mottled clays.

Leached zones are adjacent to zones of accumulation. This is a state of constant desequilibrium, but in the long run, kaolinite has a tendency to concentrate in deeper and more downslope areas. Whitened and leached horizons with abundant quartzose skeleton can develop in places beneath iron crusts, undergo compaction, and lead to collapsing of the crust, which breaks up in situ in individual blocks. Meanwhile, the crust itself through continuously repeated redistribution of iron generates pisolites, which progressively become separated from each other, forming a pebbly horizon. This horizon is expressed in the field by small depressions that allow the development of a local drainage and hence, in time, the appearance of a linear erosional cut. Other portions of geomorphic units display an equilibrium between leaching and accumulation and allow a geochemical sinking of such portions without major discontinuities. Iron crusts develop at the expense of mottled clays horizons, and the latter at the expense of saprolites. Consequently, pedologic mantles with iron crusts display a morphology of asym-

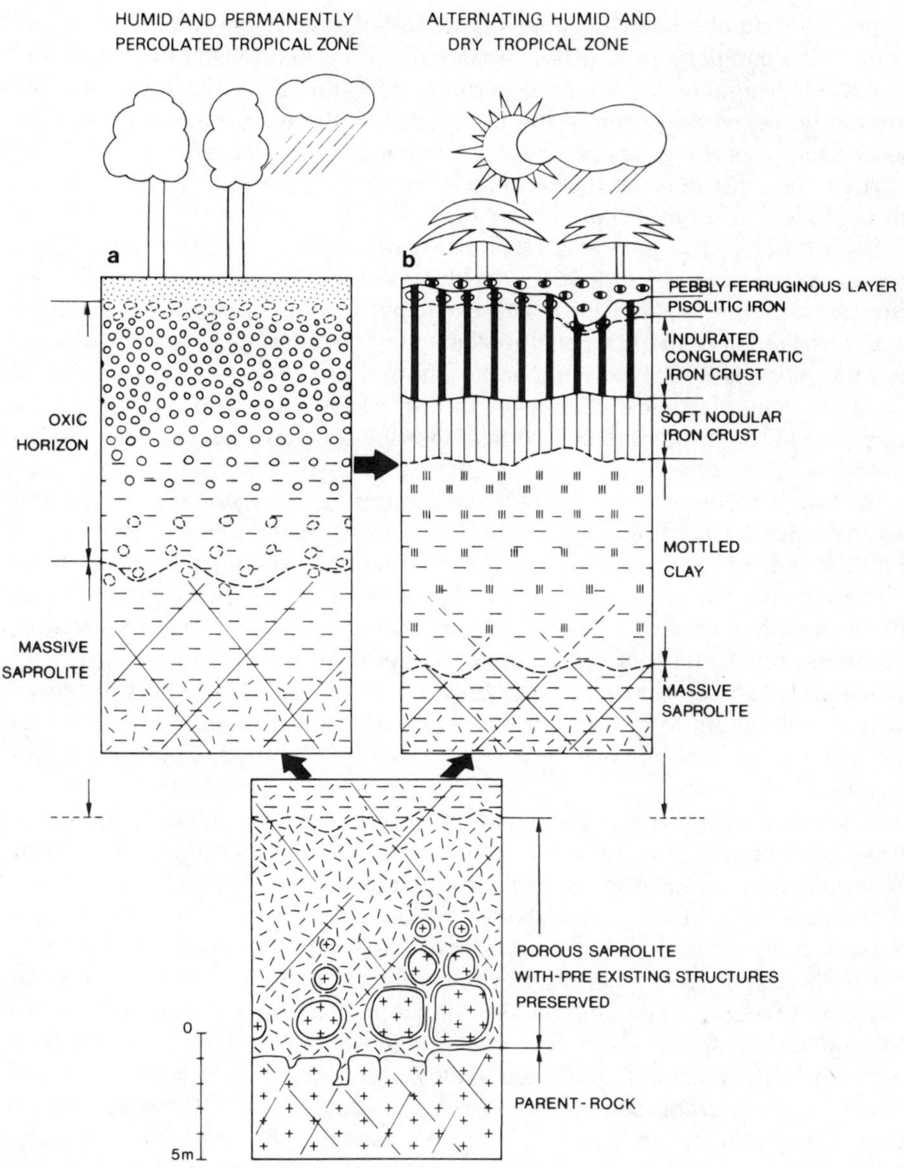

Figure 5.12. Sketch of the Evolution of the Pedological Mantle in Tropical Regions with Alternating Dry and Rainy Seasons. Sequence "b" can be original and formed at the expense of a parent rock under bioclimatic conditions of alternating seasons, or can be derived from a sequence "a" formed under more humid conditions (modified from Nahon, 1987).

metric hills (Fig. 5.13), showing the following features: roughly tabular tops with very low uniform inclination, rather steep concave slopes where the erosional cut is located, low angle slopes, slightly convex and concave when joining broad thalwegs.

PEDOLOGIC MANTLES OF TROPICAL LANDSCAPES WITH PREDOMINANT EXTENDED DRY SEASON

The surficial mantle of regions submitted to a tropical climate with extended dry season (7–9 months) was studied in detail by Boulet (1974) in Burkina Faso (former Upper Volta) in West Africa.

In the region of Garango, southern part of Burkina Faso, the pedologic mantle is differentiated over a migmatitic parent-rock. The geomorphology of this area consists of short interfluves with convex summit and gentle slope, of the order of 1–2%. Inselbergs are scattered throughout the region. Two types of sequences were recognized: the first one with predominant vertical differentiation; the second with lateral differentiation superposed on the vertical one. The first is an original sequence, the second a derived one.

The original sequence called Garango I is developed over a geomorphic unit (interfluve) away from any inselberg. This sequence is called a normal sequence characteristic of the surrounding bioclimatic conditions.

Garango I

The pedological mantle of Garango I is rather thin and reaches only a few meters' thickness. Overlying a sandy gruss with preserved structure (coarse saprolite) is an argillaceous-sandy horizon, 2 m thick upslope and 1.5 m thick downslope. At its maximum thickness, this horizon (eutrophic brown soil) consists of an argilloplasma with predominant kaolinite and smectite (Fig. 5.14) generated by weathering of parent minerals whose relicts form the essential part of the sandy and silty fraction. Microscopic studies show that, with the exception of the near-surface, parent-relicts have not been displaced and the weathering is isovolumetric.

The amount of smectite of the argillaceous sandy horizon increases appreciably downslope and particularly toward the upper part of this horizon to the extent of differentiating a vertisol in which parent relicts are affected by pedoplasmation. Within this downslope smectitic argillaceous sandy horizon, a calcitic accumulation is expressed by calcitic nodules.

Alteroplasmation that predominates in the gruss (coarse saprolite) of the upslope portion of the sequence (Fig. 5.14) is 60% smectitic and 40% kaolinitic, but becomes predominantly kaolinitc (70% at the top of the argillaceous sandy horizon).

Alteroplasmation that predominates in the gruss in the downslope portion of the sequence is 90% smectitic and is followed in the argillaceous sandy

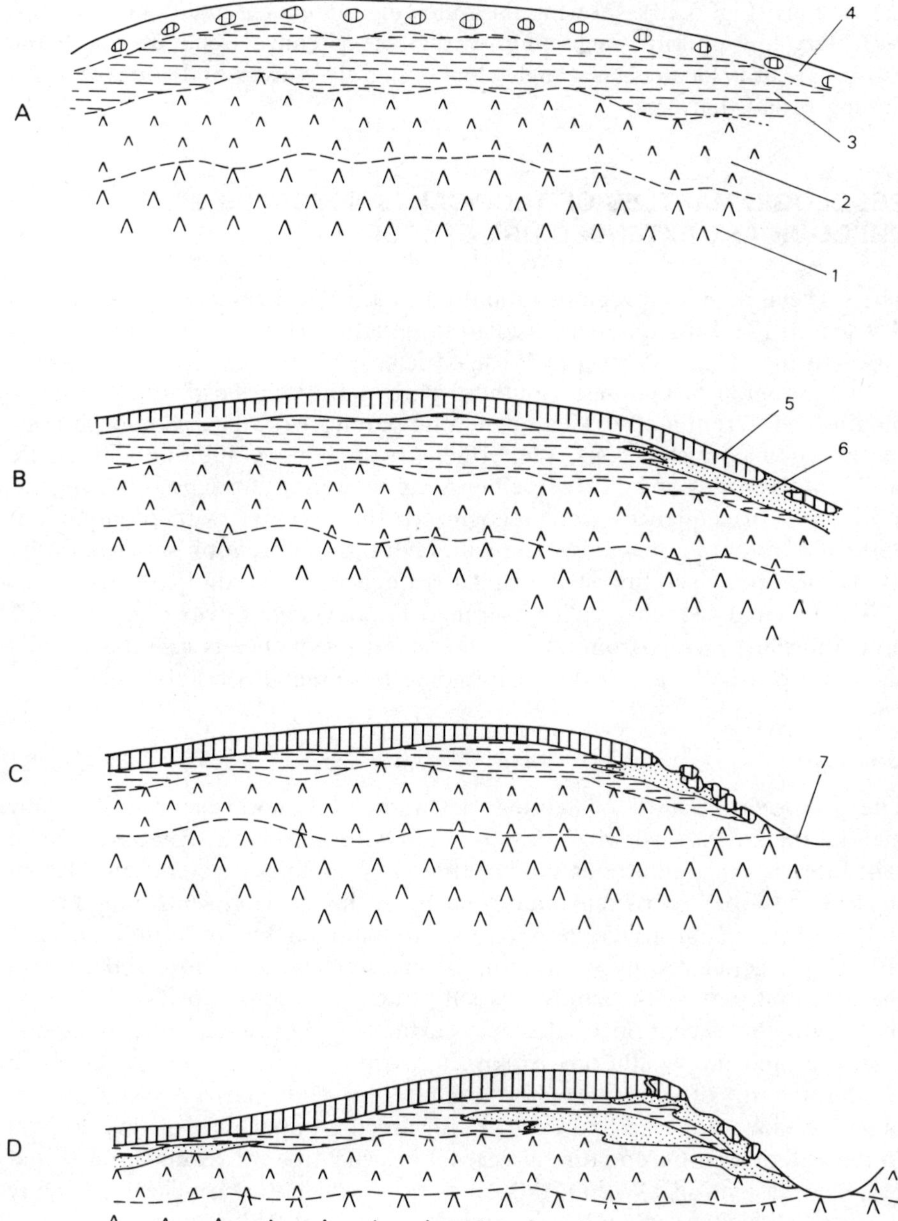

Figure 5.13. Sketch of the Evolution of a Pedologic Mantle with Ferruginous Crust. (1) Unweathered parent rock; (2) saprolite; (3) mottled clay; (4) horizon with ferruginous nodules; (5) horizon with ferruginous crust; (6) leached horizons; (7) thalweg. Stages A to D show formation and evolution through Time.

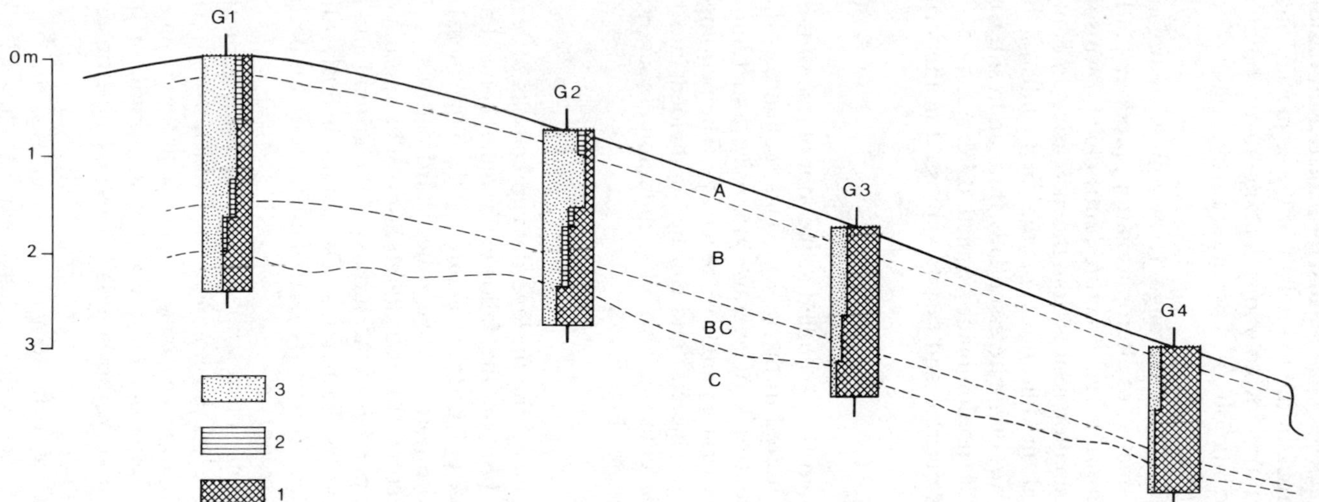

Figure 5.14. Sequence of Garango I. Distribution of Major Clay Minerals. (1) Smectite; (2) illite; (3) kaolinite. Horizon C: coarse saprolite. Horizon BC: transition. Horizon B: eutrophic brown soil upslope, vertisol downslope. Horizon A: humic sand to argillaceous sand. (Modified from Boulet, 1974. Reproduced by permission of ORSTOM Editions.)

horizon by an essentially smectitic pedoplasma in which kaolinite can reach 20%. This difference in the mineralogical nature of argillaceous plasmas, in a downslope direction, can be explained first by an increasing basic character of the migmatitic parent rock in the lower half of the sequence, and second by a discrete lateral migration of solutions leading to a confinement more favorable to the precipitation of smectite (Tardy, 1969; Paquet, 1969; Bocquier, 1973).

In the original sequence of Garango I, two units of organization are recognized. First, the s-matrix of alteroplasmation, predominantly smectitic, represent horizons with preserved structure (eutrophic brown soil). Second, the s-matrix of pedoplasmation results from the increasing amount of smectite that generates internal modifications of the first unit of organization (vertisol). Both s-matrixes originate essentially from an in situ weathering of parent minerals: It is a predominant vertical differentiation. The lateral differentiation remains discrete and poorly expressed in the structures.

Garango II

The sequence of Garango II was studied at the foot of one of the inselbergs, 90 m high, that are scattered in the region. This sequence is differentiated over a migmatitic parent-rock of average grain-size and rather acid. The slope is concavo-convex with an angle of about 2%. The pedologic mantle is thin, reaching only a few meters, and can be subdivided into three major domains: upslope argillaceous sandy with kaolinite, middle sandy, downslope argillo-sandy with smectite (Fig. 5.15).

Upslope Domain

At the foot of the inselberg, the upslope domain consists of ferralitic soils, showing, from top to bottom, the following profile: homogeneous red B horizons with fine friable polyhedric structure (~50 cm thickness), reticulated red B horizon with more porous beige zones, hard bedrock.

In a downslope direction, this reticulated horizon has an upper limit that remains approximately parallel to the surface of the ground and increases in thickness at the expense of the parent-rock. Simultaneously, the overlying homogeneous red horizon looses its ferralitic character by becoming more massive and of lighter color.

Study under petrographic microscope shows that inside the red reticulated horizon and at the base of the homogeneous red horizon, parent relicts scattered within the alteroplasma have not been displaced and, therefore, the weathering is isovolumetric. However, the dispersion of fragments of skeleton in the upper part of the homogeneous red horizon expresses the effect of pedoplasmation.

At 35 m downslope from the foot of the inselberg, the red reticulated horizon at its basal contact with the hard bedrock shows its beige zones emptied of their fine fraction with the reticulated network untouched. This

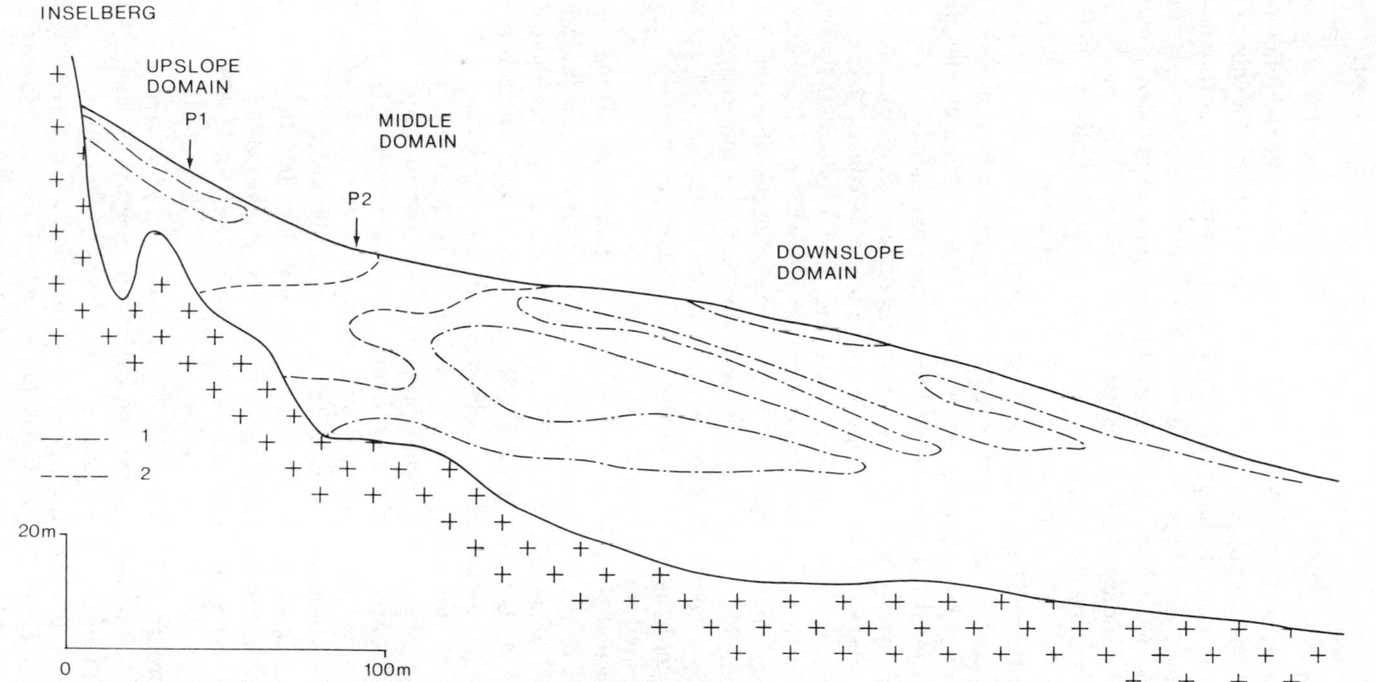

Figure 5.15. Sequence of Garango II, Location of the Three Major Domains (modified from Boulet 1974). (1) curves of isovalue at 25% clay with respect to percentage of total soil; (2) curves of isovalue at 10% clay with respect to percentage of total soil. (Reproduced with permission of ORSTOM Editions.)

is the appearance of a A′2 leached horizon at the base of the red reticulated horizon.

Middle Domain

Further downslope, horizon A′2 develops at the expense of the red reticulated one, beige zones are completely emptied with the exception of a white sand deposited at the bottom of the cavities. The red network that surrounds the emptied zones keeps its shape, but indurates. Gradually, leaching develops from the bottom upwards at the expense of the entire reticulated horizon, with an upper boundary dipping against the slope (Fig. 5.16, profile P_1). Furthermore, leaching rises and extends above the reticulated horizon reaching the surface of the ground at a distance of about 100 m from the foot of the inselberg. The leached horizon is expressed by loss of the fine fraction and appearance of a millimetric porosity and with the bottom of the pores coated with a fine white quartzose sand. This situation characterizes the middle domain of the sequence, which is essentially sandy.

At the same time as the leached horizon developed at the expense of all the other horizons of the initial pedologic mantle becomes thicker and extensive, clay accumulates in its lower part (Fig. 5.16, profile P_2). This clay, which is mixed with fine sand deposited by an almost vertical decantation within the aquifer, invades the lower portion of the leached horizon during the rainy season. This process of accumulation is first expressed by clay coatings, then by a complete filling of the porosity of the leached horizon. These coatings and fillings of clay cover and enclose the fine white sand previously deposited on the floors of the pores.

Inside the leached horizon, which developed at the expense of the reticulated one, clays clog the porous portions existing between the indurated ferruginous red framework. With gradual accumulation of clays increasing downslope (from A'_2B' to B'_2 horizons), this ferruginous framework becomes friable and discontinuous and then disappears, being replaced by a new prismatic structure (formation of a B'_2 horizon) (Fig. 5.16, profile P_2).

Downslope, another type of accumulation occurs, originating from particles in suspension and solutes migrating laterally in the aquifer. Through its repeated oscillations, the aquifer, whenever being lowered, deposits clays forming superposed tongue-shaped horizons prograding upslope. These horizons decrease gradually the reservoir capacity of the aquifer and the roof of the aquifer rises toward the surface. Consequently, the slope of the tongue-shaped horizons is steeper than that of the topography (Fig. 5.17).

Downslope Domain

The downslope domain consists in its lower part of gruss horizons where major parent minerals are weathered into asepic alteroplasma. These minerals have kept their original crystallographic orientation, indicating that weathering occurred in situ and was largely isovolumetric. Upward, the argilliplasma is very developed and grades into a pedoplasma (vertisol). The

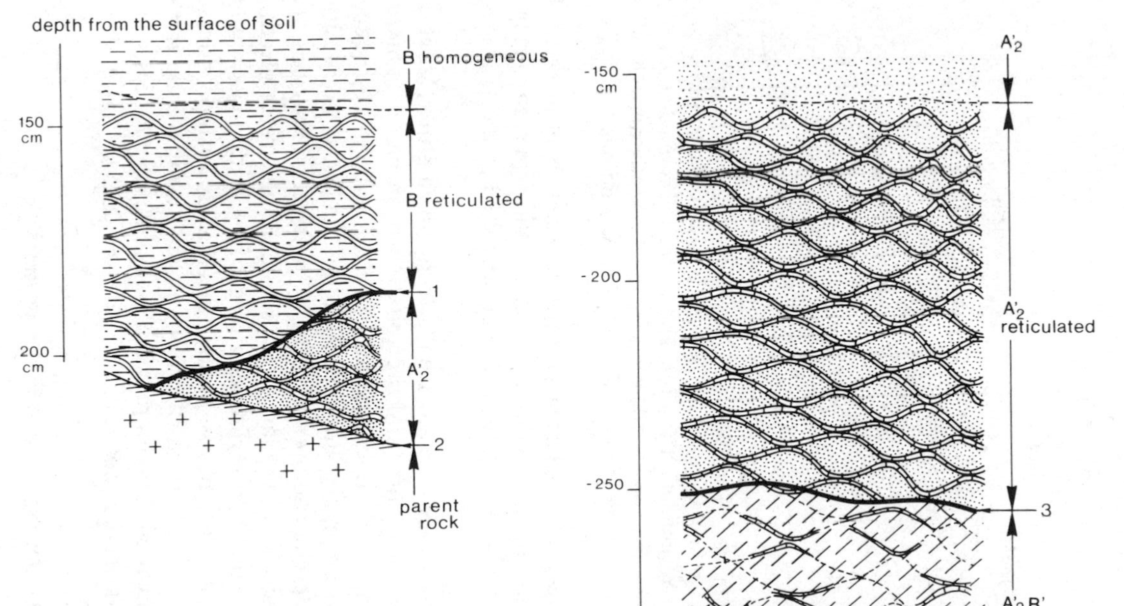

Figure 5.16. Sequence of Garango II. Development of leached horizon A'_2 in the lower part of profiles of upslope domain (P1 profile of Figure 5.15) and of middle domain (P2 profile of Figure 5.15) (1) Upper boundary of A'_2 (leached) horizon, dipping against the slope; (2) boundary with hard bed rock; (3) boundary between leached horizon (A'_2) and accumulation of clay in the lower part of A'_2 leading to formation of A'_2B'; (4) upper boundary of B'_2 horizon; (5) limit of water table. (Modified from Boulet, 1974. Reproduced with permission of ORSTOM Editions.)

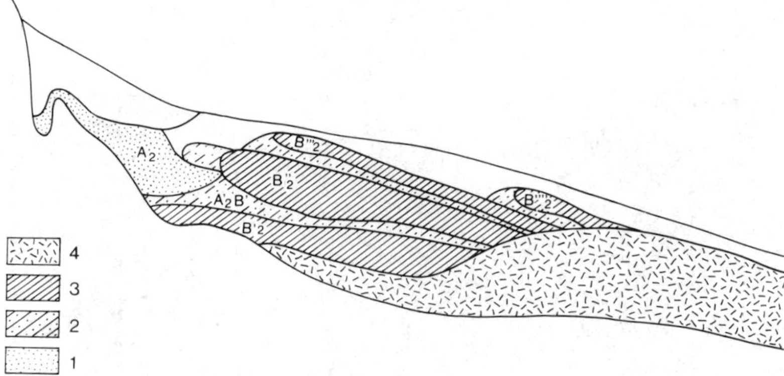

Figure 5.17. Distribution in the Garango II Sequence of Leached Horizons A'_2 (1), of Horizons of Clay Accumulation B'_2 (3), and of their Intermediates A'_2 B1 (2). Downslope argillaceous horizons (4) develop by in situ weathering. (Modified from Boulet, 1974. Reproduced by permission of ORSTOM Editions.)

general gradation between alteroplasma and pedoplasma demonstrates the autochthonous generation of the argillaceous plasma through weathering of parent minerals. This vertisol is identical to that described in the Garango I sequence.

In summary, the downslope domain results from an in situ relative accumulation through transformation and precipitation from parent minerals. This process is identical to the one described in the Garango I sequence. The tongues originate from an absolute accumulation of clays (by deposition or precipitation) from perched aquifers existing in the leached horizons. The distribution of clay minerals in the Garango II sequence illustrates the three domains recognized in a downslope direction (Fig. 5.15), with an essentially kaolinitic upslope domain and an essentially smectitic downslope domain. Kaolinite and smectite result in each case from an in situ differentiation by weathering of the parent-rock. The more complex middle domain consists in part of kaolinite, illite-smectite mixed layers, and smectite.

Comparison between Garango I and II Sequences

A comparison between the original sequences of Garango I and II separated by a distance of 2 km, reveals for both a tendency of the alteroplasmation to be kaolinitic upslope and smectitic downslope. This tendency is even more accentuated at Garango II when the upslope domain shows an almost exclusive kaolinitic alteroplasmation and the downslope domain an almost exclusive smectitic one. This situation can be explained at Garango II by the inselberg effect, which concentrates its aquifers against the foot of the inselberg, allowing formation of ferralitic soils.

However, at Garango II, a derived sequence (tongue-shaped horizons of the middle domain) that transforms the original sequence occurs (Fig. 5.18).

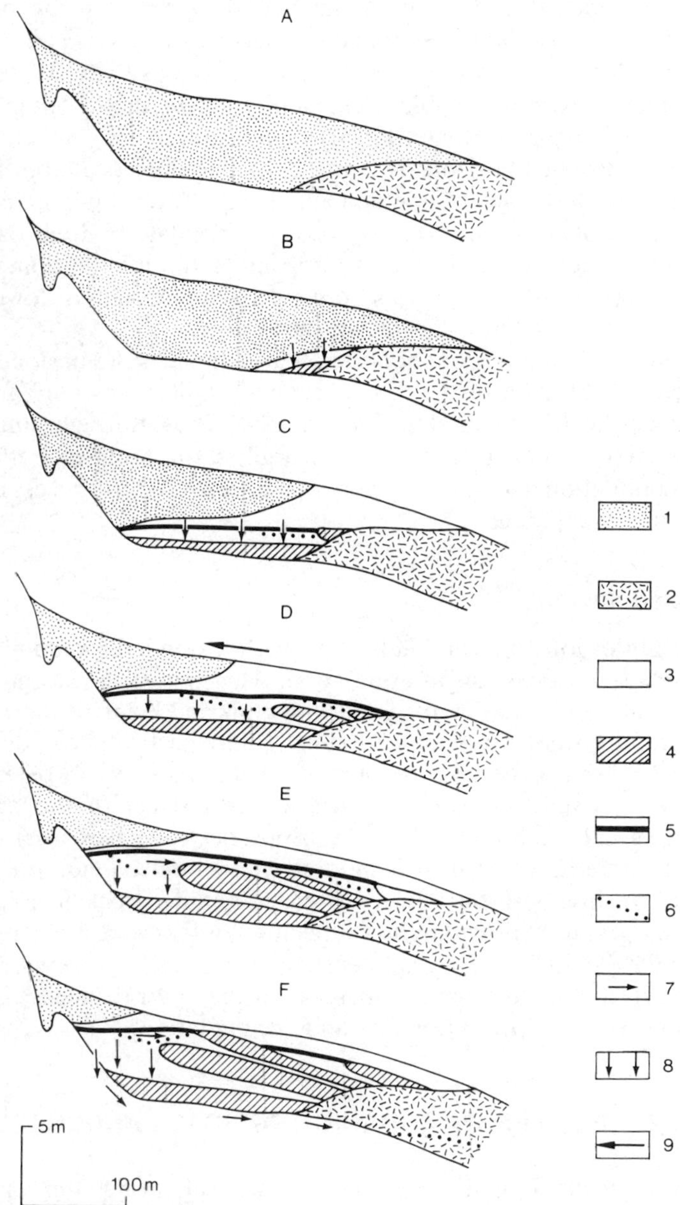

Figure 5.18. Stages of Evolution of Original and Derived Sequences at Garango II (modified from Boulet 1974). Stage A is the differentiation of the original sequence. Stages B through F illustrate progressive development of a derived sequence through transformation of the original upslope domain which tends to be destroyed. F is the present-day soil sequence. (1) Kaolinitic horizons of original cover; (2) smectitic horizons of original cover; (3) leached horizons; (4) smectitic illuvial horizons; (5) upper level of water-table; (6) lowering water table; (7) lateral transfers; (8) vertical transfers; (9) progression of the transformation. (Reproduced by permission of ORSTOM Editions.)

This sequence seems to develop at the boundary between the permeable kaolinitic domain and the less permeable smectitic one. This is how leached horizons develop at the expense of ferralitic horizons. The existence of a pedologic mantle with a variable permeability allows the development of a particular hydric regime. Boulet (1974) demonstrated the existence of two aquifers in the leached horizons A′2 and of one aquifer located in the lower part of the gruss in the downslope domain. The first two aquifers blocked by the argillaceous plug of the downslope domain regulate by their oscillations the position of the tongue-shaped B horizons of the middle domain. Only the most upslope aquifer seems to feed very slowly, through its lower part of the most downslope aquifer located in the gruss.

The rising of the top of both aquifers located induces the upslope rising in the leached A′2 horizon of the tongue-shaped argillaceous horizons, which in turn induces further rising of the two aquifers. Thus, through simultaneous and reciprocal effect between aquifers and argillaceous tongues, both aquifers and the accumulation of clay rise upslope. Once started, this process continues through self-development.

General Conclusion

Surficial mantles of tropical landscapes with extended predominant dry season display two types of differentiation. The first, original and vertical, results from alteroplasmation of parent minerals and leads to the formation of an essentially smectitic mantle, except under particular conditions (inselberg effect) where the upslope portion of this mantle can be kaolinitic. The second, derived and lateral, starts at the contact of the downslope smectitic plug and then rises within the slope destroying original organizations. Such systems of transformation may be induced not only by the presence of an inselberg but also by a coarse-grained rock (for instance a porphyroidal granite), providing low smectite content and more permeable pedologic materials. Hence an acceleration of percolation, which might be caused by external factors such as increase of the contrast between dry and rainy season, triggers rising leaching and accumulation.

PEDOLOGIC MANTLES OF SUBARID TROPICAL LANDSCAPES

In subarid tropical zones where annual rainfall is about 500 mm or less, the pedologic mantle is rather thin (~ 2 m). A sequence studied at Soffokel in the northern part of Burkina Faso (Boulet, 1974) illustrates the type of pedologic mantle developed over migmatites and gneisses.

The sequence at Soffokel displays a morphology of broad interfluves (~ 2 km), with rounded summits and low angle (< 1%) slightly convexo-concave slopes. Vertic subarid brown soils cover all slopes with the following succession of horizons from bottom to top (Fig. 5.19): (1) gruss (coarse

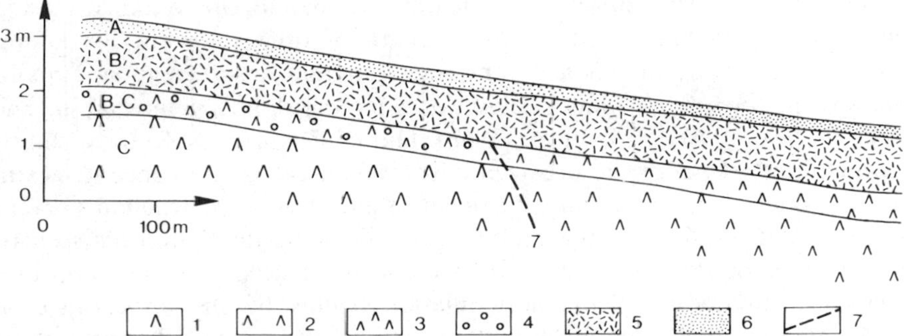

Figure 5.19. Sequence of Soffokel (modified from Boulet, 1974). (1) C horizons, gruss developed over gneiss with biotite, epidote, and garnet; (2) C horizons, gruss developed over gneiss with quartz and biotite; (3) BC horizons; (4) calcitic nodules in BC horizons; (5) B horizons with prismatic structure; (6) A horizons; (7) boundary between the two types of gneisses. (Reproduced with permission of ORSTOM Editions.)

saprolite) transitional to the unweathered rock; (2) BC horizons, transitional with the underlying gneiss and containing oblique alternating bands of sands and sandy clays with polyhedric structure about 40 cm thick; (3) more argillaceous brown B horizons, with prismatic structure about 1 m thick; (4) argillo-sandy A horizons with cubical structure, about 25 cm thick, brown or red in color.

Petrographic study of gruss (coarse saprolite) and BC horizons shows undisplaced parent relicts inside an asepic argilliplama indicating in situ alteroplasmation. This alteroplasma grades into an oriented pedoplasma (vo-squel-masepic) which forms the B horizons. In this case, alteroplasma and pedoplasma are essentially smectitic. Toward the upper part of the profile, kaolinite reaches 10–20% of the clay fraction.

The sequence of Soffokel rests over a parent-rock, which changes in composition: gneiss with biotite, epidote (pistachite) and garnet in an upslope position, and very quartzose gneiss with biotite in a downslope position. This difference of mineralogical composition (more basic upslope) is expressed in the field by the generation of calcitic nodules located in the upslope position of BC horizons, namely, where the parent-rock consists of more basic minerals capable through their weathering to liberate calcium necessary for the accumulation of calcite. Such a concentration emphasizes the close relationship between parent-rock and vertical pedologic differentiation. This is a real case of lithodependence. Furthermore, uniformity of pedologic organizations all along the slope indicates an absence of lateral differentiation.

Whereas at Soffokel, vertical differentiation is marked by development of calcitic nodules above the most basic parent-rocks, numerous examples occur in tropical subarid and Mediterranean areas where lateral differentiation of calcitic accumulations are known without any change in the nature of parent-rock. This concentration of calcium carbonate occurs either when the pedologic system has operated for a length of time sufficient for lateral migration to occur, or when parent rocks contain a sufficient amount of minerals capable of providing calcium, or when the slope is appreciable. The works of Gile (1961), Gile et al. (1966), and of Ruellan (1968) have shown in which kind of pedologic organizations calcium carbonate concentrates in subarid zones. But, in examples from northern Morocco, Ruellan (1971a) demonstrated the lateral differentiation of calcitic accumulation (calcretes). Indeed, in a sequence developed over the same shale parent-rock and in a downslope direction, calcitic accumulation begins by the appearance of friable nodules that become indurated; then these nodules become more numerous and coalesce into laminated crusts, and eventually into a more compact calcitic slab. This lateral differentiation occurs also from bottom to top of the most downslope profiles.

Petrographic studies of such sequences show that calcite accumulates through epigenetic replacement of most parent minerals and solutions weathered minerals that preceded calcite accumulation. The final result is that a very solid parent-rock can be replaced by calcite. Therefore, calcite accumulation in subarid climates can play the role of a real weathering tool.

Furthermore, if calcite accumulation develops within previously differentiated pedologic organizations, it cuts across and destroys them during replacement. Thus, in arid regions that border humid ones (Central America, Australia, Africa), one frequently encounters isolated buttes that are relicts of pedologic mantles formed during earlier more humid climates. These relicts are particularly obvious whenever they represent indurated mantles. Such is the case in the Sahara, where conglomeratic ferruginous crusts occur as relicts of earlier climates with extended rainy seasons. The example presented is located in northern Saharan Mauritania, near Fdérik in the Achouil valley, and shows the relationships between old ferruginous crusts and associated calcretes (Nahon, 1976).

Isolated buttes with ferruginous crust are connected by a break in slope to the most recent morphology represented by calcretes. Profiles of recent geomorphic units consist, from bottom to top, as follows: (1) weathered shale; (2) accumulation of friable to indurated calcitic nodules (these nodules replace locally the minerals of the weathered shale without disturbing the schistose structure); (3) calcitic slab resulting from the increasing number of nodules and their coalescence (only patches of unreplaced weathered shales occur in places with their original schistosity undisturbed); (4) very indurated calcitic slab containing scattered clasts of ferruginous crust (Fig. 5.20).

During the gradual change, in an upslope direction, from the recent morphology with isolated buttes to the older ferruginous crusts, one can see calcretes invading completely the downslope and basal part of the sequence with ferruginous crust (Fig. 5.20) the weathered shale is partially replaced whereas calcite precipitates inside all fractures and cracks of the ferruginous crust to the extent of isolating clasts. The boundary separating the developing calcrete from the untouched ferruginous crust dips against the slope.

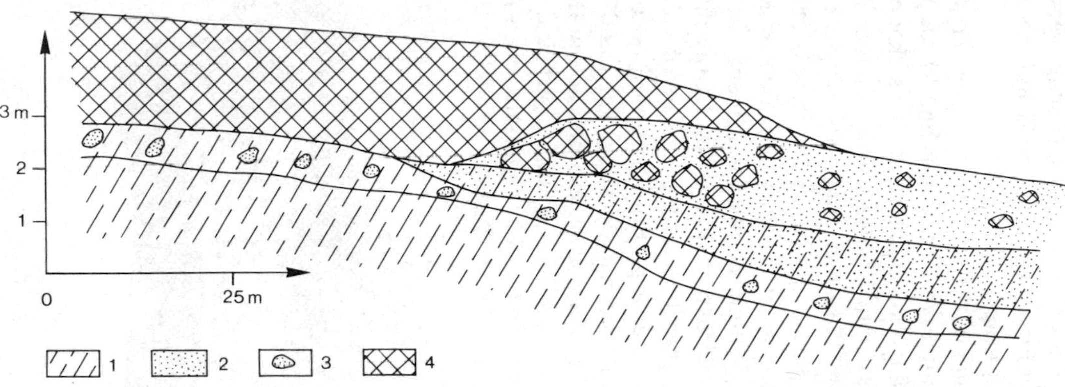

Figure 5.20. Sequence of Achouil in Mauritania. (1) Weathered shale; (2) calcitic nodules; (3) calcrete slab with preserved original schistose structure, or with scattered blocks and pebbles of ferruginous crust; (4) ferruginous crust.

Petrographic examination of the margin of blocks of ferruginous crust engulfed in the calcrete showns that the gradual reduction in size of these blocks occurs simultaneously with the development of the calcrete. An open crack most frequently corresponds to the boundary between the block of crust and the margin of the surrounding calcrete, which consists of an association of sparitic and micritic crystals (Fig. 5.21). Individual crystals of sparite are adjacent to each other and perpendicular to the external boundary of the ferruginous block. These sparite crystals show very fine ferruginous concentric lines (several tens for a single crystal 0.20 mm in size), discontinuous but always parallel to each other.

Along the external margin of the open crack, single crystals of sparite are always coated with a discontinuous ferruginous line whose shape is similar to that of the facing wall of the block of crust. This demonstrates that the ferruginous line originates from a pulling away of the external film of a block of crust. Away from the open crack, the single crystals of sparite within the calcrete are replaced by a micritic cement at the boundary of which the ferruginous lines show a diffused margin and are replaced by diffused ferruginous halos. Within the micritic cement, the lines have disappeared and only shapeless diffused halos persist. X-ray diffraction analyses indicates that most of the micritic cement consists of calcite with 25% of ferroan calcite. This shows that a portion of ferric iron of the ferruginous lines (hematite, goethite) has been reduced to Fe^{2+} and integrated in the crystalline lattice of calcite. Furthermore, examination of the micrite cement by Mossbaüer spectrometry shows that goethite particles, several hundreds of Å in size, persist inside the calcite. They represent relicts of the ferruginous lines pulled away from the blocks of crust.

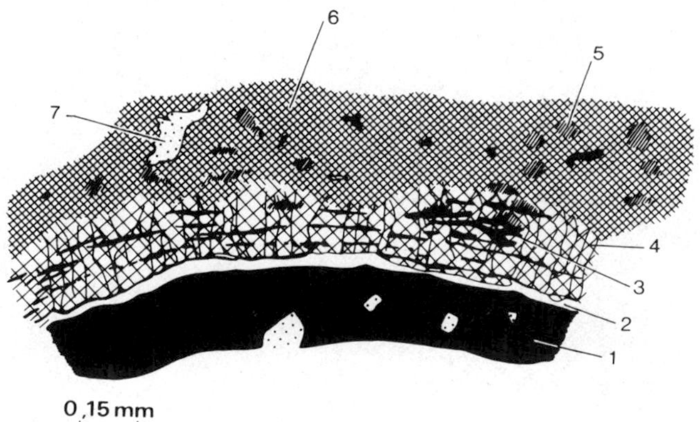

Figure 5.21. Petrographic Aspect of the Margin of Blocks of Ferruginous Crust Within the Calcrete (modified from Nahon, 1976). (1) Zone of ferruginous crust; (2) open crack; (3) ferruginous lines; (4) sparite crystals (calcrete); (5) diffused halos of goethite in micrite of calcrete; (6) micrite of calcrete; (7) quartz grain. (Reprinted by permission of Sciences Géologiques.)

The structures described above can be explained as follows. Alkaline solutions enter first open cracks of the blocks of iron crust and precipitate calcite inside them. During each dry season, contraction of the calcitic cement pulls away a ferruginous film, creating a new open crack. The process is repeated several times, as demonstrated by the numerous ferruginous lines that show individual crystals of sparite do not result from a single episode of crystallization but from a series of successive crystallizations. Finally, the ferruginous lines are incorporated into the micritic cement to the calcrete through reprecipitation of calcite. Iron diffuses in halos and oxides are thus pulverized into minute particles or partially dissolved and integrated into the calcite as Fe^{2+}.

The rising of the calcrete within the old sequences with ferruginous crusts gradually leads through the above process to their replacement. This process operates in an upslope and upward direction within the sequences.

Thus, the differentiation of pedologic sequences with calcretes is widespread in subarid zones. The calcite accumulation that clogs porosity downslope rises into upslope portions of sequences to the extent of invading fossilized pedologic organizations of earlier more humid climates, even destroying them to the extent of leasing only residual buttes.

Whereas, as seen in the sequences of Garango, rising tongues of smectite at the expense of kaolinite was preceded by leaching, here, on the other hand, rising of calcretes at the expense of ferruginous crusts occurred without intermediate leaching. Pedologic organizations generated by more humid bioclimatic conditions are in both cases destroyed and replaced. Original and vertical sequences are wiped out by derived and lateral sequences. In the first case, the cause is self-development of the pedologic system; in the second case, changes of climatic conditions are the responsible factors.

PEDOLOGIC MANTLE IN TEMPERATE CLIMATE

Pedological mantles of zones of temperate climates were studied more extensively than those of tropical zones. Among the numerous published examples, the one presented here is a mantle in which lateral variations along the slope are clearly differentiated between upslope podzolics, a middle domain with brown soils, and a downslope domain with brown soils showing at depth an argillaceous horizon with prismatic structure (planosols). Description begins with the median domain because its soil horizons represent the original sequence devoid of the transformations that characterize the other domains.

The example (Boulet et al., 1987) is located in Brittany, France, in the forest of Paimpont, 40 km west of Rennes, under a temperate oceanic climate (average yearly rainfall of 760 mm). The pedologic cover developed at the expense of a Paleozoic sandstone.

Middle Domain

Brown soils consist of horizons 4, 3, and 2 (Fig. 5.22) developed under a surficial humic horizon 1. The soil grades from a silty sandy composition with microaggregated structure (horizon 2) to an argillaceous silty composition with polyhedric structure and ferruginous mottles (horizon 4). Under the petrographic microscope, horizons 2 and 3 are characterized by caps over the coarsest grains and the presence of argilanes along the margins of certain voids, both features typical of leached horizons.

Horizon 4 shows a strong and rapid increase of argilanes that clog voids. Through swelling and contraction, these argilanes are gradually integrated into the argillaceous matrix, which takes on a polyhedric structure typical of an illuvial argillaceous horizon, particularly in the lower part of horizon 4, where illuviation is the most important.

In summary, the middle domain consists of a leached brown soil developed over a sandstone with a tendency toward argillaceous illuviation at depth. Relicts of the parent sandstone occur throughout the profile. The sandstone is weathered and very friable.

Upslope Domain (Plateau)

In the direction of the plateau, the brown leached soil disappears in a wedgelike fashion and grades into a podzol consisting of two horizons (7, 8): first a dark brown horizon Bh rich in organic matter and less argillaceous than the leached brown soil; second a white sandy silty A_2 podzol of leached type.

The Bh horizon appears beneath the leached brown soil (2) overlying the parent sandstone. This horizon enriched in organic matter and impoverished in clay isolates, therefore, the brown soil from the parent-rock at this particular place. The structures of Bh intersect those of the brown soil, indicating that the latter developed first from the sandstone and subsequently changed laterally into a podzol.

Downslope Domain

In a downslope direction, the various horizons of the leached brown soils gradually change. Microaggregated and leached horizons 2 and 3 show an increasing content of argilanes and grade into a horizon 2′ with polyhedric structure. Argillaceous illuviation, which appears in the lower part of horizon 4, develops downslope, leading to horizon 5 where the clogging by clay becomes complete, forming an argillaceous brown and compact horizon with prismatic structure called planosol.

The study of the hydric regime during rainy seasons shows that drainage is essentially vertical on the plateau, leading to the formation of the podzol. In the middle part, microaggregated porous structures of the brown soils

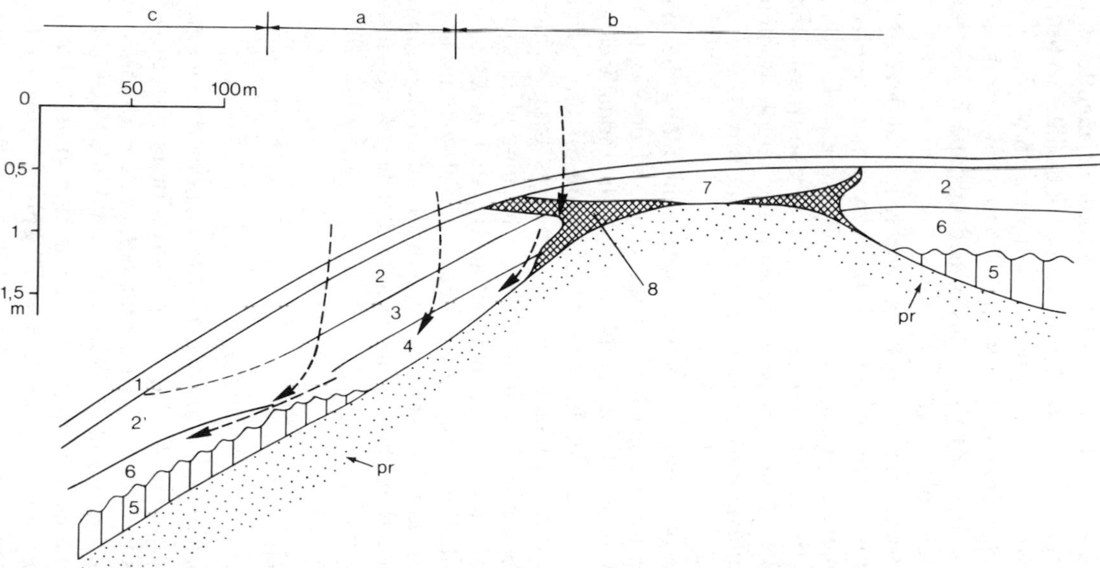

Figure 5.22. Organization of the Sequence of Soils at Paimpont (Brittany, France), (modified from Boulet et al. 1987). (a) middle domain; (b) upslope domain (plateau); (c) downslope domain. (1) humic surficial horizon; (2) microaggregated silty-sandy horizons; (2′) silty-argillaceous horizons; (3) transitional horizons between 2 and 4; (4) illuvial argillaceous-silty horizons; (5) brown argillaceous horizons with prismatic structure; (6) leached transitional horizons between 2′ and 5; (7) white, sandy-silty horizon of A2 type, devoid of clay; (8) horizon Bh, enriched in organic matter, impoverished in clay. (7, 8) podzol; (2, 3, 4) leached brown soils; (2′, 6, 5) planosols; (pr) parent rock (sandstone). (↑) direction of drainage of solutions. (Reproduced with permission of ORSTOM Editions.)

also allow a vertical drainage. However, argillaceous illuviation increasing with depth interrupts vertical penetration of solutions in favor of a lateral drainage, which operates in the upper part of the illuviated horizons (lower part of horizons 4 and 5). Therefore, the flow of the aquifer is downslope and lateral within horizon 6, which results from the leaching of both the upper part of horizon 5 and the lower part of horizon 2′.

In conclusion, the study of structures of the pedologic mantle shows a podzol on the plateau and a planosol downslope both intersecting and destroying leached brown soils with preservation of a few relicts. This situation means that an original sequence of leached brown soils existed over the entire landscape before being transformed upslope into a podzol and downslope into a planosol.

These two pedologic transformation systems display lateral development. The first transformation is centrifugal and advances from the center of the plateau toward its periphery changing the brown soil into a podzol through complete elimination of the argillaceous matrix. The second transformation is centripetal and advances from downslope toward the middle domain by upslope accumulation (illuviation) of clay changing the brown soil into a planosol.

The vertical dynamics of drainage reaches a maximum along the margin of the plateau, allowing intense leaching of the clay and development of podzols. In the middle domain, clay accumulation, which gradually clogs the porosity of the brown soils at depth, leads to a "tilting of the drainage" (Boulet et al., 1987, p. 195). Indeed, drainage changes from vertical to lateral and feeds the downslope domain in the clay, forming planosols. The latter, being impervious, modify the lateral drainage that flows upslope, allowing a rising accumulation of clay. In essence, the self-evolution of the pedologic system induces transformation systems with lateral development. Thus, through self-organization, the original system of leached brown soils is gradually destroyed.

CONCLUSIONS

Investigations of pedologic mantles that are no longer limited to identifying and classifying soils, but, instead, that are concerned with analysis of vertical and lateral organizations have revealed a complete hierarchy of microscopic and megascopic levels of organization. Two motors are responsible for the generation and the dynamic evolution of organization units, that is, of pedologic differentiation: first, *self-development*, which seems to be the essential motor and, second, *change of external factors* (anthropic, climatic, tectonic). If the evolution of these factors has not been sufficiently long or advanced, it is difficult, at any scale, to detect the effects of such an evolution on the organizations of pedologic mantles. In the opposite situation, pedologic mantles display a polyphased differentiation.

Therefore, pedologic mantles correspond to sequences of organization

(Boulet et al., 1987) or to biogeodynamic systems (Bocquier, 1973), which can be understood only through the study of these constitutive horizons (Boulet et al., 1982).

Pedologic horizons, regardless of size, are defined by their various units or organization and by their spatial limits, namely, their envelopes, called fronts. These horizons, like original or derived units of organization, evolve and develop from one another.

Original Horizons and Their Dynamic Evolution

Original horizons develop in their lower part at the expense of indurated parent-rocks essentially through *alteroplasmation*. The front of alteroplasmation has a very irregular spatial distribution. This front results from reactivity of parent minerals to percolating or infiltrating fluids, that is, from mineralogical composition and texture of the parent-rock, as well as from different accessibility of minerals to chemical weathering (particularly their fissuring and fracturing), and hence from the velocity and the nature of percolating solutions.

Although the state of fracturing or fissuring of a mineral does strongly influence its rate of weathering (Colin et al., 1985), only chemical composition and texture of the parent-rock are considered here, namely, nature and structure of parent minerals. Goldich (1938) showed that chemical weathering of parent minerals is differential, and consequently minerals of basic or ultrabasic rocks weather more easily that those of acid rocks. Under comparable hydrodynamic and climatic conditions, initial original horizons are therefore thicker over basic rocks than over acid ones. At the scale of the parent-rock, and also disregarding its fracturation, alteroplasmation fronts are less irregular in the case of basic rocks, and more irregular in the case of acid ones in which very reactive minerals (Na-Ca plagioclases) and weakly reactive ones (quartz) occur.

Ortoleva et al. (1987b, p. 1010) discuss evolutive reaction fronts (Fig. 5.23) resulting from interactions between an aqueous solution flowing through a rock and its constitutive minerals as follows:

> Reactive water that flows through a porous rock establishes a moving reaction front within it. The front can spontaneously become regularly scalloped even in rocks having initially uniform texture. Suppose that X is an aqueous species brought in by the water and A is a mineral disseminated in the rock, and assume that X and A react through the irreversible process $X_{(aq)} + A_{(s)} \rightarrow$ aqueous products. Assume the other minerals in the rock do not react with the incoming aqueous solution under the prevailing conditions. An A-dissolution front moves slowly downflow driven by the input of X dissolved in the water and divides the rock into two regions: in the region downflow from the front, mineral A is still unreacted whereas in the region upflow from it, A has already been dissolved out. The front itself is a transition zone whose thickness depends on the kinetics of reaction and on the characteristics of the transport mechanisms involved.

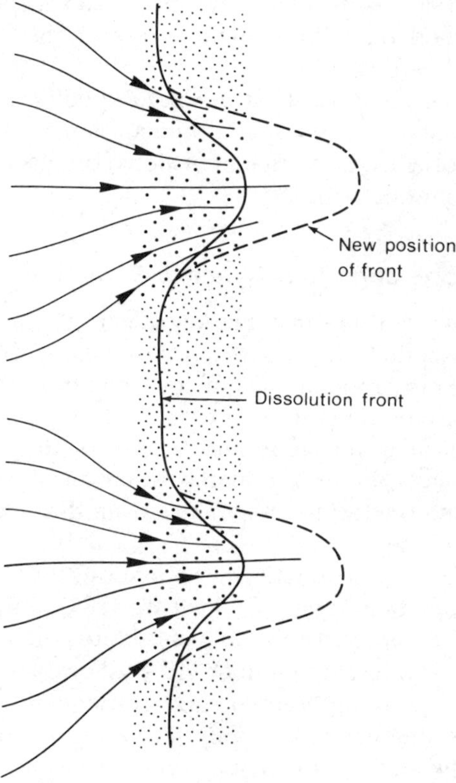

Figure 5.23. Evolutive Dissolution Front (Heavy Line) in a Porous Rock with Incoming Water Flow Arrows (modified from Ortoleva et al., 1987a, Fig. 3, p. 988. (Reproduced by permission of the American Journal of Science.)

The fingerlike shape of the weathering front results, according to Ortoleva et al. (1987b, p. 1010), from a "reactive-infiltration feedback." Its development is self-regulated and described as follows:

> The fingers developed by the reaction front as it sweeps through the rock increase progressively in length but only up to a limit. The longer a finger becomes, the longer the water reaching its tip has had to pick up solutes by diffusion from the sides. Hence the fluids reaching the tip are increasingly equilibrated, as the finger length-to-width ratio increases. This phenomenon prevents fingers from growing indefinitely and forces them to reach a given length-to-width ratio dictated by the system parameters (composition of the inlet water, imposed rate of flow, and composition of the unaltered rock).

Evolutive reaction fronts, and particularly alteroplasmation fronts, operate through reactive-infiltration feedbacks, which represent a self-development process of the biogeodynamic system. This process, also called *self-*

organization (Ortoleva et al., 1987a,b), appears common to numerous geochemical systems. This self-organization is defined by the above authors as the spontaneous and autonomous change of an unpatterned to a patterned system without intervention of periodical external factors.

Alteroplasmation fronts advance as percolating and infiltrating solutions react with parent minerals. These reactions lead to alteroplasma A′, which expresses the nature of the first dissolved minerals A while preserving in its crystalline structure a chemical memory. However, solutions reach continuously alteroplasmation fronts, dissolving other parent minerals B and also reacting with the earlier weathering products A′. Thus a new alteroplasma A′B′ is generated, which, in turn, preserves a chemical memory of minerals A and B, and so on. Weathering minerals A′, B′, to X′ are precipitated under equilibrium conditions, except if prevented from doing so by kinetic reasons, however, equilibria are local and fugaceous because they are repeatedly challenged by continuously renewed percolating solutions. The horizon that develops while weathering fronts lower appears as a reaction-transport geochemical process in constant desequilibrium.

For self-organization to exist, the system must be maintained out of equilibrium and at least two active processes in the system must be coupled (Nicolis and Prigogine, 1977). As mentioned by Ortoleva et al. (1987a, p. 980), "because several types of reaction-transport loops can operate in geochemical systems, and because these systems are most often out of equilibrium, geochemical self-organization should be expected to be common place."

A quantitative modeling of the self-organization process in reaction-transport systems was proposed by Ortoleva et al. (1987a, b) in order to account for structures and organizations resulting from self-development of each system.

Original horizons are those in which alteroplasmation develops: Coarse and fine saprolite are differentiated in the lower part of the pedologic mantle. These original horizons are characterized by preservation of volumes and of general structures of the parent-rock. Furthermore, chemical memory with respect to parent minerals is the most accentuated in these horizons.

Original horizons appear as geochemical systems evolving in three dimensions with their proper dynamics. They are separated from the underlying parent-rock and other horizons by fronts that are sometimes cleanest. These horizons are *spatial concentration discontinuities*. A simplified approach to their mathematical modeling was recently proposed by Guy (1984). This model is a geometrical solution to the infiltration model of Korzhinski (1970) in order to visualized the evolution in time and space of concentration profiles as well as the appearance and growth of discontinuity fronts.

Original horizons express structural and chemical filiations of the parent-rock from which they originate. In this case, *pedologic differentiation is vertical*, and results from *self-development of a geochemical system* through repeated interactions between percolating fluids and both parent and weathering minerals.

Horizons with Limited Transformation

Above original horizons with preservation of the structure of parent-rock, other horizons develop with internal redistributions leading to new structural organizations in which relative accumulation predominates.

Weathering of parent minerals continues in these horizons through reaction between percolating fluids and remaining parent minerals. But, alteroplasma being increasingly abundant, reactions occur more and more with secondary minerals of the alteroplasma. Two consequences result from this situation. First, geochemical reactions such as dissolutions, very short-distance transfers, and reprecipitations allow internal redistributions of certain elements, forming streaks, mottles, and nodules (oxides, oxyhydroxides, and carbonates). Second, physicogeochemical reactions such as redistributions of plasma generate a rearrangement under repeated swelling and shrinking of clay minerals, leading to pedoplasma with limited crystallo-chemical modifications of minerals. Thus, horizons are formed with mottles, bands, or nodules of iron and alumina oxides, and argillaceous horizons with polyhedric and prismatic structures. In this case also, processes of self-development are widespread. They are the continuation of those that allowed the differentiation of original pedologic associations. Self-organization leads here to individualization of pedologic structures that gradually destroy preserved structures of the parent-rock characteristic of original horizons. Furthermore self-development processes allow the generation of several levels of organization apparently independent from each other but nevertheless reacting upon each other. This is exemplified by the formation of ferruginous nodules and associated minerals.

Alteroplasmation in humid climate favors formation of kaolinite and iron oxyhydroxides. The spatial distribution of these minerals in the alteroplasma generates different porosities. Endlessly repeated reactions lead to local dissolution of oxyhydroxides, and through ionic iron transfer to their reprecipitation at a short distance in small pores of the pedoplasma. The small size of the pores favors precipitation of hematite, which itself, by liberating protons, allows dissolution of the host kaolinite. Silica and most of alumina migrate at a short distance and reprecipitate as another generation of kaolinite, which, in turn, clogs a portion of the porosity of the pedoplasma. In this kaolinite, iron can concentrate again as hematite, and so on. However, self-development does not stop at this stage. Iron accumulates as hematite mottles, then as nodules, until the latter are sufficiently indurated to generate a clearly defined pedologic feature within the pedoplasma. Then reactivity to phases of wetting and drying affecting the pedoplasma allows the appearance of contraction voids between surrounding kaolinitic plasma and ferruginous nodule, both of which no longer react in the same way. At this stage the nodule can evolve independently in a centripetal manner and becomes surrounded by a banded cortex. This cortex, in which minerals of ABAB composition alternate (for instance different types of aluminous goethite), is explained by a simple mathematical model in which each mineral selectively accelerates its own formation through the essential role played by autocatalytic effect of surfaces (Gruffat and Guy, 1984). In these nodules, centripetal cortification takes place under the effect of diffusion by mineralogical reconstruction of alternating bands. It is a kind of restructuring by autocannibalism.

In these horizon with limited transformation, everything acts on everything. Feedbacks occur continuously and at all scales. Evolving mineral series and evolving structural series endlessly follow each other under the effect of reactions between fluids and minerals. Equilibria reached at each step are so fugacious and local that each wetting and each percolation puts them in jeopardy. If each forming mineral stage can be expressed by a thermodynamic equilibrium diagram, the entire evolution results from constant preservation of disequilibrium at the boundary between successive organizations. The controlling factor is again self-organization of the geochemical system.

Horizons with limited transformation also keep a chemical memory of the parent-rock, not only in relicts (grains of residual skeleton), but also in newly constructed crystalline frameworks. (Leprun, 1979). However, this memory becomes increasingly more diffuse after successive dissolutions and reprecipitations.

For these horizons with limited transformation, pedologic differentiation is vertical and essentially results from self-development of the geochemical system.

Horizons with Advanced Transformation

Self-development of horizons with limited transformation can lead to horizons with advanced transformation. Reactions-transports of the system remain essentially vertical but differentiation becomes more pronounced for a chemical element, in particular, and for one or two minerals. This type of accumulation is called "monomineralic" (Millot, 1979). It is a spatial concentration discontinuity that takes place in general in the upper part of sequences and hence in the upper part of the previously described systems, but can also appear in their lower part. Representative of this situation are horizons with aluminous crusts (bauxites), with manganesiferous crusts, or ferruginous crusts. In the latter, for instance, repeated and relative accumulation of iron as indurated nodules leads progressively to a horizon in which nodules become juxtaposed and weld to each other, eventually forming nodular ferruginous crust. In these horizons, and particularly in ferruginous crusts, chemical memory is preserved for a few weakly mobile trace elements (Mazaltarin, 1989).

Original Sequences with Vertical Differentiation

The three types of pedologic structural units described above occur in a same geomorphic unit in three dimensions and constitute a sequence with vertical differentiation. Lateral variations occur and were demonstrated many times, particularly in humid zones or in relation to specific morphological conditions. But these variations or lateral distributions remain limited. These sequences express parent-rocks from which they formed, and are eluvial, lithodependent, and vertical. The evolving dynamics controlling these sequences lead to a succession of relative accumulations, absolute accumulations, and absolute subtractions. Accumulations are essentially vertical, whereas subtractions are vertical or lateral.

Relative accumulations take place in situ by reconstruction of weakly mobile chemical elements into weathering minerals, and the latter into organizations. Migrations of these elements and their reprecipitation occur over short distances within a given horizon. Absolute accumulations take place from an overlying horizon toward an underlying one during lowering of the pedologic sequence in the landscape. This dynamics of lowering, and hence of absolute transfer, is vertical. Thus the overlying horizon E3 (Fig. 5.24) feeds horizon E2, which in time feeds horizon E3 by means of its residual minerals and organizations. It also feeds horizon E1 by means of its transferred elements in dissolved or particulate state. Horizon E1 acts in the same fashion at the expense of the parent rock and for the benefit of horizon E2.

Absolute subtractions are vertical and allow transfer of the most mobile elements (alkalis and alkaline earths, silica in part) at depth where they can reach deep aquifers. Absolute subtractions can also be lateral, but in the investigated sequences, they remain of limited extent, feeding shallower aquifers located downslope of the sequences, often in original horizons.

In summary, water-mineral interactions are responsible for continuous lowering the sequences with vertical differentiation through self-development of the geochemical system.

Horizons with Lateral Transformation

Horizons of sequences with vertical differentiation evolve through self-organization of the geochemical system consisting of endlessly repeated interactions between fluids and minerals. Constant disequilibrium of the reaction-transport system could continue indefinitely. However, the numerous examples discussed above show that during the history of a pedologic mantle, variations of external factors (mainly climatic or tectonic) can induce changes in the hydrodynamic evolution of that mantle. The immediate consequence is a profound modification of the geochemical system, which through its self-evolution controls organization of the mantle. Any change of external conditions leads to a change of internal ones. The generalized

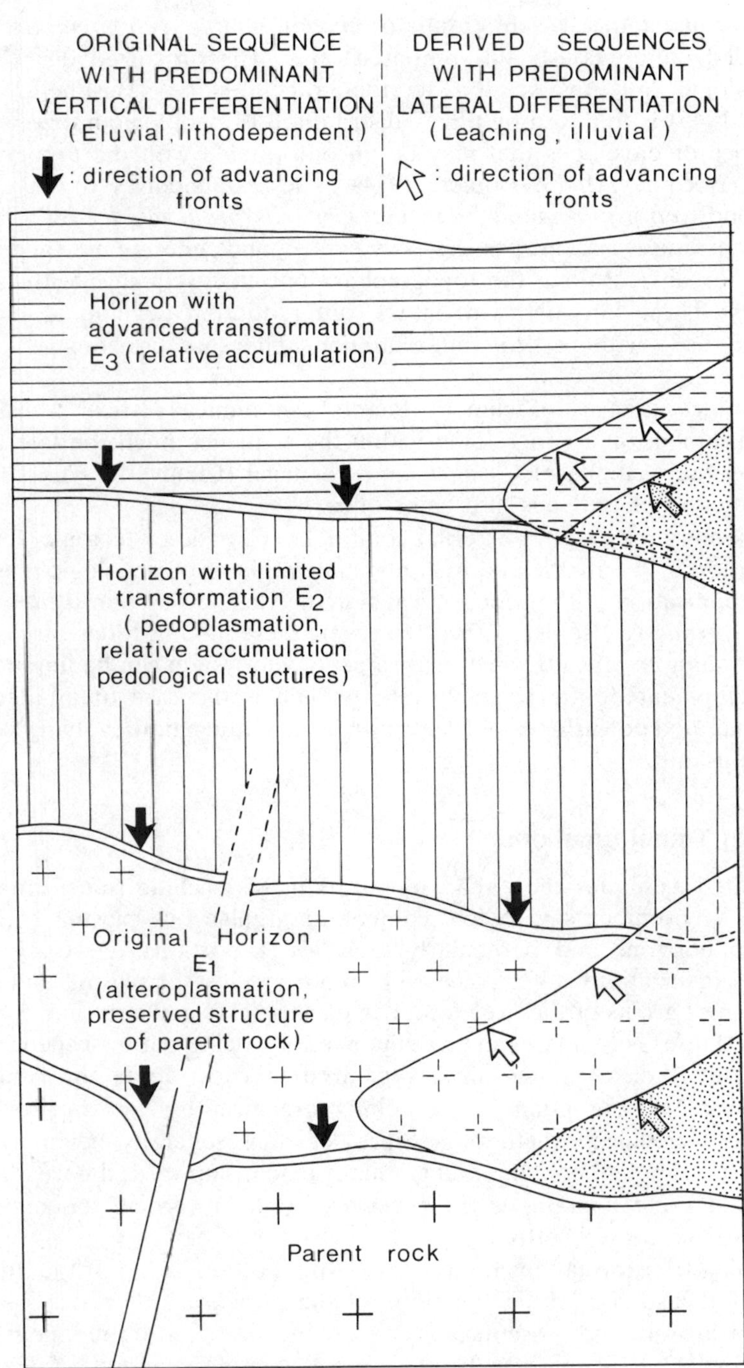

Figure 5.24. Evolution Sketch of Horizons in Original and Derived Sequences.

disequilibrium tends to differentiate *unconformable pedologic horizons* superposed on previously differentiated ones. The organized distribution of pedologic horizons related to topographic slope (Bushnell, 1942; Greene, 1945) is followed by a new distribution of pedologic horizons with boundaries or envelopes that may be unconformable with the topographic surface (Bocquier, 1973; Boulet, 1974). These boundaries or envelopes correspond to *transformation fronts that generally propagate laterally*. Thus, new organizations appear, progressively evolve, and generate new pedologic horizons organized along the topographic slope in a sequence with lateral dynamics. These horizons with lateral differentiation develop at the expense of those with vertical differentiation. They are horizons of lateral transformation.

Sometimes changes of climatic factors can induce vertical transformations, although their progression within the sequence might be lateral, as shown by the example of Casamance, in Senegal (Chauvel, 1976), for the transformation of red soils into beige ones.

In summary, changes in external conditions, climatic or tectonic, lead to modification of the hydric and hydrodynamic regimes of pedologic mantles, hence modification of the geochemical system. This new system is *functional and transformative* (Boulet, 1974). However, these modifications of a hydric regime leading to a transformation of a pedologic system can be initiated by self-development of the original system. Such is the case of argillaceous plugs generated downslope, which eventually generate a new evolving system further upslope.

Leaching Transformations

Investigated examples show that organizations of leaching transformations begin in environments with low content of argillaceous plasma (original kaolinitic horizons and particularly their lower portion). Destruction of original organizations takes place with important loss of plasma and reorganization of grains of the skeleton. The plasma (clay and very fine fraction of the skeleton) is to a large extent eliminated by being put in suspension or in solution. Skeletal grains are reorganized in caps, striae, or lamellae. Furthermore, skeletal grains, while being reorganized by leaching, continue to undergo chemical weathering. Since leaching generally begins in the lower part of original horizons and gradually extends upward, the weathering gradient of skeletal grains is the reverse of that in vertical sequences: It increases from top to bottom.

Leaching transformations not only rise from bottom to top of the original horizon but also rise within the slope of the horizon. This progression of leaching transformations occurs along a front inclined against the topographic slope (Boulet, 1974); it has been called the *rising invasion of leaching* (Boulet and Paquet, 1972).

The engine that starts lateral leaching transformations is a blocking of vertical percolation and appearance of perched microaquifers within original horizons. The assumed cause is often climatic or tectonic, but it can also be the result of self-development of the vertical sequence, which eventually generates inside the original horizon enough plasma (even kaolinitic) to stop any vertical percolation and induce a lateral migration of fluids.

In summary, regardless of the type of engine, a leaching pedologic horizon develops at the expense of the original one from downslope to upslope and upward, gradually destroying it.

Accumulating Transformations

Leaching transformation organizations can become receptive structures for new accumulations, which may be argillaceous (smectitic or kaolinitic), ferruginous, calcitic, and so on. Their nature depends on the climatic zone where the pedologic mantle is located. For an accumulation of material to occur in leached organizations, perched aquifers must be maintained for a sufficient period of time in a given sequence; if not, water would freely flow toward the outlets of drainage areas (Boulet, 1974). This requires the occurrence of plugs or barriers downslope. Boulet (1974) gave a possible inventory of these barriers creating total or partial imperviousness downslope of the sequence. These barriers can result either from an impervious smectitic domain developed over parent-rocks (basic) different from those upslope (acid), or from a slowing down and choking of solutions due to a change of slope and a related accelerated precipitation of clay downslope.

The result is a progressive blocking of aquifers downslope and either precipitation of neoformed minerals or deposition of detrital particles in suspension. The result of this process is to further reinforce the upper part of the downslope barrier and to allow a front of accumulation or a front of illuviation to progressively rise upslope and upward within the leached horizon while developing in it an accumulation horizon. This illuviation front often dips against the topographic slope and is tongue-shaped or staircase-shaped for smectite accumulation (Boulet, 1974; Bocquier, 1973), and only staircase-shaped in the case of calcitic accumulation (Nahon, 1976).

In the initial phase of accumulation, earlier leaching organizations can be recognized. But accumulation becomes increasingly transforming either because of repeated phases of wetting-drying in the case of clays, or because of epigenetic phases in the case of calcite or iron. The result is that, progressively, leaching organizations are destroyed by accumulating ones.

Lateral Eluvial–Illuvial Transformations: Derived Sequences with Lateral Differentiations

Leaching and accumulating transformations build a geochemical system with lateral dynamics leading to a sequence with lateral differentiation. This sequence is transformative and develops at the expense of certain horizons or entire sequences with vertical differentiation. Boulet (1974) showed that such a sequence can be intercalated laterally between two weathering domains.

Initiation of lateral dynamics can result for a change of climatic or tectonic conditions, but also from a simple self-development of a sequence with vertical differentiation without any change of external constraints. At any rate, any generalized disequilibrium induced by a change of external conditions once started will self-catalyse. This happens particularly at the places where external conditions are at their maximum (bottom of slope, center of plateau, etc.). In other words, it is a self-organization of the newly modified pedologic system.

In summary, pedologic mantles represent geochemical systems in constant search for equilibrium. They are functional and living systems.

CHAPTER 6

CONCLUSIONS

THE PEDOLOGIC COVER: A BIOGEOCHEMICAL SYSTEM

The study of the organization of soils and alterites reveals the direction of evolution and the genetic and historical relationships between their constitutive minerals and how, subsequently, geochemical processes could be visualized.

Differentiation followed by evolution of the pedologic cover (soils and alterites) belong to the same biogeochemical system in which all factors interact with each other and at all scales.

Meteoric solutions whose nature and abundance are controlled by original climatic factors percolate and react with the minerals of the rocks they permeate, generating weathering minerals (alteroplasmas, crystalliplasmas). As soon as they are formed, these minerals also react with percolating solutions from season to season. In this manner, microscopic and macroscopic structures of soils and alterites are generated.

These structures do not last but evolve in time and space by self-development of the system and/or by effect of changing external factors (climatic, tectonic, and, in the future, probably human).

The pedologic cover appears as a complex biogeochemical system in continuous operation except in areas where water is in a solid state (glacial regions) or absent (completely desert regions).

A fundamental question can be raised at the end of this book: What is the contribution of this biogeochemical system to the global geochemical system of the planet? A few answers are available, but too many problems remain as yet unanswered to allow the presentation of a coherent and general interpretation. The main unsolved points pertain, in particular, to the age of the pedologic cover and to the factors that control its genesis and development.

THE PEDOLOGIC COVER: REFLECTION OF THE PRESENT AND THE PAST

Monophased Pedologic Covers of the Present

Pedologic covers with vertical differentiation are the best reflections of the climatic conditions under which they developed. Historically, the differentiation of such pedologic covers can be considered as monophased. The analysis of original vertical sequences in the equatorial and tropical zones of West Africa by Boulet (1974) allowed him to present an evolutionary sketch of the pedologic cover developed on migmatites from humid to semiarid climates (Fig. 6.1). This sketch displays the general nature and distribution of the major products of weathering generated under these climates and their related morphology.

The pedologic cover of semiarid regions is thin and does not exceed a few tens of centimeters; active erosion can expose the underlying parent-rock; the major weathering minerals are smectites and calcite, uniformely distributed but invisible to the naked eye. Vegetation is sparse; the hydrographic network is very spread out; the flat landscape is predominantly controlled by erosion.

With increasing rainfall, smectitic weathering covers are gradually replaced by kaolinitic ones, at first in well-drained upslope areas, then over the entire landscape. The relief becomes increasingly accentuated as an effect of the denser hydrographic system, of the more abundant vegetation, and of the deepening of the weathering front with respect to surficial erosion. Consequently, pedologic covers become thick saprolites. The occurrence of calcite is less and less surficial and, in the most humid zones, this mineral is eliminated from the profiles, whereas oxyhydroxides of iron and alumina are differentiated and accumulate.

Polyphased Pedologic Covers of the Past

More difficult is the analysis of polyphased pedologic covers considered during geological time in which humid climates alternated with arid ones.

The most interesting examples belong to zones tectonically stable during the last several million years, such as the West African craton, where the effect of the various climates is expressed in old pedologic covers.

In such old covers, and particularly in zones corresponding to climatic transitions, sequences with lateral differentiation are well developed. These conditions occurred either because the pedologic cover lasted long enough to start a dynamics of self-evolution, or because generalized disequilibrium was initiated by the climatic change. Thus, sequences developed under

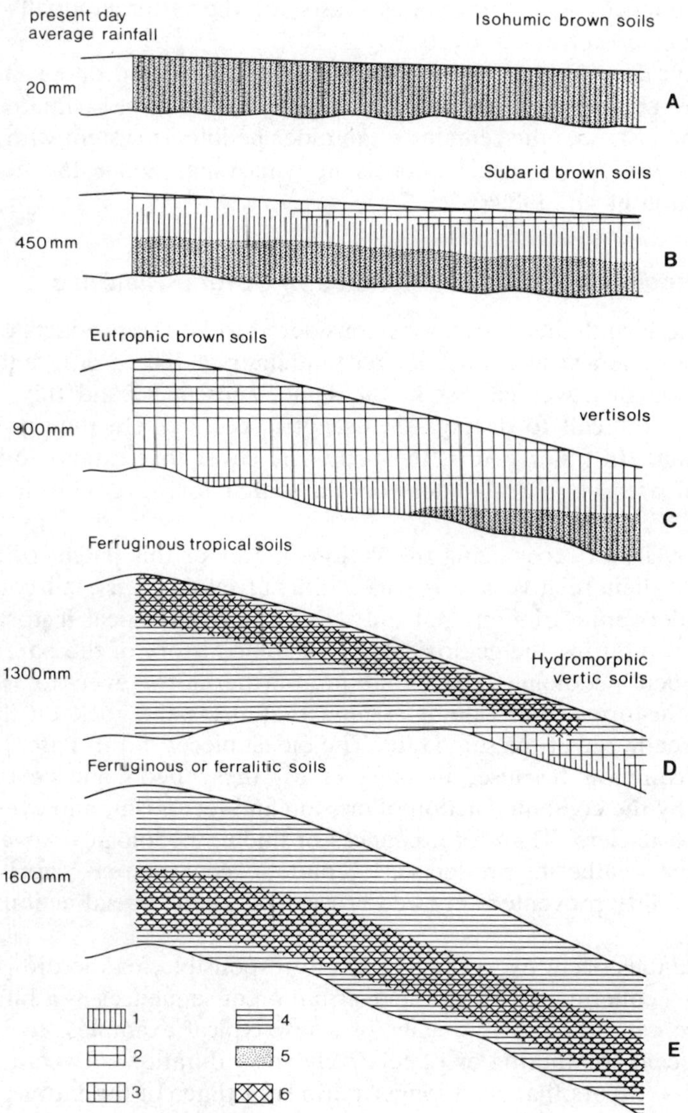

Figure 6.1. Latitudinal Variation of the Pedologic Cover from the Semiarid Tropical Zone (A) to the Humid Equatorial Zone (E) (modified from Boulet, 1974). (1) predominant smectites; (2) predominant kaolinite; (3) smectite-kaolinite mixture with predominant smectite; (4) kaolinite-smectite mixture; (5) accumulation of calcium carbonate; (6) accumulation of iron oxyhydroxides. (Reproduced by permission of ORSTOM Editions.)

humid climates can be replaced by other pedologic sequences during a climatic change toward more arid conditions: Smectitic covers rise and wipe out kaolinitic ones; calcitic covers associated with smectites invade and destroy sequences with ferruginous crusts, or the latter gradually replace microaggregated ferralitic covers.

Similarly, the reverse replacement of sequences formed under more arid climates by pedologic sequences representing more humid climates is also known. For instance, the remains of an older pedologic system with ferruginous crust was found inside latosols in Amazonia, under the equatorial forest (Nahon et al., 1989a).

The Pedologic Cover: An Interference of Several Histories

If certain soils and alterites may be considered to be the products of recent weathering, others span a much longer time interval. These different aspects of the pedologic cover coexist in the same landscapes, and this situation makes them difficult to distinguish, except in cases where they are clearly differentiated. In other words, the pedologic cover that extends over most continental areas is a real patchwork in search for an equilibrium never reached.

The capacity of recognizing the various pieces of this patchwork and of establishing their relative age is still a difficult task because all controlling factors underwent variations not only during the geological history of the cover but also during the geological and tectonic history of the parent-rocks that preceded pedologic differentiation. Furthermore, even for identical geological histories, one cannot assume that the same rock or the same minerals weathered at the same rate. The oldest pieces of this patchwork are the least common because, in most cases, these pedologic covers were destroyed by the combined action of erosion and tectonism, and even by the abrasion of glaciers. Therefore, chances of finding pedological cover relicts of these past weathering products are remote, except in areas where relative tectonic stability prevented renewed erosion or where glacial action did not take place.

Uncertainties occur everywhere and are responsible for the difficulties in presenting a coherent sketch of the distribution of sequences as a function of time. Once can attempt, by means of a few typical examples, to illustrate cases of recent weathering or of relatively short duration of weathering, as well as a few others that span longer periods of time. In the latter case, the understanding of past climatic parameters requires a comparison with recent data of the global geochemical evolution.

Weathering Rates and Young Pedological Covers

Results of laboratory experiments and observations of natural processes, as reported in previous chapters, indicate that estimated weathering rates can vary by a factor of 10. A kaolinitic saprolite is assumed to differentiate at rates ranging from 4–40 mm per thousand years. On the average, weathering of a basic rock is about 2.5 times faster than that of an acidic one.

These figures give an idea of the length of time necessary for the formation of a soil or of an alterite, but they do not take into account the following three factors: (1) flow rate of reactant fluids and their chemical composition, which change in time or at least while the soil is being formed; (2) rate of surficial erosion; and (3) occurrence of microorganisms, bacteria, or organic acid compounds. The proposed average rates of weathering can be either decreased or increased by the effect of one or several of these factors. The ages proposed for certain covers result from relative inferences, since no technique is yet available for precise and absolute dating of soils and alterites.

The following cases illustrate the effects of relatively recent weathering that continues at Present. Bouchard and Pavich (1989) stated that in the Grapé Peninsula, Quebec, a kaolinitic saprolite, 2–3 m thick, formed over a variety of rocks during an interglacial period that lasted 70,000 years. Lateritic weathering, about 10 m thick, of the Pliocene Bone Valley phosphates and Citronelle Formation of Florida (Altschuler et al., 1956; Altschuler et al., 1963), lateritic weathering of Eocene and Oligocene phosphates of Senegal, about 20 m thick (Flicoteaux and Tessier, 1971), that of Miocene basic volcanic rocks of the region of Dakar, about 10 m thick (Nahon and Lappartient, 1977), and that of Miocene sediments and underlying basement of the Appalachian Piedmont, about 15 m thick (Pavich and Obermeier, 1985) lead us to assume that weathering began after the Miocene and even after the Pliocene and continues at Present.

Furthermore, a study by Pavich (1989) of the metamorphic rocks of the Appalachian Piedmont between Alabama and Maryland and their saprolites, which are 10–20 m thick, shows of the basis of measurements of cosmogenic Be^{10} that both soil and alterite are at least younger than 5 MY and continue at Present. This opinion confirms the studies of Nahon (1986), who assumed that a complete lateritic profile developed over basement rocks requires a span of time ranging from 1–6 MY and that such profiles are active to-day.

Weathering Rates and Old Pedologic Covers

Relicts of Mesozoic and Cenozoic weathering mantles that underwent fossilization have been described in Europe, the United States, and Africa (Tessier, 1965; Valeton, 1972, Bardossy, 1982). These mantles, essentially bauxitic and ferruginous, are fossil alterites integrated within local stratigraphic columns. They represent excellent paleoclimatic and paleogeographic indicators.

More difficult is the problem of exposed alterites, with thicknesses ranging from tens to hundreds of meters, which form high residual plateaus, dissected and overlain by thick bauxites. The relative ages of these residual deposits are established by means of correlations with sediments buried in adjacent basins whose mineralogical and chemical features indicate an intense leaching of the continent undergoing lateritization. Typical examples are described in West Africa by King (1962) and Michel (1973). These alterites were interpreted as expressing a succession of older climates around the margins of the Atlantic Ocean during the past 150 MY (Tardy et al., 1988). Using a new technique based on the isotopic composition of the oxygen of minerals generated within certain thick alterites of Eastern Australia, Bird and Chivas (1988) concluded that some weathering covers considered earlier of Cretaceous and Cenozoic ages could be of pre-late Mesozoic age or older. Although those authors are not sure about the age of residual thick alterites, and particularly about those overlain by bauxites, they agree on the fact that they represent ancient weathering profiles expressing old climates. They would be the oldest pieces of the patchwork forming the pedologic cover.

In summary, the major portion of the pedologic cover appears in agreement with the present-day morphology, which means that its differentiation occurred during the last thousands years, or at the most during the last million years, and continues today. This has been the opinion of this author throughout all the studies he has undertaken. No absolute criteria are available as yet for dating this major portion of the pedologic cover. The same situation applies to the residues, in the shape of buttes and dissected plateaus, of the oldest and restricted portions of that same cover. Therefore, the question remains: What was the nature of the past climates that formed the ancient covers, and how could certain residues of them, buried or not, have escaped subsequent erosions, dissolutions, and geochemical reworkings? The answer lies in the reconstruction of the most important climatic variations during the Phanerozoic. Current research in global geochemistry is precisely oriented toward the investigation of such problems.

THE PAST PEDOLOGIC HISTORY SEEN IN THE LIGHT OF GLOBAL EVOLUTION

Global tectonics provided not only a knowledge of the position of the continents during the successive periods of the Phanerozoic age, but also afforded data on the average variations through time of the composition of the oceans, of sea level, of the elevation of continents and the nature of their surfaces, of the composition of the atmosphere, and so on.

This variety of average and global data can be used to reconstruct past climates and related weathering processes that were acting on exposed lands. The studies analyzed below, which pertain to past geological times, are keyed to present-day conditions and allow us, at the same time, to abstract several facts at different scales discussed in previous chapters.

Past Climates and Continental Areas

Tardy et al. (1989) calculated the Phanerozoic global runoff and concluded that the most humid periods occurred during the Cambrian (500 MY), the Silurian-Devonian (400 MY) and the Cretaceous (100 MY) age while the driest occurred during the Permo-Triassic (200 MY) ages, and is comparable to the Present. The surface area of continents was continuously modified during the Phanerozoic age, either by the formation of supercontinents or by the division into several smaller continents of variable size (Ziegler et al., 1982).

Tardy et al. (1989) proposed a holospheric distribution of continental areas versus oceanic areas for the Present and for an average reconstruction during the Phanerozoic age. On the basis of present-day data, these authors concluded that rainfall is a function of the surface of a continent and its latitude. A decrease of rainfall occurs toward the interior of continents that is more effective in zones of higher rainfall than in those of lower rainfall. Obviously, the more widespread a continent, the more arid its internal portion. On the basis of data from Ronov et al. (1980), Tardy (1986) proposed a relative curve of continental surfaces during the Phanerozoic age with respect to present ones. The curve shows that during most of the Cretaceous age exposed lands corresponded to 70–75% of present ones, whereas during the Paleocene and the Eocene ages, the figure was 85%.

Time Variation of The CO_2 Factor

The BLAG model (Berner et al., 1983) attempted to reconstruct for the last 100 millions years the variation of the average CO_2 content of the atmosphere and its possible effect on average temperatures. This model was recently improved by Lasaga et al. (1985) by taking into account the imputs of organic carbon (Fig. 6.2). One can see that the CO_2 content of the atmosphere was in the middle Cretaceous (−90 MY) 13 times higher than at present and that a peak occurred during the middle Eocene age. The greenhouse effect that resulted from these peaks in CO_2 produced warmer and probably more humid climates (Fig. 6.3), an interpretation supported by independent paleontological and paleogeographic data (Frakes, 1976; Savin, 1977).

The major factor responsible for the above-mentioned changes of the CO_2 content of the atmosphere and related increases of the average temperature is assumed to be a global renewal of volcanic activity.

The effects on the weathering of rocks of warmer and more humid climates suggested by the BLAG model were discussed by Berner and Barron (1984). Increase of CO_2 in the atmosphere leads to an increase of plant growth (see Lemon, 1983, *in* Berner and Barron, 1984). This greater productivity of plants on continents “means higher plant respiration, and higher respiration, in turn, suggests higher root production of CO_2 in soils; greater CO_2 production would infer higher soil acidity and faster weathering” (Berner and Barron, 1984, p. 1187).

Sinking of Oxidation Front

Decomposition of organic matter in soils consumes oxygen; similarly, weathering reactions of minerals, whenever increased, use more oxygen. This greater consumption of oxygen than during normal weathering conditions

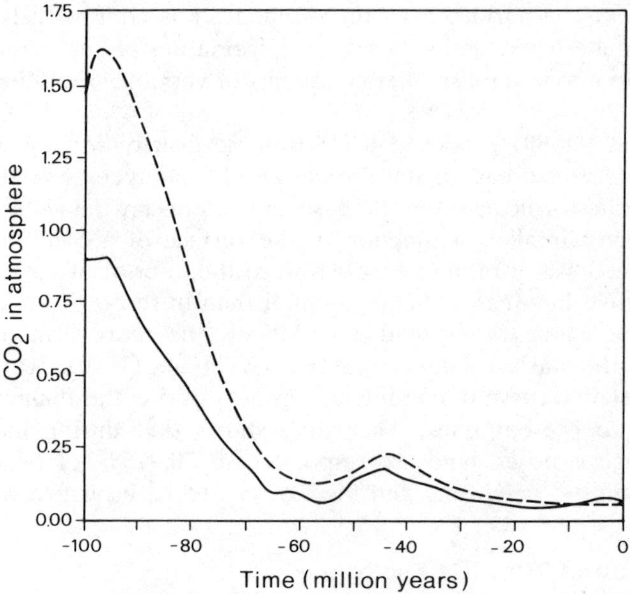

Figure 6.2. Variation of Atmospheric CO_2 as a Function of Time during the Last 100 Million Years. Units for the amount of CO_2 are 10^{13} moles. The dashed curve is to that of the BLAG model of 1983 (modified from Lasaga et al., 1985). (Copyright by the American Geophysical Union.)

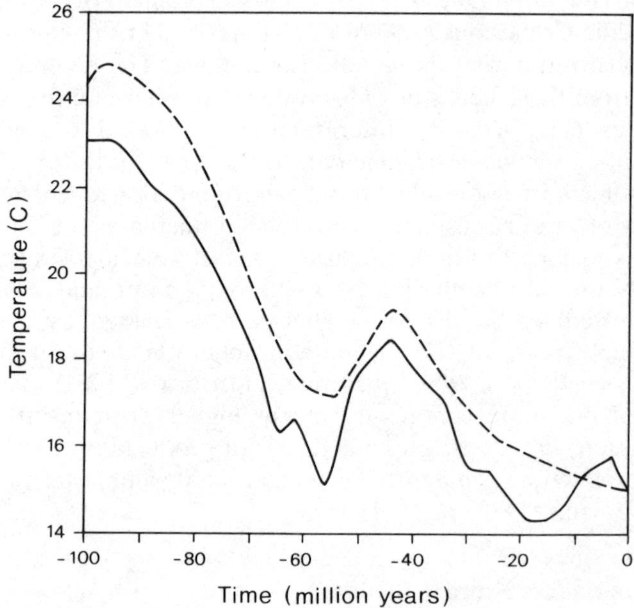

Figure 6.3. Variation of the Average Temperature as a Function of Time during the Last 100 Million Years. The dashed curve is to that of the BLAG model of 1983 (modified from Lasaga et al., 1985). (Copyright by the American Geophysical Union.)

could have been responsible for the generation in soils and alterites of slightly more reducing conditions that would enhance the formation of bauxites over extensive surfaces (Nahon, 1976, 1986) and of thicker profiles than at Present.

Could this greater consumption of oxygen in the soil also act as a feedback on the oxygen content of the atmosphere? This oxygen consumption involved mainly the upper part of weathering profiles and soils and decreased gradually with the sinking of the weathering front. The rate of sinking of the front is a logarithmic function of time and of the thickness of the weathering mantle. Therefore, each change of oxygen content in the atmosphere required the weathering mantle to consume, at the same rate and during a sufficiently long time, an appropriate amount of oxygen. This would imply a continuous regeneration by uplift and erosion of the epidermis of the continent and hence an equally active oxidizing weathering. In other words, the loss of atmospheric oxygen through weathering reactions would be controlled more by isostatic compensation or tectonic uplifting than directly by the oxidation process (Berner and Canfield, 1989).

The Temperature Factor

The increase of average temperature at the surface of continents favors weathering rates. Meybeck (1980) demonstrated these conditions once more by measuring the silica content of stream waters that increases with the rise of the average temperature of watersheds. This strenghthens the idea that during the Cretaceous age and, to a lesser extent, during the Eocene age, weathering covers were thicker and the formation of bauxites enhanced with respect to conditions during the Quaternary.

INTERACTION BETWEEN CONTROLLING FACTORS OF WEATHERING

Weathering and Structural Discontinuities

The global weathering approach requires us first, to distinguish the products of a recent evolution from those of an evolution spanning a much greater length of time, and then to identify the continental surfaces undergoing weathering.

At the smallest scales, rates of weathering were shown to be related to the specific surface of the mineral in contact with the solution and also to the number of defects exposed on that surface, which represent the points of excess energy where weathering is enhanced. At a much greater scale, the surface of continents exposed to weathering should be taken into consideration and also its defects, namely, discontinuities of tectonic or sedimentological nature along which weathering is also enhanced. However, on a global scale, an exceedingly large exposed surface, as in a supercontinent, can generate a water deficit in the center and hence a slowing down of geochemical weathering. On the other hand, when continental surfaces are reduced, the climate over continents is controlled by large anticyclonic zones that allow a more humid climate (Tardy, 1986).

Weathering and Isostasy

The weathering cover at the surface of continents has a thickness ranging from 1 m to almost 200 m. Unaltered silicate rocks, when submitted to weathering, have specific gravities of the order of 2.5 (2.58 for a syenite, 2.61 for a hornfels, 2.80 for a dunite). Specific gravities of the corresponding saprolites are, respectively, only 1.54, 1.40 and 1.58 (Millot, 1964). Therefore, the material eliminated through the weathering processes over continental surfaces leads to an isostatic compensation, which itself enhances the sinking of the weathering front. If the effect of erosion that controls surfaces is added, the uplift is appreciable. For instance, in the Appalachian Piedmont since the end of the Cenozoic age, continuous isostatic uplift, in response to weathering and surficial erosion, is in the order of 10–20 m per million years (Pavich, 1985). These figures are naturally much lower than the maximum ones given for areas undergoing active isostasy (3 cm per year) or active epeirogenesis (0.3 cm per year) given by Perrodon (1972), Howell and Von Huene (1981), and Blatt et al. (1980). In these areas, erosion is the predominant factor. If these regions of high isostatic uplift are located in a tropical humid zone, chemical weathering is strongly reactivated and leads to the development of thick and young profiles. Tectonics becomes then the controlling factor.

Water Flushing and Reactivity of Rocks to Weathering

Past or present rainfall on continental surfaces has a critical effect on the development rates of soils and alterites. However, calculations by Berner (1978) showed that flushing by water, if the flux is continuous and important, reaches a limit beyond which it is no longer the controlling factor of the rate of weathering. When percolating waters are close to saturation (in the case of weathering of carbonate rocks), the flux of fluids appears as the major controlling factor. On the other hand, solutions weathering silicate rocks remain generally undersaturated with respect to these rocks, and reactivity of parent minerals becomes the predominant controlling factor with respect to flushing (Berner, 1978).

Obviously, the proportions of various rocks relative to the continental surface have to be taken into account in the preparation of global budgets of weathering, not only with respect to the control of rates of weathering, but also to the control of CO_2 budgets. On this subject, Berner and Barron (1984, p. 1185) wrote:

'Carbon dioxide is removed during carbonate weathering but this CO_2 is soon returned to the atmosphere upon subsequent precipitation of $CaCO_3$ in the oceans; with silicate weathering some of the CO_2 is permanently lost to rocks, as Ca derived from the weathering of silicate minerals and CO_2 derived from the atmosphere are buried together as $CaCO_3$.

Therefore, it is important to consider the flux of solutions not only for itself, but with respect to the reactivity of minerals in contact with this flux (Fouillac et al., 1977).

The different reactivities of minerals, either through their crystallo-chemical structure or through their variable amount of defects, lead to weathering products variable in nature and in quantity. At a higher scale, the results are soils and alterites of different nature and thickness, eventually expressed at the surface by a morphology of buttes and thalwegs.

When the flux of percolating solutions becomes the major factor, either because the nature of the rock allowed it, as mentioned above, or because in time porosities were sufficiently open, weathering products become uniform and express the rates of flux. At the surface, runoff and erosion act on the same type of material, and the morphology has a tendency to become uniform. Dissolution rates are, in fact, a logarithmic function of both time and depth beneath the soil surface. Furthermore, for a given flux, the chemical composition of the percolating water changes (downwards and downslope of the pedologic cover) so that the controlling factors of weathering rates can change within a given profile. Thus, the dissolution features displayed by a given mineral are a function of the degree of saturation of the solution in contact with it (Brantley et al., 1986; Lasaga, 1983).

Endless Feedbacks and Evolutive Weathering

The flux of solutions should be considered with respect to the reaction rate of a given mineral, which acts on the composition of the solution, which, in turn, influences the dissolution of the mineral. But these feedbacks among factors controlling weathering do not stop at this level. The evolution of the organizations of soils and alterites during precipitation of secondary minerals and their reaction with percolating fluids eventually initiates, through self-development, new organizations of transformation that locally modify structure and texture of the pedologic cover. Changes of porosities affect the circulation and activity of water, which, in turn, modify the nature of the chemical reactions of weathering (Mattigod and Kittrick, 1980; Tardy and Novikoff, 1988). Therefore, for a given continental surface, a given parent-rock and the same rainfall, weathering rates evolve while the pedologic cover undergoes differentiation. Feedbacks between the various factors controlling weathering reactions are constant. Such interactions, which self-develop within soils and alterites, eventually react on vegetation, morphology, and pedoclimate.

Weathering rates of minerals and rocks are strongly modified by the action of living organic matter, plants, and animals, which is an intrinsic part of soils and alterites. This action occurs in numerous direct or indirect ways, such as respiration of plants and their roots, activity of micro- and macro-organisms, decay and mineralization of organic matter, formation of acids, and so on. Although this action was not discussed in this book, it is important to stress its essential role of catalyst in local and global weathering processes.

REFERENCES

Aagaard, P., and Helgeson, H. C., 1982. Thermodynamic and kinetic constraint on reaction rates among minerals and aqueous solutions. I. Theoretical considerations. *Am. J. Sci.*, 282, pp. 237–285.

Alexander, L. T., and Cady, J. G., 1962. Genesis and hardening of laterite in soils. *U.S. Dept. Agric., Soil Cons. Serv., Bull.*, Washington, 1282, p. 90.

Altschuler, Z. S., 1965. Precipitation and recycling of phosphate in the Florida land-pebble phosphate deposits. *Prof. Pap. U.S. Geol. Surv.*, Washington, 525, B, pp. 91–95.

Altschuler, Z. S., 1973. The weathering of phosphate deposits. Geochemical and environmental aspects. In: *Environmental Phosphorus Handbook*, Griffith D., Beeton A., Spencer J. M., Mitchell, D. T. Ed., John Wiley & Sons, Inc., New York, pp. 33–96.

Altschuler, Z. S., and Boudreau, C. E., 1949. A mineralogical and chemical study of the leached zone of the Bone Valley formation. A progress report. *Trace Elements Inv. U.S. Geol. Surv.*, Washington, 102, p. 67.

Altschuler, Z. S., Clarke, R. S., and Young, E. G., 1958. Geochemistry of uranium in apatite and phosphorite. *Prof. Pap. U.S. Geol. Surv.*, Washington, 314, D, pp. 45–90.

Altschuler, Z. S., Dwornik, E. J., and Kramer, H., 1963. Transformation of montmorillonite to kaolinite during weathering. *Science*, 141, 3576, pp. 148–152.

Altschuler, Z. S., Jaffe, E. B., and Cuttita, F., 1956. La zone des phosphates d'aluminium de la formation Bone Valley et ses dépôts d'uranium. *Actes Conf. internat. util. énerg. atom. fins pacif. VI: Géologie de l'uranium et du thorium*, Genève, 1955, pp. 576–588.

Ambrosi, J. -P., 1984. Pétrologie et géochimie d'une séquence de profils latéritiques cuirassés ferrugineux de la région de Diouga, Burkina Faso. Thesis, Univ. Poitiers, unpublished, p. 215.

Ambrosi, J. -P., and Nahon, D., 1986. Petrological and geochemical differentiation of lateritic iron crust profiles. *Chem. Geol.*, 57, pp. 371–393.

Ambrosi, J. -P., Nahon, D., and Herbillon, A. J., 1986. The epigenetic replacement of kaolinite by hematite in laterite. Petrographic evidence and the mechanisms involved. *Geoderma*, 37, pp. 283–294.

Amouric, M., Baronnet, A., Nahon, D., and Didier, P., 1986. Electron microscopic investigations of iron oxyhydroxides and accompanying phases in lateritic iron crust pisolites. *Clays and Clay Miner.*, 34, pp. 45–52.

Arnaud, G., 1945. Les ressources minières de l'Afrique occidentale. *Bull Direction des Mines et de la Géologie, Afrique Occidentale Française (A.O.F.)*, 8, 1, and *Mines et Charbon.* Paris, p. 100.

Aylmore, L. A. G. and Quirk, J. P., 1960. Domain or turbostratic structure of clays. *Nature*, 187, pp. 1046–1048.

Aylmore, L. A. G., and Quirk, J. P., 1962. The structural status of clay systems. *Clays and Clay Miner.*, 9, pp. 104–130.

Aylmore, L. A. G., and Quirk, J. P., 1971. Domains and quasicrystalline regions in clay systems. *Soil Sci. Soc. Amer. Proc.*, 35, pp. 652–654.

Bardossy, G., 1982. *Karst Bauxites.* Elsevier, Amsterdam, p. 441.

Baron, G., Caillère, S., Lagrange, R., and Pobeguin, Th., 1959. Etude du Mondmilch de la grotte de la Clamouse et de quelques carbonates et hydrocarbonates alcalino-terreux. *Bull. Soc. Fr. Minéral. Crist.*, 82, pp. 150–158.

Barshad, I., 1948. Vermiculite and its relation to biotite as revealed by base exchange reactions, X-ray analyses, differential thermal curves and water content. *Am. Mineral.*, 33, pp. 655–678.

Barshad, I., 1966. The effect of variation in precipitation on the nature of clay mineral formation in soils from acid and basic igneous rocks. *Int. Clay Conf. Proc.* Jerusalem, 1, pp. 167–173.

Barshad, I., and Fawzy, F. M., 1968. Oxidation of ferrous iron in vermiculite and biotite alters fixation and replaceability of potassium. *Science*, 162, pp. 1401–1402.

Basham, I. R., 1974. Mineralogical changes associated with deep weathering of gabbro in Aberdeenshire. *Clay Min.*, 10, pp. 189–202.

Bassett, W. A., 1960. Role of hydroxyl orientation in mica alteration. *Bull. Geol. Soc. Amer.*, 71, 4, pp. 449–556.

Baumgartner, A., and Reichel, E., 1975. *The World Water Balance. Mean Annual Global, Continental and Maritime Precipitation, Evaporation and Run-off.* Elsevier, Amsterdam, p. 179.

Baver, L. D., 1948. *Soil Physics.* Chapman-Hall, Ltd., London, p. 498.

Beauvais, A., and Nahon, D., 1985. Nodules et pisolites de dégradation des profils d'altération manganésifères sous conditions latéritiques. Exemples de Côte d'Ivoire et du Gabon. *Bull. Sci. Géol.*, 38, 4, pp. 359–381.

Bech, J., Nahon, D., Paquet, H., Ruellan, A., and Millot, G., 1980. Sur l'extension géographique et climatique des phénomènes d'épigénie par la calcite dans les encroûtements calcaires. Exemple de la Catalogne. *C. R. Acad. Sci. Paris*, 291, pp. 371–376.

Bennema, J., Jongerius, A., and Lemos, R. C., 1970. Micromorphology of some oxic and argillic horizons in South Brazil in relation to weathering sequences. *Geoderma*, 4, pp. 333–355.

Berner, R. A., 1971. *Principles of Chemical Sedimentology*. McGraw-Hill, New York, p. 240.

Berner, R. A., 1978. Rate of control of mineral dissolution under earth surface conditions. *Am. J. Sci.*, 278, pp. 1235–1252.

Berner, R. A., and Barron, E. J., 1984. Comments on the BLAG Model: Factors affecting atmospheric CO_2 and temperature over the past 100 million years. *Am. J. Sci.*, 284, pp. 1183–1192.

Berner, R. A., and Canfield, D. E., 1989. A new model for atmospheric oxygen over Phanerozoic Time. *Am. J. Sci.*, 289, pp. 333–361.

Berner, R. A., and Holdren, G. R. Jr., 1977. Mechanism of feldspar weathering. Some observational evidence. *Geology*, 5, pp. 369–372.

Berner, R. A., and Holdren, G. R. Jr., 1979. Mechanism of feldspar weathering. II. Observations of feldspars from soils. *Geochim. Cosmochim. Acta*, 43, pp. 1173–1186.

Berner, R. A., Holdren, G. R. Jr., and Schott, J., 1985. Protective surface layers on dissolving silicates. Comments on the paper "Study of the weathering of albite at room temperature and pressure with a fluidized bed reactor" by L. Chou and R. Wollast (Geochim. Cosmochim. Acta 48, 2205–2217, 1984). *Geochim. Cosmochim. Acta*, 49, pp. 1657–1658.

Berner, R. A., Lasaga, A. C., and Garrels, R. M., 1983. The carbonate-silicate geochemical cycle and its effect on atmospheric carbon dioxide over the past 100 million years. *Am. J. Sci.*, 283, pp. 641–683.

Berner, R. A., and Schott, J., 1982. Mechanism of pyroxene and amphibole weathering. II. Observations of soil grains. *Am. J. Sci.*, 282, pp. 1214–1231.

Berner, R. A., Sjöberg, E. L., Velbel, M. A., and Krom, M. D., 1980. Dissolution of pyroxenes and amphiboles during weathering. *Science*, 207, pp. 1205–1206.

Bésairie, H., 1943. Les latéritoïdes phosphatées du plateau de Thiès. Unpublished report, Arch. Dir. Mines. A.O.F., Dakar, mimeographed., p. 13.

Bigotte, G., and Bonifas, G., 1968. Faits nouveaux sur la géologie de la région de Bakouma (Préfecture du M'Bomou, République Centrafricaine). *Chroniques Mines Rech. Min.*, Paris, 370, pp. 43–46.

Bird, M. I., and Chivas, A. P., 1988. Oxygen isotope dating of the Australian regolith. *Nature*, 6156, pp. 513–516.

Bisdom, E. B. A., 1967. The role of micro-crack systems in the spheroidal weathering of an intrusive granite in Galicia (NW Spain). *Geologie en Mijnbow*, 46, pp. 333–340.

Black, C. A., 1943. Phosphate fixation by kaolinite and other clays as affected by pH, phosphate concentration, and time of contact. *Soil Sci. Soc. Amer. Proc.*, 7, pp. 123–133.

Blatt, H., Middleton, G., Murray R., 1980. *Origin of Sedimentary Rocks* (2nd ed). Prentice Hall, Englewood Cliffs, New Jersey, p. 782.

Blokhuis, W. A., Slager, S., and Van Schagen R. H., 1970. Plasmic fabrics of two Sudan vertisols. *Geoderma*, 4, 2, pp. 127–137.

Blum, A. E., and Lasaga, A. C., 1987. Monte Carlo simulations of surface reaction rate laws. In: *Aquatic Surface Chemistry: Chemical Processes at the Particle-Water Interface*, Stumm W., Ed., John Wiley & Sons, Inc., New York, pp. 255–292.

Bocquier, G., 1973. Genèse et évolution de deux toposéquences de sols tropicaux du Tchad. Interprétation biogéodynamique. *Mém. Office Rech. Sci. Tech. Outre-Mer (ORSTOM)*, Paris, 62, p. 350.

Bocquier, G., Boulangé, B., Ildefonse, P., Nahon, D., and Muller, D., 1984. Transfers, accumulation modes, mineralogical transformations and complexity of historical development in lateritic profiles. *Int. Seminar on Lateritization Proccesses*, 2, *Proc.*, Sao Paulo, Brazil, pp. 331–343.

Bocquier, G., Muller, J. P., and Boulangé, B., 1984. Les latérites: Connaissances et perspectives actuelles sur les mécanismes de leur différenciation. *Livre Jubilaire du Cinquantenaire de l'Association Française pour l'Etude du Sol*, Paris, pp. 123–128.

Bocquier, G., and Nalovic, L., 1972. Utilisation de la microscopie électronique en pédologie. *Cah. Office Rech. Sci. Tech. Outre-Mer (ORSTOM), sér. Pédologie, X*, 4, pp. 411–434.

Bouchard, M., and Pavich, M. J., 1989. Characteristics and significance of pre-Wisconsinan saprolites in the northern Appalachians. *Zeit. Geomorph. N. F.*, 72, pp. 125–137.

Boulangé, B., 1984. Les formations bauxitiques latéritiques de Côte d'Ivoire. Les faciès, leur transformation, leur distribution et l'évolution du modelé. *Trav. Docum. Office Rech. Sci. Tech. Outre-Mer (ORSTOM)*, Paris, 175, p. 341.

Boulangé, B., and Bocquier, G., 1983. Le rôle du fer dans la formation des pisolites alumineux au sein des cuirasses bauxitiques latéritiques. *Mém. Sci. Géol.*, Strasbourg, 72, 2, pp. 29–36.

Boulangé, B., Paquet, H., and Bocquier, G., 1975. Le rôle de l'argile dans la migration et l'accumulation de l'alumine de certaines bauxites tropicales. *C.R. Acad. Sci. Paris*, 280, D, pp. 2183–2186.

Boulet, R., 1974. Toposéquences de sols tropicaux en Haute-Volta. Equilibre et déséquilibre pédobioclimatique. *Mém. Office Rech. Sci. Techn. Outre Mer, (ORSTOM)*, 85, pp. 272.

Boulet, R., Bellier, G., and Humbel, F. X., 1987. Différenciation toposéquentielle et transformations morphologiques d'un sol brun de Bretagne. *Cah. Office Rech. Sci. Tech. Outre-Mer (ORSTOM), sér, Pédologie*, 23, 3, p. 187–196.

Boulet, R., Chauvel, A., Humbel, F. X., and Lucas, Y., 1982. Analyse structurale et cartographie en pédologie. I. Prise en compte de l'organisation bidimensionnelle de la couverture pédologique: les études de toposéquences et leurs principaux apports à la connaissance des sols. *Cah. Office Rech. Sci. Tech. Outre-Mer (ORSTOM), Sér. Pédologie*, 4, pp. 309–321.

Boulet, R., Chauvel, A., and Lucas, Y., 1984. Les systèmes de transformation en pédologie. *Livre jubilaire du Cinquantenaire de l'Association Française pour de l'Étude du Sol*, Paris, pp. 167–179.

Boulet, R., Humbel, F. X., and Lucas, Y., 1982. Analyse structurale et cartographie en pédologie. II. Une méthode d'analyse prenant en compte l'organisation tridimensionnelle des couvertures pédologiques. *Cah. Office Rech. Sci. Tech. Outre-Mer (ORSTOM), Sér. Pédologie*, 4, pp. 323–339.

Boulet, R., and Nahon, D., 1970. Observations pédologiques dans la région de Nouadhibou, République Islamique de Mauritanie. *Rapp. Prov. de mission, Cent. ORSTOM, Dakar, et Lab. Géol. Fac. Sci.*, Dakar, mimeographed, unpublished, p. 21.

Boulet, R., and Paquet, H., 1972. Deux voies différentes de la pédogenèse en Haute-Volta. Convergence finale vers la montmorillonite. *C.R. Acad. Sci., Paris*, 275, 12, pp. 1203–1206.

Bourrié, G., and Pédro, G., 1979. La notion de pF, sa signification physicochimique et ses implications pédogénétiques, signification physicochimique. Relations entre le pF et l'activité de l'eau. *Sci. Sol*, 4, pp. 313–322.

Brantley S. L., Crane, S. R., Crerar, D. A., Hellmann, R., and Stallard, R., 1986. Dissolution at dislocation etch pits in quartz. *Geochim. Cosmochim. Acta*, 50, pp. 2349–2361.

Bresson, L. M., 1981. Ion micromilling applied to the ultramicroscopic study of soils. *Soil Sci. Soc. Amer. J.*, 45, pp. 568–573.

Brewer, R., 1960. The petrographic approach to the study of soils. *Int. Cong. Soil. Sci., 7, Proc.*, Madison, Wisconsin, 1, pp. 1–13.

Brewer, R., 1964. *Fabric and Mineral Analysis of Soils* (First Printing). John Wiley & Sons, Inc., New York, p. 470.

Brewer, R., 1976. *Fabric and Mineral Analysis of Soils* (Second Printing). Robert E. Krieger Publishing Company, Huntington, New York, p. 482.

Brewer, R., and Sleeman, J. R., 1960. Soil structure and fabric: Their definition and description. *J. Soil Sci.*, 11, pp. 172–185.

Brewer, R., and Sleeman, J. -R., 1964. Glaebules: Their definition, classification, and interpretation: *J. Soil Sci.*, 15, pp. 66–78.

Brimhall, G. H., and Dietrich, W. E., 1987. Constitutive mass balance relations between chemical composition, volume, density, porosity, and strain in metasomatic hydrochemical systems: Results on weathering and pedogenesis. *Geochim. Cosmochim. Acta*, 51, 3, pp. 567–587.

Brinkman, R., 1970. Ferrolysis, a hydromorphic soil forming process. *Geoderma*, 3, pp. 199–206.

Brinkman, R., 1979. *Ferrolysis, a Soil Forming Process in Hydromorphic Conditions.* Agric. Res. Rep., Center for Agricultural Publishing and Documentation, Wageningen, The Netherlands, p. 887.

Brinkman, R., Jongmans, A. C., Miedema, R., and Maaskant, P., 1973. Clay decomposition in seasonally wet, acid soils: Micromorphological, chemical and mineralogical evidence from individual argillans. *Geoderma*, 10, pp. 250–270.

Brückner, W. D., 1957. Laterite and bauxite profiles of West Africa as an index of rhythmical climatic variations in the tropical belt. *Eclog. Geol. Helv.*, Lausanne, 50, 2, pp. 239–256.

Bruand, A., 1985. Contribution à l'étude de la dynamique de l'organisation de matériaux gonflants. Application à un matériau argileux d'un sol argilo-limoneux de l'Auxerois. Thesis, Univ. Paris 7, unpublished, p. 225.

Bryan, W. A., 1952. Soils nodules and their significance. In: *Sir Douglas Mawson Anniversary Volume*, University of Adelaïde, Australia, pp. 43–53.

Buol, S. W., and Eswaran, H., 1978. Micromorphology of oxisols. *Int. Meet. Soil Micromorph., 5, Grenade, Proc.*, pp. 325–347.

Busenberg, G., and Clemency, C. V., 1976. The dissolution kinetics of feldspar at 25°C and 1 atm CO_2 partial pressure. *Geochim. Cosmochim. Acta*, 40, pp. 41–49.

Bushnell, T. M., 1942. Some aspects of the soil catena concept. *Soil. Sci. Soc. Amer. Proc.*, 7, pp. 466–476.

Butel, P., 1982. Formes et mécanismes de l'accumulation carbonatée dans les sols de la plaine poitevine. Thesis, Univ. Poitiers, unpublished, p. 123.

Butler, B. E., 1955. A system for the description of soil structure and consistence in the field. *J. Austr. Inst. Agr. Sci.*, 21, pp. 239–249.

Caillère, S., and Hénin, S., 1951. Etude de quelques altérations de la phlogopite à Madagascar. *C.R. Acad. Sci. Paris*, 233, pp. 1383–1385.

Camara, L., 1982. Comportement hydrique et propriétés de gonflement macroscopique de mélanges d'argile. Thesis, Univ. Paris VI, unpublished, p. 155.

Capdecomme, L., 1952. Sur les phosphates alumineux de la région de Thiès (Sénégal). *C.R. Acad. Sci. Paris*, 235, pp. 187–189.

Capdecomme, L., and Kulbicki, G., 1954. Argiles des gîtes phosphatés de la région de Thiès (Sénégal). *Bull. Soc. fr. Minér. Crist.*, 87, pp. 500–518.

Carroll, D., 1958. Role of clay minerals in the transportation of iron. *Geochim. Cosmochim. Acta*, 14, pp. 1–27.

Cartledge, G. H., 1928. Studies on the periodic system. I. The ionic potential as a periodic function. *J. Amer. Chem. Soc.*, 50, pp. 2855–2863.

Cathcart, J. B., 1966. Economic geology of the Fort Meadle quadrangle, Polk and Hardee counties, Florida. *Bull. U.S. Geol. Surv.*, 1207, p. 97.

Cathcart, J. B., Blade, L. V., Davidson, D. F., and Ketner, K. B., 1953. The geology of the Florida land-pebble phosphate deposits. *XIXe Congr. Internat. Géol., Alger 1952, XI*, 11, pp. 77–91.

Cathcart, J. B., and Mc Greevy, L. J., 1953. Results of geologic exploration by core drilling, 1953, land-pebble phosphate district Florida. *Bull. U.S. Geol. Surv.*, 1046, K, pp. 221–297.

Chauvel, A., 1976. Recherches sur la transformation des sols ferrallitiques dans la zone tropicale à saisons contrastées. *Trav. Documents Office Rech. Sci. Tech. Outre-Mer (ORSTOM)*, 62, p. 532.

Chauvel, A., Bocquier, G., and Pédro, G., 1977. Géochimie de la surface et des formes du relief. III. Le mécanisme de la disjonction des constituants des couvertures ferrallitiques et l'origine de la zonalité des couvertures sableuses dans les régions intertropicales de l'Afrique de l'Ouest. *Bull. Sci. Géol.*, Strasbourg, 30, 4, pp. 255–263.

Chauvel, A., Bocquier G., and Pédro G., 1978. La stabilité et la transformation de la microstructure des sols rouges ferrallitiques de Casamance (Sénégal). Analyse microscopique et données expérimentales. *Int. Meet. Soil Micromorph., 5, Grenade, Proc.*, II, pp. 779–813.

Chauvel, A., Lucas, Y., and Boulet, R., 1987. On the genesis of the soil mantle of the region of Manaus, Central Amazonia, Brazil. *Experientia*, 43, pp. 234–241.

Chauvel, A., and Pédro, G., 1978. Sur l'importance de l'extrême dessiccation des sols (ultradessiccation) dans l'évolution pédologique des zones tropicales à saisons contrastées. *C.R. Acad. Sci., Paris*, 226, pp. 1581–1584.

Chou, L., and Wollast, R., 1984. Study of the weathering of albite at room temperature and pressure with a fluidized bed reactor. *Geochim. Cosmochim. Acta*, 48, pp. 2205–2217.

Chou, L., and Wollast, R., 1985. Steady-state kinetics and dissolution mechanisms of albite. *Am. J. Sci.*, 235, pp. 963–993.

Churchman, G. J., 1980. Clay minerals formed from micas and chlorites in some New Zealand soils. *Clay Min.*, 15, pp. 59–76.

Clauer, N., and Tardy, Y., 1971. Distinction par la composition isotopique du strontium contenu dans les carbonates, entre le milieu continental des vieux socles cristallins et le milieu marin. *C.R. Acad. Sci.*, 273, pp. 2191–2194.

Cole, C. V., and Jackson, M. L., 1951. Solubility equilibrium constant of $Al(OH)_2H_2PO_4$ relating to a mechanism of phosphate fixation in soils. *Soil Sci. Soc. Amer. Proc.*, 15, pp. 84–89.

Coleman, R., 1944. Phosphorus fixation by coarse and fine clay fraction of kaolinitic and montmorillonitic clays. *Soil Sci.*, 58, pp. 71–78.

Coleman, S. M., and Dethier, D. P., 1986. *Rates of Chemical Weathering of Rocks and Minerals.* Academic Press, Inc., San Diego, p. 603.

Colin F., Noack, Y., Trescases, J. J., and Nahon, D., 1985. L'altération latéritique débutante des pyroxénites de Jacuba, Niquelandia, Brésil. *Clay Min.*, 20, pp. 93–113.

Colin, F., Parron, C., Bocquier, G., and Nahon, D., 1980. Nickel and chromium concentrations by chemical weathering of pyroxenes and olivines. In: '*Metallogeny of Mafic and Ultramafic Complexes*, UNESCO Intern. Symposium, Proc. Athens 1980, 2, pp. 56–66.

Correns, C. W., and Von Engelhardt, W., 1938. Neue Untersuchungen über die Vermitterung des Kalifeldspates. *Chemie der Erde*, 12, pp. 1–22.

Curmi, P., 1979. Altération et différenciation pédologique sur granite en Bretagne. Etude d'une toposéquence. Thesis, Ecole Nat. Supér. Agron. Rennes, 2, unpublished, p. 176.

Curmi, P., and Fayolle, M., 1981. Caractérisation microscopique de l'altération dans une arène granitique a structure conservée. In: *Submicroscopy of Soils and Weathered Rocks*, Bisdom, E.B.A., Ed., Centre for Agricultural Publishing and Documentation, Wageningen, The Netherlands, ch. 13, pp. 249–270.

Decarreau, A., Colin, F., Herbillon, A., Manceau, A., Nahon, D., Paquet, H., Trauth-Badaud, D., and Trescases J. -J., 1987. Domain segregation in Ni-Fe-Mg smectites. *Clays Clay Miner.*, 35, pp. 1–10.

Delvigne, J., 1965. La formation des minéraux secondaires en millieu ferrallitique. *Mém. Office Rech. Sci. Tech. Outre-Mer (ORSTOM)*, Paris, 13, p. 177.

Delvigne, J., and Boulangé, B., 1973. Micromorphologie des hydroxydes d'aluminium dans les niveaux d'altération et dans les bauxites. In: *Int. Working-Meeting on Soil Micromorphology, IVth, Proc.*, Rutherford, G. K., Ed., Kingston, Ontario, pp. 665–681.

Delvigne, J., and Martin, H., 1970. Analyse à la microsonde électronique de l'altération d'un plagioclase en kaolinite par l'intermédiaire d'une phase amorphe. *Cah. Office Rech. Sci. Tech. Outre-Mer (ORSTOM), sér. Géologie*, 2, pp. 259–295.

Dethier, D. P., 1986. Weathering rates and the chemical flux from catchments in the Pacific Northwest, U.S.A. In: *Rates of Chemical Weathering of Rocks and Minerals*, Coleman, S. M., and Dethier, D. P., Eds., Academic Press, Inc., San Diego, ch. 21, pp. 503–530.

D'Hoore, J., 1954. L'accumulation des sesquioxydes libres dans les sols tropicaux. *Publ. Inst. Nation. Etude Agron. Congo (INEAC)*, 62, p. 132.

Didier, P., 1983. Paragénèses à oxydes et hydroxydes de fer et d'alumine dans les cuirasses ferrugineuses. Thesis, Univ. Poitiers, unpublished, p. 150.

Didier, P., Fritz, B., Nahon, D., and Tardy, Y., 1983. Fe^{3+}-kaolinites, Al-goethites, Al-hematites in tropical ferricretes. In: *Petrology of Weathering and Soils*, Nahon, D., and Noack, Y., Eds., *Mém. Sci. Géol.*, Strasbourg, 71, pp. 35–44.

Dowty, E., 1980. Crystal-chemical factors affecting the mobility of ions in minerals. *Am. Mineral.*, 65, pp. 179–182.

Drever, J. I., 1982. *The Geochemistry of Natural Waters.* Prentice Hall, Inc., Englewood Cliffs, New Jersey, p. 388.

Drosdoff, M., and Nikiforoff, C. C., 1940. Iron-manganese concretions in Dayton soils. *Soil Sci.*, 49, p. 333.

Duchaufour, P., 1970. *Précis de Pédologie*, 3rd Ed., Masson, Paris, p. 481.

Ducloux, J., Dupuis, T., Butel, P., and Nahon, D., 1984. Carbonates de calcium amorphe et cristallisés dans les encroûtements calcaires des milieux tempérés. Comparaison des séquences minérales naturelles et expérimentales. *C. R. Acad. Sci., Paris*, 298, pp. 147–149.

Duplay, J., 1984. Analyses chimiques ponctuelles d'argiles. Relations entre variations de composition dans une population de particules et température de formation. *Bull. Sci. Géol.*, Strasbourg, 37, p. 307–317.

Duplay, J., 1988. Géochimie des argiles et géothermométrie des populations monominérales de particules. Thesis, Univ. Strasbourg, unpublished, p. 222.

Du Preez, J. W., 1954. Notes on the occurrence of oolites and pisolites in Nigerian laterites. *Congr. Géol. Int.*, 19, Alger, 1952, 21, pp. 163–169.

Dupuis, T., Ducloux, J., Butel, P., and Nahon, D., 1984. Etude par spectrographie infrarouge d'un encroûtement calcaire sous galet. Mise en évidence et modélisation expérimentale d'une suite minérale évolutive à partir de carbonate de calcium amorphe. *Clay Min.* 19, pp. 605–614.

Durand, R., 1979. La pédogenèse en pays calcaire dans le Nord-Est de la France. *Mém. Sci. Géol.*, Strasbourg, 55, p. 198.

Eggleton, R. A., 1975. Nontronite topoaxial after hedenbergite. *Am. Mineral.*, 60, pp. 1063–1068.

Eggleton, R. A., 1986. The relation between crystal structure and silicate weathering rates. In: *Rates of Chemical Weathering of Rocks and Minerals*, Coleman S. M. and Dethier D. P., Eds., Academic Press, Inc., San Diego, CA, ch. 1, pp. 21–39.

Eggleton, R. A., and Boland, J. N., 1982. The weathering of enstatite to talc through a series of transitional phases. *Clays Clay Miner.*, 30, pp. 11–20.

Escande, M. A., 1983. Echangeabilité et fractionnement isotopique de l'oxygène des smectites magnésiennes de synthèse. Etablissement d'un géothermomètre. Thesis, Univ. Orsay, unpublished, p. 150.

Eswaran, H., and Bin, W. C., 1978. A study of a deep weathering profile on granite in Peninsular Malaysia. III. Alteration of feldspars. *Soil, Sci., Soc. Am. J.*, 42, pp. 154–158.

Eswaran, H., and Heng, Y. Y., 1976. The weathering of biotite in a profile on gneiss in Malaysia. *Geoderma*, 16, pp. 9–20.

Eswaran, H., Lim, C. H., Sooryanarayanan, V., and Daud, N., 1977. Scanning electron microscopy of secondary minerals in Fe-Mn glaebules. *Int. Meet. Soil Micromorph., 5, Proc., Granada, Spain*, pp. 851–885.

Fanning, D. S., and Keramidas, V. Z., 1977. Mica. In: *Minerals in Soil environments*, Dixon J. B., Ed., Soil Society of America, Madison, Wisconsin, pp. 195–258.

Farmer, V. C., Russell, J. D., Fardy, J. Mc., Newman, A. C. D., Ahlrichs, J. L., and Rimsaite, J. Y. H., 1971. Evidence of loss of protons and octahedral iron from oxidized biotites and vermiculites. *Mineral. Mag.*, 38, 294, pp. 121–137.

Fayolle, M., 1979. Caractérisation analytique d'un profil d'argile à silex de l'ouest du bassin de Paris. Thesis, Univ. Paris VII, unpublished, p. 162.

Fieldes, M., and Swindale, L. D., 1954. Chemical weathering of silicates in soil formation. *New Zealand Journ. Sci. Tech.*, 36, pp. 140–154.

Fies, J. C., 1978. Porosité du sol: Étude de son origine texturale. Thesis, Univ. of Strasbourg, unpublished, p. 139.

Fischer, W. R., and Schwertmann, U., 1975. The formation of hematite from amorphous Fe(III) hydroxide. *Clays Clay Miner.*, 23, pp. 33–37.

Flach, K. W., Cady, J. G., and Nettleton, W. D., 1968. Pedogenetic alteration of highly weathered parent materials. *Int. Cong. Soil Sci., 9, Proc., Adelaide*, 4, pp. 343–351.

Flicoteaux, R., 1982. Genèse des phosphates alumineux du Sénégal Occidental. Etapes et guides de l'altération. *Mém. Sci. Géol.*, Strasbourg, 67, p. 229.

Flicoteaux, R., Nahon, D., and Paquet, H., 1977. Genèse des phosphates alumineux à partir des sédiments argilo-phosphatés du Tertiaire de Lam-Lam (Sénégal). Suite minéralogique, permanences et changements de structure. *Bull. Sci. Géol.*, Strasbourg, 30, pp. 153–174.

Flicoteaux, R., and Tessier, F., 1971. Précisions nouvelles sur la stratigraphie des formations du plateau de Thiès (Sénégal occidental) et sur leurs altérations. Conséquences paléogéographiques. *C.R. Acad. Sci. Paris*, 272, pp. 364–366.

Fouillac, C., Michard, G., and Bocquier, G., 1977. Une méthode de simulation de l'évolution des profils d'altération. *Geochim. Cosmochim. Acta*, 41, pp. 207–214.

Frakes, L. A., 1976. *Climates throughout Geologic Time.* Elsevier, Amsterdam, p. 310.

Frankel, J. -J., and Bayliss, P., 1966. Ferruginized surface deposits from Natal and Zululand, South Africa. *J. Sedim. Petrol.*, 36, pp. 193–201.

Frederickson, A. F., 1951. Mechanism of weathering. *Bull. Geol. Soc. Am.*, 62, pp. 221–232.

Fripiat, J. J., 1960. Application de la spectroscopie infra-rouge à l'étude des minéraux argileux. *Bull. Groupe Fr. Argiles*, 12, pp. 25–43.

Fritsch, E., 1984. Les transformations d'une couverture ferrallitique en Guyanne française. *Mém. Off. Rech. Sci. Tech. Outre Mer (ORSTOM)*, Paris, p. 190.

Fritz, B., 1975. Etude thermodynamique et simulation des réactions entre minéraux et solutions. Application à la géochimie des altérations et des eaux continentales. *Mém. Sci. Géol.*, Strasbourg, 41, p. 153.

Fritz, B., and Tardy, Y., 1976. Séquence des minéraux secondaires dans l'altération des granites et roches basiques; modèles thermodynamiques. *Bull. Soc. Géol. France.*, 7, pp. 7–12.

Furrer, G., and Stumm, W., 1986. The coordination chemistry of weathering. I.

Dissolution kinetics of δ-Al_2O_3 and BeO. *Geochim. Cosmochim. Acta*, 50, pp. 1847–1860.

Furukawa, H., Handawella, J., Kyuma, K., and Kawaguchi, X., 1976. Chemical, mineralogical and micromorphological properties of glaebules in some tropical low land soils. *S.E. Asian Stud.*, Japan, 14, 3, pp. 365–388.

Gac, J. Y., 1979. Géochimie du bassin du lac Tchad. Thesis, Univ. of Strasbourg, unpublished, p. 251.

Garcia Hernandez, J. E., 1981. Interprétation cinétique de la géochimie d'altération de la silice à basse température (25° C). *Publication Instit. Nation. Recherch. Agron. (INRA)* Versailles, p. 213.

Gardner, L. R., 1970. A chemical model for the origin of gibbsite from kaolinite. *Am. Miner.*, 55, pp. 1380–1389.

Garrels, R. M., and Mackenzie, F. T., 1971. *Evolution of Sedimentary Rocks.* W. W. Norton, New York, p. 397.

Gavaud, M., 1968. Projet de corrélation pédologique dans le bassin du lac Tchad. O.R.S.T.O.M.-U.N.E.S.C.O., Rep., Paris, mimeographed, unpublished, p. 117.

Giese, R. F. Jr., 1979. Hydroxyl orientations in 2/1 phyllosilicates. *Clays Clay Miner.*, 27, 3, pp. 213–223.

Gile, L. H., 1961. A classification of Ca horizons in soils of a desert region, Dona Ana Country, New Mexico. *Soil Sci. Soc. Amer. Proc.*, 25, pp. 52–61.

Gile, L. H., Peterson, F. F., and Grossman, R. B., 1966. Morphological and genetic sequences of carbonate accumulation in desert soils. *Soil Sci.*, Baltimore, Maryland, 101, pp. 347–360.

Gilkes, R. J., 1973. The alteration products of potassium depleted oxybiotite. *Clays and Clay Miner.*, 21, pp. 303–313.

Gilkes, R. J., and Suddhiprakarn, A., 1979. Biotite alteration in deeply weathered granite I and II. *Clays and Clay Miner.*, 27, pp. 349–367.

Gilkes, R. J., and Young, R. C., 1974. Artificial weathering of oxidized biotite. IV. The inhibitory effect of potassium on dissolution rate. *Soil Sci. Soc. Amer. Proc.*, 38, pp. 529–532.

Gilkes, R. J., Young, R. C., and Quirk, J. P., 1972. The oxidation of octahedral iron in biotite. *Clays and Clay Miner.*, 20, pp. 303–315.

Glossaire de Pédologie, 1969. Office Rech. Sci. Techn. Outre-Mer (ORSTOM) Editions, Paris, p. 82.

Goldich, S. S., 1938. A study in rock-weathering. *J. Geol.*, 46, pp. 17–58.

Goldschmidt, V. M., 1934. Drei Vorträge über Geochemie. *Geol. Fören. Förhandl.*, 56, pp. 385–427.

Goldschmidt, V. M., 1937. The principles of the distribution of chemical elements in minerals and rocks. *Journ. Chem. Soc.*, London, pp. 655–672.

Goudie A., 1973. *Duricrusts in Tropical and Subtropical Landscapes.* Oxford Univ. Press, New York, London, p. 174.

Graham, E. R., 1949. The plagioclase feldspars as an index to soil weathering. *Soil Sci. Amer. Proc.*, 14, pp. 300–302.

Grandstaff, D. E., 1977. Some kinetics of bronzite orthopyroxene dissolution. *Geochim. Cosmochim. Acta*, 41, pp. 1097–1103.

Grandstaff, D. E., 1978. Changes in surface area and morphology and the mechanism of forsterite dissolution. *Geochim. Cosmochim. Acta*, 42, pp. 1899–1901.

Grandstaff, D. E., 1981. The dissolution rate of forsterite olivine from Hawaiian beach sand. *Int. Symp. on Water-rock Interaction, 3, Proc.*, Edmonton, Can., pp. 72–74.

Grandstaff, D. E., 1983. The dissolution rate of forsteritic olivine from Hawaiian beach sand. In: *Rates of Chemical Weathering of Rocks and Minerals*, by Coleman, S. M., and Dethier, D. P. Eds., Academic Press Inc., San Diego, ch. 3, pp. 41–59.

Greene, H. 1945. Classification and use of tropical soil. *Soil Sci. Soc. Amer. Proc.*, 10, pp. 392–396.

Grim, R. E., 1968. *Clay Mineralogy*. McGraw-Hill, New York, p. 384.

Grimaldi, M., 1981. Contribution à l'étude du tassement des sols: Evolution de la structure d'un matériau limoneux soumis à des contraintes mécaniques et hydriques. Thesis, Ecole Nat. Sup. Agron., Rennes, unpublished, p. 220.

Gruffat, J. J., and Guy, B., 1984. Un modèle pour les précipitations alternantes de minéraux dans les roches métasomatiques: L'effet autocatalytique des surfaces. *C.R. Acad. Sci. Paris*, 299, 14, pp. 961–964.

Grüner, J. W., 1934. The structure of vermiculites and their collapse on dehydratation. *Am. Miner.*, 19, pp. 557–575.

Guy, B., 1984. Contribution to the theory of infiltration metasomatic zoning; the formation of sharp fronts: A geometrical model. *Bull. Miner.*, 107, pp. 93–105.

Haines, W. B., 1923. The volume changes associated with variation of water content in soil. *J. Agric. Sci.*, 13, pp. 296–310.

Hall, R. D., and Martin, R. E., 1986. The etching of hornblende grains in the matrix of alpine tills and periglacial deposits. In: *Rates of Chemical Weathering of Rocks and Minerals*, Coleman, S. M., and Dethier, D. P., Ed., Academic Press, Inc., San Diego, ch. 6, pp. 101–128.

Hay, R. L., 1959. Origin and weathering of late Pleistocene ash deposits on St. Vincent B.N.I. *J. Geol.*, 67, pp. 65–87.

Helgeson, H. C., 1968. Evaluation of irreversible reactions in geochemical processes involving minerals and aqueous solutions. I. Thermodynamic relations. *Geochim. Cosmochim. Acta*, 32, pp. 853–877.

Helgeson, H. C., 1969. Thermodynamics of hydrothermal systems at elevated temperatures and pressures. *Am. J. Sci.*, 267, pp. 729–804.

Helgeson, H. C., 1970. Description and interpretation of phase relations in geochemical processes involving aqueous solutions. *Am. J. Sci.*, 268, pp. 415–438.

Helgeson, H. C., 1971. Kinetics of mass transfer among silicates and aqueous solutions. *Geochim. Cosmochim. Acta*, 35, pp. 421–469.

Helgeson, H. C., Brown, T. H., Nigrini, A., and Jones, T. A., 1970. Calculation of mass transfer in geochemical processes involving aqueous solutions. *Geochim. Cosmochim. Acta*, 34, pp. 569–592.

Helgeson, H. C., Garrels, R. M., and Mackenzie, F. T., 1969. Evaluation of irreversible reactions in geochemical processes involving minerals and aqueous solutions. II. Applications. *Geochim. Cosmochim. Acta*, 33, pp. 455–481.

Helgeson, H. C., Murphy, W. M., and Aagaard, P., 1984. Thermodynamic and kinetic constraints on reaction rates among minerals and aqueous solutions. II. Rate constants, effective surface area and the hydrolysis of feldspar. *Geochim. Cosmochim. Acta*, 48, pp. 2405–2432.

Hemwall, J. B., 1957. The role of soil clay minerals in phosphorus fixation. *Soil Sci.*, 83, pp. 101–108.

Hendricks, D. M., and Whittig, L. D., 1968. Andesite weathering. I. Mineralogical transformation from andesite to saprolite. II. Geochemical changes from andesite to saprolite. *J. Soil Sci.*, 19, pp. 135–153.

Hénin, S., and Pédro, G., 1979. Rôle de l'hétérogénéité minéralogique du milieu sur les modalités de l'altération. *Sci. Sol.*, 2–3, pp. 209–221.

Herbillon, A., 1961. Contribution à l'étude des gels d'alumine. Synthèse et genèse de la gibbsite. Thesis, Univ. Louvain, Belgique, unpublished, p. 89.

Herbillon, A., and Gastuche, M. C., 1962. Etude des complexes kaolinite-hydroxyde d'aluminium. Synthèse et genèse des trihydrates cristallisés. *Bull. Gr. Fr. Argiles*, 13, 8, pp. 87–94.

Herbillon, A., and Nahon, D., 1988. Laterites and lateritization processes. In: *Iron in Soils and Clay Minerals*, Stucki, J. W., Goodman, B. A., and Schwertmann, U., Eds., NATO Advanced Science Institutes Series, D. Reidel Publish., Dordrecht, The Netherlands, 217, pp. 779–796.

Holdren, G. R. Jr., and Berner R. A., 1979. Mechanism of feldspar weathering. I. Experimental studies. *Geochim. Cosmochim. Acta*, 43, pp. 1161–1171.

Holdren, G. R. Jr., and Speyer, P. M., 1985a. Reaction rate-surface area relationships during the early stages of weathering. I. Initial observations. *Geochim. Cosmochim. Acta*, 49, pp. 671–681.

Holdren, G. R. Jr., and Speyer, P. M., 1985b. pH dependent changes in the rates and stoichiometry of dissolution of an alkali feldspar at room temperature. *Am. J. Sci.*, 285, pp. 994–1026.

Holdren, G. R. Jr., and Speyer, P. M., 1986. Stoichiometry of alkali feldspar dissolution at room temperature and various pH values. In: *Rates of Chemical Weathering of Rocks and Minerals*, Coleman, S. M., and Dethier, D. P., Eds., Academic Press, Inc., San Diego, ch. 4, pp. 61–81.

Hotz, P. E., 1961. Weathering of peridotite, Southwest Oregon. *U.S. Geol. Surv. Prof. Paper*, 424-D, pp. 327–331.

Howeler, R. H., and Bouldin, D. R., 1971. The diffusion and consumption of oxygen in submerged soils. *Soil Sci. Soc. Amer. Proc.*, 35, pp. 202–208.

Howell, D. G. and Von Huene, R. 1981. Tectonic and sedimentation along active continental margins. In Depositional systems of active continental margins; short course notes (R. G. Douglas *et al*, Eds). Soc. Econ. Paleont. Mineral, San Francisco, pp. 1–13.

Hudcova, O., 1970. The reaction of kaolinite with phosphates. *Conf. Clay Miner. Proc.*, 5, Prague, pp. 21–26.

Humbert, R., 1948. The genesis of laterite. *Soil. Sci.*, 65, p. 281.

Hurlbut Jr., C. S., and Klein, C., 1977. *Manual of Mineralogy, after James D. Dana*, 19th ed. John Wiley & Sons, Inc., New York, p. 532.

Iler, R. K., 1979. The chemistry of silica: solubility, polymerization, colloid and surface properties, and biochemistry. John Wiley & Sons, Inc., New York, p. 866.

Jackson, M. L., 1959. Frequency distribution of clay minerals in major great soil group as related to the factors of soil formation. *Clays and Clay Miner.*, pp. 133–143.

Jackson, M. L., 1968. Weathering of primary and secondary minerals in soils. *Int. Cong. Soil Sci., 9, Proc.*, Adelaide, 4, pp. 281–292.

Jackson, M. L., Hseung, Y., Corey, R. B., Evans, E. J., and Van Den Heuvel, R. C., 1952. Weathering of clay size minerals in soils and sediments. II. Chemical weathering of layer silicates. *Soil Sci. Soc. Am. Proc.*, 16, pp. 3–6.

Jackson, M. L., Tyler, S. A., Willis, A. L., Bourbeau, G. A., and Pennington, R. P., 1948. Weathering sequence of clay-size minerals in soils and sediments. *J. Phys. Coll. Chem.*, 52, pp. 1237–1260.

Jones, H. A., 1965. Ferruginous oolites and pisolites. *J. Sedim. Petrol.*, Tulsa, Oklahoma, 35, 4, pp. 838–845.

Joshi, M. S., and Paul, B. K., 1977. Surface structures of trigonal bipyramid faces of natural quartz crystals. *Am. Mineral.*, 62, pp. 122–126.

Kehres, A., 1983. Isothermes de déshydratation des argiles; énergies d'hydratation; diagrammes de pores; surfaces internes. Thesis, Univ. Toulouse, unpublished, p. 163.

Keller, W. D., Balgord, W. D., and Reesman, A. L., 1963. Dissolved products of artificially pulverized silicate minerals and rocks: Part I. *J. Sedim. Petrol.*, 33, pp. 191–204.

King, L., 1962. *The Morphology of the Earth. A Study and Synthesis of World Scenery*. Oliver and Boyd. Edinburgh, p. 699.

Kittrick, J. A., and Jackson, M. L., 1954. Electron microscope observation of the formation of aluminum phosphate crystals with kaolinite as the source of aluminum. *Science*, New York, 120, pp. 508–509.

Kittrick, J. A., and Jackson, M. L., 1955. Rate of phosphate reactions with soil minerals and electron microscope observations on the reaction mechanism. *Soil Sci. Soc. Amer. Proc.*, 19, pp. 292–295.

Kittrick, J. A., and Jackson, M. L., 1956. Electron microscope observations of the reaction of phosphate with minerals, leading to a unified theory of phosphate fixation in soils. *J. Soil Sci.*, 7, pp. 81–88.

Korzhinski, D. S., 1965. The theory of system with perfectly mobile components and processes of mineral formation. *Am. J. Sci.*, 263, pp. 193–205.

Korzinski, D. S., 1970. *Theory of Metasomatic Zoning*. Clarendon Press, Oxford, p. 162.

Kounestron, O., Robert, M., and Berrier, J., 1977. Nouvel aspect de la formation des smectites dans les vertisols. *C.R. Acad. Sc., Paris*, 284, pp. 733–737.

Kovda, V. A., 1933. Principles of soil classification. *Trudy Soviet Sekta MAP*, Moscow, 2, pp. 7–22.

Kovda, V. A., Zimovets, B. A., and Amchislavskaya, A. G., 1958. The hydromorphic accumulation of compounds of silica and sesquioxides in the soils of the Amur region. *Pochvovedenic*, 5, pp. 1–11.

Krauskopf, K. B., 1967. *Introduction to Geochemistry*. MacGraw-Hill, Inc., New York, p. 617.

Krinsley, D. H., and Doornkamp, J. C., 1973. *Atlas of Quartz Sand Surface Textures*. Cambridge Univ. Press, Cambridge U. K., p. 91.

Kubiena, W. L., 1938. *Micropedology*. Collegiate Press Inc., Ames, Iowa, p. 242.

Labeyrie, L., and Juillet, A., 1982. Oxygen isotopic exchangeability of diatom value silica: Interpretation and consequences for paleoclimatic studies. *Geochim. Cosmochim. Acta*, 46, pp. 967–975.

Lacroix, A., 1914. Les latérites de la Guinée et les produits d'altération qui leur sont associés. *Nouvelles Archives Museum Histoire Nat. Paris*, 5, pp. 255–356.

Lafeber, D., 1962. Aspects of stress-induced differential movements of fabric elements in mineral soils. *Conf. Austr. Road Res. Board, 1, Proc.*, pp. 1059–1067.

Lagache, M., 1965. Contribution à l'étude de l'altération des feldspaths dans l'eau entre 100° et 200° C, sous diverses pressions de CO_2 et application à la synthèse des minéraux argileux. *Bull. Soc. Fr. Minér. Crist.*, 88, pp. 223–253.

Lagache, M., 1976. New data on the kinetics of the dissolution of alkali feldspars at 200° C in CO_2 charged water. *Geochim. Cosmochim. Acta*, 40, pp. 157–161.

Langford-Smith, T., 1978. *Silcrete in Australia*. Dept. Geography, Univ. New-England, Sydney, Australia, p. 304.

Lapparent, J. de, 1909. Etude comparative de quelques porphyroïdes françaises. *Bull. Soc. Fr. Minéral.*, 32, pp. 174–304.

Lappartient, J., 1970. La latérite récente des environs de Dakar (République du Sénégal). *Travaux Lab. Sci. Terre, St-Jérôme*, Marseille, A, 3, p. 57.

Lasaga, A. C., 1981. Rate laws of chemical reactions. In: *Kinetics of Geochemical Processes, Reviews in Mineralogy*, vol. 8, Lasaga, A. C., and Kirkpatrick R. J., Eds., Mineralogical Society of America, Washington, D.C., pp. 1–67.

Lasaga, A. C., 1983. Kinetics of silicate dissolution. *Int. Symp. on Water-rock Interaction, 4, Proc.*, Misasa (Japan), pp. 269–274.

Lasaga, A. C., 1984. Chemical kinetics of water-rock interactions. *J. Geophys. Res.*, 89, B6, pp. 4009–4025.

Lasaga, A. C., Berner, R. A., and Garrels, R. M., 1985. An improved geochemical model of atmospheric CO_2 fluctuations over the past 100 million years. In: *The Carbon Cycle and Atmospheric CO_2: Natural Variations Archean to Present*, Geophysical Monograph 32, AGU Ed., pp. 397–411.

Lasaga, A. C., and Blum, A. E., 1986. Surface chemistry, etch pits and mineral-water reactions. *Geochim. Cosmochim. Acta*, 50, pp. 2363–2379.

Lawrence, J. R., and Taylor, H. P., 1971. Deuterium and oxygen 18 correlation: Clay minerals and hydroxides in Quaternary soils compared to meteoric waters. *Geochim. Cosmochim. Acta*, 35, pp. 993–1003.

Lawrence, J. R., and Taylor, H. P., 1972. Hydrogen and oxygen isotope systematics in weathering profiles. *Geochim. Cosmochim. Acta*, 36, pp. 1377–1393.

Lemon, E. R., 1983. CO_2 and Plants. Am. Assoc. Adv. Sci. Selected Symposium 84, Washington, D.C., p. 280.

Leneuf, N., 1959. L'altération des granites calco-alcalins et des granodiorites en Côte d'Ivoire forestière et les sols qui en sont dérivés. Thesis, Univ. of Paris, unpublished, p. 210.

Leprun, J. C., 1979. Les Cuirasses ferrugineuses des pays cristallins d'Afrique Occidentale sèche. Genèse, transformation, dégradations. *Mém. Sci. Géol.*, Strasbourg, 58, p. 224.

Leprun, J. C., and Nahon, D., 1973. Cuirassements ferrugineux autochtones sur deux types de roches. *Bull. Soc. Géol. France*, Paris, XV, 3–4, pp. 356–361.

Le Ribault, L., 1977. *L'exoscopie des Quartz*. Masson, Paris, p. 150.

Lietard, O., 1977. Contribution à l'étude des propriétés physicochimiques, cristallographiques et morphologiques des kaolins. Thesis, Inst. National Polytechnique de Lorraine, unpublished, p. 377.

Lin, F. C., and Clemency, C. V., 1981a. Dissolution kinetics of phlogopite. I. Closed system. *Clays and Clay Miner.*, 29, pp. 101–106.

Lin, F. C., and Clemency, C. V., 1981b. The kinetics of dissolution of muscovites at 25°C, and 1 atm CO_2 partial pressure. *Geochim. Cosmochim. Acta*, 45, pp. 571–576.

Lin, F. C., and Clemency, C. V., 1981c. The dissolution kinetics of brucite, antigorite, talc, and phlogopite at room temperature and pressure. *Am. Mineral.*, 66, pp. 801–806.

Locke, W. W., 1986. Rates of hornblende etching in soils on glacial deposits, Baffin Island, Canada. In: *Rates of Chemical Weathering of Rocks and Minerals*, Coleman, S. M., and Dethier, D. P., Eds., Academic Press Inc., San Diego, ch. 7, pp. 129–144.

Loughnan, J. C., 1969. *Chemical Weathering of the Silicate Minerals*. Elsevier, New York, p. 154.

Low, P. F., and Black, C. A., 1948. Phosphate induced decomposition of kaolinite. *Soil Sci. Soc. Amer. Proc.*, 12, pp. 180–184.

Low, P. F., and Black, C. A., 1950. Reactions of phosphate with kaolinite. *Soil Sci.*, 70, pp. 273–290.

Lucas, Y., Boulet, R., and Chauvel A., 1988. Intervention simultanée des phénomènes d'enfoncement vertical et de transformation latéral dans la mise en place de systèmes de sols de la zone tropicale humide. *C. R. Acad. Sci., Paris*, 306, pp. 1395–1400.

Lucas, Y., Boulet, R., and Veillon, L., 1987. Systèmes sols ferrallitiques-podzols en région Àmazonienne. In: *Podzols et Podzolisation*, Righi, D., and Chauvel, A., Eds., AFES (Plaisir). INRA, Paris, pp. 53–65.

Lucas, Y., and Chauvel, A., 1990. Soil formation in tropically weathered terranes. In: *Handbook of Exploration, Geochemistry, Soil, Laterite and Saprolite Geochemistry in Mineral Exploration of Tropically Weathered Terranes*, Butt, C., and Zeegers, H., Eds., Elsevier, Amsterdam, New York, in press.

Lucas, Y., Chauvel, A., and Ambrosi, J. P., 1986. Processes of aluminum and iron accumulation in latosols developed on quartz rich sediments from Central Amazonia, Manaus, Brazil. *Int. Meet. Geochem. Earth Surface and Processes Min. Formation, 1, Proc.*, Granada, Spain, pp. 289–299.

Lucas, Y., Chauvel, A., Boulet, R., Ranzani, G., and Scatolini, F., 1984. Transição latossolos-podzoils sobre a Formação Barreias, região de Manaus, Amazonia. *R. Bras. Ci. Solo*, 8, pp. 325–335.

Lucas, Y., Kobilsek, B., and Chauvel, A., 1988. Structure, genesis and present evolution of Amazonian bauxites developed on sediments. *Int. Cong. ICSOBA, 6, São Paulo*, Brazil, Abstract.

Luce, R. W., Bartlett, R. W., and Parks, G. A., 1972. Dissolution kinetics of magnesium silicates. *Geochim. Cosmochim. Acta*, 36, pp. 35–50.

Maignien, R., 1958. Le cuirassement des sols en Guinée, Afrique Occidentale. *Mém. Serv. Carte Géol. Alsace Lorraine*, Strasbourg, 16, p. 223.

Manceau, A., 1989. Cristallochimie des éléments de transition dans les géomatériaux supergènes. Thesis, Univ. Paris 7, unpublished, p. 241.

Manceau, A., and Calas, G., 1986. Nickel-bearing clay minerals. II. Intracrystalline distribution of nickel: An X-ray absorption study. *Clay Min.*, 21, pp. 341–360.

Mattigod, S. V., and Kittrick, J. A., 1980. Temperature and water activity as variables in soil mineral activity diagrams. *Soil Sci. Soc. Am. J.*, 44, pp. 149–154.

Mazaltarim, D., 1989. Géochimie des cuirasses ferrugineuses et bauxitiques de l'Afrique Occidentale et Centrale. Thesis, Univ. Louis Pasteur, Strasbourg, mimeographed, unpublished, 262 p.

Mejsner, J., 1978. Alteration products from chlorite. I. The Taro Valley mineral Series (Italy). *Archivum Mineral.*, 34, pp. 5–16.

Mering, J., 1962. Gonflement, dispersion et hydratation des argiles. *Bull. Groupe Fr. Argiles*, 13, pp. 115–123.

Meunier, A., 1980. Les mécanismes de l'altération des granites et le rôle des microsystèmes. Etude des arènes du massif granitique de Parthenay (Deux-Sèvres). *Mém. Soc. Géol. France*, 140, pp. 80.

Meybeck, M., 1980. Pathways of major elements from land to ocean through rivers. In: *River inputs to Ocean Systems; UNEP, IOC, SCOR Workshop Proceedings*, Martin, J. M., Burton, J. D., and Eisma, D. Eds. United Nations Press, New York, p. 384.

Michel, P., 1973. Les bassins des fleuves Sénégal et Gambie. Etude géomorphologique. *Mém. Office Rech. Sci. Techn. Outre-Mer, (ORSTOM)*, Paris, 1, 2, 3, 63, p. 752.

Millot, G., 1949. Relations entre la constitution et la genèse des roches sédimentaires argileuses. *Bull. Géol. Appl. Prospection Minière*, 2, pp. 1–352.

Millot, G., 1964. *Géologie des Argiles*. Masson et Cie. Ed., Paris, p. 499.

Millot, G., 1979. Rôle des épigénies dans l'enrichissement des gîtes minéraux météoriques: quatre exemples. Présentation et enseignements. *Bull. Sci. Géol.*, Strasbourg, 32, 4, pp. 139–145.

Millot, G., and Bonifas, M., 1955. Transformations isovolumétriques dans les phénomènes de latéritisation et de bauxitisation. *Bull. Serv. Carte Géol. Alsace Lorraine*, 8, pp. 3–10.

Millot, G., Lucas, J., and Paquet, H., 1965. Evolution géochimique par dégradation et aggradation des minéraux argileux dans l'hydrosphère. *Geol. Rundsch.*, 55, pp. 1–20.

Millot, G., Nahon, D., Paquet, H., Ruellan, A., and Tardy, Y., 1977. L'épigénie calcaire des roches silicatées dans les encroûtements carbonatés en pays subaride, Anti Atlas, Maroc. *Bull. Sci. Géol.* Strasbourg, 30, 3, pp. 129–152.

Mitsuchi, M., 1976. Characteristics and genesis of nodules and concretions occurring in soils of the R. Chinit area, Kompong Thom Province, Cambodia. *Soil Sci. Plant Nutr.* 22, 4, pp. 409–421.

Mogk, D. W., and Locke, W. W., 1988. Application of Auger Electron Spectroscopy (A.E.S.) to naturally weathered hornblende. *Geochim. Cosmochim. Acta*, 52, pp. 2537–2542.

Moniz, A. C., and Jackson, M. L., 1970. Quantitative mineralogical analysis of Brasilian soils derived from basic rocks and slate. Madison Wisconsin, unpublished Soil Science report, 212, p. 75.

Monnier, G., Stengel, P., and Fies, J. C., 1973. Une méthode de mesure de la densité apparente de petits agglomérats terreux. Application à l'analyse des systèmes du porosité du sol. *Ann. Agron.*, 24, pp. 533–545.

Mortland, M. M., 1958. Kinetics of potassium release from biotite. *Soil Sci. Soc. Am. Proc.*, 22, pp. 503–508.

Moss, A. J., Walker, P. H., and Hutka, J., 1973. Fragmentation of granitic quartz in water. *Sedimentology*, 20, pp. 489–511.

Moss, A. J., and Green, P., 1975. Sand and silt grains: Predetermination of their formation and properties by microfractures in quartz. *J. Geol. Soc. Aust.*, 22, pp. 485–495.

Mpiana, K., 1980. Contribution à l'étude des profils bauxitiques de Côte d'Ivoire et du Cameroun. Relations entre les microstructures et la minéralogie. Thesis, Univ. Aix-Marseille 3, unpublished, p. 187.

Muljadi, D., Posner, A. M., and Quirk, J. P., 1966. The mechanism of phosphate adsorption by kaolinite, gibbsite and pseudoboehmite. Part II. The location of the adsorption sites. *J. Soil Sci.*, 17, pp. 230–237.

Muller, D., Bocquier, G., Nahon, D., and Paquet, H., 1980. Analyse des différentiations minéralogiques et structurales d'un sol ferrallitique à horizons nodulaires du Congo, *Cah. Office Sci. Tech. Outre-Mer (ORSTOM), sér. Pédologie*, 18, 2, pp. 87–109.

Muller, J. -P., 1987. Analyse pétrologique d'une formation latéritique meuble du Cameroun. Thesis, Univ. Paris 7, unpublished, p. 174.

Murphy, W. M., 1985. 'Thermodynamic and kinetic constraints on reaction rates among minerals and aqueous solutions'. Thesis, University of California-Berkeley, unpublished.

Nahon, D., 1971. Contribution à l'étude de la genèse des cuirasses ferrugineuses quaternaires sur grès: Exemple du massif de Ndias, Sénégal occidental, *Rapp. Dept. Géol. Fac. Sci.*, Dakar, 31, p. 81.

Nahon, D., 1976. Cuirasses ferrugineuses et encroûtements calcaires au Sénégal occidental et en Mauritanie. Systèmes évolutifs: Géochimie, structures, relais et coexistence. *Mém. Sci. Géol.*, Strasbourg, 44, p. 232.

Nahon, D., 1986. Evolution of iron crusts in tropical landscapes. In: *Rates of Chemical Weathering of Rocks and Minerals*, Coleman, S. H., and Dethier D. P., Eds., Academic Press Inc., San Diego, ch. 9, pp. 169–191.

Nahon, D., 1987. Microgeochemical environments in lateritic weathering. In: *Geochemistry and Mineral Formation in the Earth Surface, Int. Meeting Proc., Granada, Spain, 1986*, Rodriguez-Clemente, R., and Tardy, Y., Eds., Consejo Superior de Investigaciones Cientificas (CSIC), Madrid, pp. 141–156.

Nahon, D., Beauvais, A., Boeglin, J. L., Ducloux, J., and Nziengui-Mapangou, P., 1983. Manganite formation in the first stage of the lateritic manganese ores in Africa. *Chem. Geol.*, 40, pp. 25–42.

Nahon, D., Beauvais, A., Nziengui-Mapangou, P., and Ducloux, J., 1984. Chemical weathering of Mn-garnets under lateritic conditions in northwest Ivory Coast (West Africa). *Chem. Geol.*, 45, pp. 53–71.

Nahon, D., and Bocquier, G., 1983. Petrology of elements transferred in weathering and soil systems. *Mém. Sci. Géol.*, Strasbourg, 72, pp. 111–121.

Nahon, D., and Colin, F., 1982. Chemical weathering of orthopyroxenes under lateritic conditions. *Am. J. Sci.*, 282, pp. 1232–1243.

Nahon, D., Colin, F., and Tardy, Y., 1982. Formation and distribution of Mg, Fe, Mn-smectites in the first stages of the lateritic weathering of forsterite and tephroïte. *Clay Min.*, 17, pp. 1–9.

Nahon, D., Ducloux, J., Butel, P., Augas, G., and Paquet, H., 1980. Néoformation d'aragonite, première étape d'une suite minéralogique évolutive dans les encroûtements calcaires. *C. R. Acad. Sc. Paris*, 291, pp. 725–727.

Nahon, D., Herbillon, A., and Beauvais, A., 1989b. The epigenetic replacement of kaolinite by lithiophorite in a manganese-lateritic profile, Brazil. *Geoderma*, 44, pp. 247–259.

Nahon, D., Janot, C., Karpoff, A. M., Paquet, H., and Tardy, Y., 1977. Mineralogy, petrography and structures of iron crusts (Ferricretes) developed on sandstones in the western part of Senegal. *Geoderma*, 19, pp. 263–277.

Nahon, D., Janot, C., Paquet, H., Parron, C., and Millot G., 1979. Epigénie du quartz et de la kaolinite dans les accumulations ferrugineuses superficielles. *Bull. Sci. Géol.*, Strasbourg, 32, pp. 165–180.

Nahon, D., and Lappartient, J. R., 1977. Time factor and geochemistry in iron crusts genesis. *Catena*, 4, pp. 249–254.

Nahon, D., Melfi, A., and Conte, C. N., 1989a. Présence d'un vieux système de cuirasses ferrugineuses latéritiques en Amazonie du Sud. Sa transformation in situ en latosols sous la forêt équatoriale actuelle. *C. R. Acad., Sci., D, Paris*, 308, pp. 755–760.

Nahon, D., and Millot, G., 1977. Géochimie de la Surface et formes du relief. V. Enfoncement géochimique des cuirasses ferrugineuses par épigénie du manteau d'altération. Influence sur le paysage *Bull. Sci. Géol.*, Strasbourg, 30, pp. 283–288.

Nahon, D., Paquet, H., and Delvigne, J., 1982. Lateritic weathering of ultramafic rocks and the concentration of nickel in the Western Ivory Coast. *Econ. Geology*, 77, pp. 1159–1175.

Nahon, D., Paquet, H., Ruellan, A., and Millot, G., 1975. Encroûtements calcaires dans les altérations des marnes éocènes de la falaise de Thiès, Sénégal. Organisation morphologique et minéralogique. *Bull. Sci. Géol.*, Srasbourg, 28, 1, pp. 29–46.

Nakajima, Y., and Ribbe, P. H., 1980. Alterations of pyroxenes from Hokkaido, Japan, to amphibole, clays and other biopyriboles. *Neues Jahrb. Mineral. Monatsh.*, 6, pp. 258–268.

Nalovic, L., 1971. Comportement du fer en présence des éléments de transition dans la nature. *C. R. Acad. Sci. Paris*, 273, pp. 1664–1667.

Newman, A. C. D., 1969. Cation exchanges properties of micas. 1. The relation between mica composition and potassium exchange in solutions of different pH. *J. Soil Sci.*, 20, pp. 357–373.

Nicolis, G., and Prigogine, I., 1977. *Self-organization in Nonequilibrium Systems*. John Wiley & Sons, Inc., New York, p. 491.

Noack, Y., Decarreau, A., and Manceau, A., 1986. Spectroscopic and oxygen isotopic evidence for low and high temperature origin of talc. *Bull. Minéral.*, 109, pp. 253–263.

Norrish, K., 1954. The swelling of montmorillonite. *Disc. Faraday Soc.*, 18, pp. 120–134.

Norrish, K., 1973. Factors in the weathering of mica to vermiculite. *Int. Clay Minerals Conf. Proc.*, Division de Ciencias, C.S.I.C., Madrid 1972, Serratosa, J. M., Ed., pp. 417–432.

Novikoff, A., 1974. Altération des roches dans le massif du Challu (République Populaire du Congo). Formation et évolution des argiles en zone ferrallitique. Thesis, Univ. Strasbourg, unpublished, p. 297.

Nriagu, J. O., 1976. Phosphate-clay mineral relations in soils and sediments. *Can. J. Earth Sci.*, 13, pp. 717–736.

Nriagu, J. O., and Dell, C. I., 1974. Diagenetic formation of iron phosphates in Recent lake sediments. *Am. Mineral.*, 59, pp. 934–946.

Nziengui Mapangou, P., 1981. Pétrologie comparée de deux gîtes supergènes manganésifères. Gisements de Ziémougoula (Côte d'Ivoire) et Moanda (Gabon). Thesis, Univ. Poitiers, unpublished, p. 110.

Ohashi, Y., and Burnham, C. W., 1972. Electrostatic and repulsive energies of the M1 and M2 cation sites in pyroxenes. *J. Geophys. Res.*, 77, pp. 5761–5766.

O'neil, J. R., and Kharaka, V. K., 1976. Hydrogen and oxygen isotope exchange between clay mineral and water. *Geochim. Cosmochim. Acta*, 40, pp. 241–246.

Ortoleva, P., Merino E., Moore, C., and Chadam, J., 1987a. Geochemical self-organization I. Reaction-transport feedbacks and modeling approach. *Am. J. Sci.*, 287, pp. 979–1007.

Ortoleva, P., Chadam, J., Merino, E., and Sen, A., 1987b. Geochemical self-organization, II. The reactive infiltration instability. *Am. J. Sci.*, 287, pp. 1008–1040.

Owens, J. P., Altschuler, Z. S., and Berman, R., 1960. Millisite in phosphorite from Homeland, Florida. *Am. Mineral.*, 45, pp. 547–561.

Papike, J. J., and Cameron, M., 1976. Crystal chemistry of silicate minerals and geophysical interest. *Reviews of Geophysics and Space Physics*, 14, pp. 37–80.

Papike, J. J., Ross, M., and Clark, J. R., 1969. Crystal-chemical characterization of clinoamphiboles based on five new structure refinements. *Mineral. Soc. Am., Spec. Pap.*, 2, pp. 117–136.

Paquet, H., 1969. Evolution géochimique des minéraux argileux dans les altérations et les sols des climats méditerranéens et tropicaux à saisons contrastées. *Mém. Serv. Carte Géol., Alsace Lorraine*, 30, p. 212.

Paquet, H., Duplay, J., and Nahon, D., 1981. Variations in the composition of phyllosilicates monoparticles in a weathering profile of ultrabasic rocks. Proc. VIIth Intern. Clay Conf., Bologna-Pavia 1981. *Dev. in Sedimentology*, 35, 1982, Elsevier, Amsterdam Publ. Co., pp. 595–603.

Parks, G. A., 1972. Free energies of formation and aqueous solubilities of aluminum hydroxides and oxide hydroxides at 25° C. *Am. Mineral.*, 57, pp. 1163–1189.

Pavich, M. J., 1985. Appalachian Piedmont morphogenesis: weathering, erosion and Cenozoic uplift. In: *Tectonic Geomorphology*, Morisawa, M., and Hack, J. T., Eds., 15th Annual Geomorphology Symposium Proceed. Binghamton, N.Y., 1984, Allen and Unwin, London, Sydney, pp. 299–319.

Pavich, M. J., 1986. Processes and rates of saprolite production and erosion on a foliated granitic rock of the Virginia Piedmont. In: *Rates of Chemical Weathering of Rocks and Minerals*, Coleman, S. M., and Dethier, D. P., Eds., Academic Press Inc., San Diego, ch. 23, pp. 551–590.

Pavich, M. J., 1989. Regolith residence time and the concept of surface age of the Piedmont "Peneplain." *Annual Geomorphology Symp., 20, Proc.* In press.

Pavich, M. J., and Obermeier, S. F., 1985. Saprolite formation beneath Coastal Plain sediments near Washington, D. C. *Geol. Soc. Amer. Bull.*, 96, pp. 886–900.

Pédro, G., 1968. Distribution des principaux types d'altération chimique à la surface du globe. *Rev. Géogr. Phys. Géol. Dyn.*, 10, 5, pp. 457–470.

Pédro, G., 1979. Les minéraux argileux. *In: 'Pédologie'*, vol. 2, Bonneau M., and Souchier, B., Eds., Masson, Paris, ch. 3, pp. 38–57.

Pédro, G., 1983. Structuring of some basic pedological processes. *Geoderma*, 31, pp. 289–299.

Pédro, G., 1984. La genèse des argiles pédologiques *Bull. Sci. Géol.*, Strasbourg, 37, pp. 333–348.

Pédro, G., Chauvel, A., and Melfi, A. J., 1976. Recherches sur la constitution et la genèse des *'Terra Roxa estructurada'* du Brésil. Introduction à une étude de pédogenèse ferrallitique. *Ann. Agron.*, 27, 3, pp. 265–294.

Pédro, G., and Delmas, A. B., 1980. Regards actuels sur les phénomènes d'altération hydrolitique. *Cah. Office Rech. Sci. Tech. Outre-Mer (ORSTOM), sér. Pédologie*, 18, 3–4, pp. 217–234.

Pédro, G., and Tessier, D., 1983. Importance de la prise en compte des paramètres texturaux dans la caractérisation des argiles. *Proc. 5th Meeting of the European Clay Groups, Praha 1983, Charles University, Praha 1985*, pp. 417–428.

Perrodon, 1972. Réflexions sur la comparaison de quelques vitesses de phénomènes géologiques. *C. R. Somm. Soc. Géol. France*, 2, pp. 50–52.

Peterson, J. B., 1944. The effect of montmorillonitic and kaolinic clays on the formation of platy structures. *Soil Sci. Soc. Am. Proc.*, 9, p. 37.

Petit, J. C., Della Mea, G., Dran, J. C., Schott, J., and Berner, R. A., 1987. Mechanism of diopside dissolution from hydrogen depth profiling. *Nature*, 325, 6106, pp. 705–707.

Petrovic, R., Berner, R. A., and Godhaber, M. B., 1976. Rate control in the dissolution of alkali feldspars. I. Study of residual feldspar grains by X-ray photoelectron spectroscopy. *Geochim. Cosmochim. Acta*, 40, pp. 537–548.

Pons, C. H., 1980. Mise en évidence des relations entre la texture et la structure dans les systèmes eau-smectites par diffusion aux petits angles du rayonnement X synchrotron. Thesis, Univ. Orléans, unpublished, p. 175.

Pouget, M., and Rambaud, D., 1980. Quelques types de cristallisation de calcite dans les sols à croûte calcaire (steppes algériennes). Apport de la microscopie électronique. *'Comptes Rendus Réunion du Groupe d'Etudes sur les Carbonates'*, Univ. Bordeaux, France, pp. 371–379.

Prewitt, C. T., and Burnham, C. W., 1966. The crystal structure of jadeite $NaAlSi_2O_6$. *Am. Mineral.*, 51, pp. 956–975.

Proust, D., and Velde, B., 1978. Beidellite crystallization from plagioclase and amphibole precursors: Local and long-range equilibrium during weathering. *Clay Min.*, 13, pp. 199–209.

Quirk, J. P., 1968. Particle interaction and soil swelling. *Israel Journal of Chemistry*, 6, 3, pp. 213–234.

Quirk, J. P., 1978. Some physico-chemical aspects of soil structural stability, a Review. In: *Modification of Soil Structure*, Emerson, W. W., Bond, R. D., and Dexter, A. R., Eds. Wiley-Interscience, New York, pp. 3–16

Rausell-Colom, J. A., Sweatman, T. R., Wells, C. B., and Norrish, K., 1965. Studies in the artificial weathering of mica. In: *Experimental Pedology*, Hallsworth, E. G., and Crawford, D. V., Eds., Butterworths, London, pp. 40–72.

Richards, L. A., 1947. Pressure membrane apparatus construction and use. *Agric. Eng.*, 28, pp. 451–454.

Rimsaite, J. H. Y., 1967. Studies of rock forming micas. *Bull. Geol. Surv. Can.*, 149, p. 82.

Rimstidt, J. D., and Barnes, H. L., 1980. The kinetics of silica-water reactions. *Geochim. Cosmochim. Acta*, 44, pp. 1683–1699.

Rimstidt, J. D., and Dove, P. M., 1986. Mineral-solution reaction rates in a mixed flow reactor: Wollastonite hydrolysis. *Geochim. Cosmochim. Acta*, 50, pp. 2509–2516.

Robert, M., 1972. Transformation expérimentale de glauconite et d'illite en smectite. *C. R. Acad. Sci. Paris*, 275, pp. 1319–1322.

Rode, A. A., Yarilova, Y. A., and Rashevskaya, I. M., 1960. Certain genetic characteristics of dark soils. *Soviet Soil Sci.*, 8, pp. 799–809.

Ronov, A. B., Khain, V. E., Balukhovsky, A. N., and Seslavinsky, K. B., 1980. Quantitative analysis of Phanerozoic sedimentation. *Sedimentary Geol.*, 25, pp. 311–325.

Ross, G. J., and Kodama, H., 1974. Experimental transformation of a chlorite into vermiculite. *Clays Clay Miner.*, 22, pp. 205.

Rothbauer, R., 1971. Untersuchung eines $2M_1$-Muskovits mit Neutronenstrahlen. *Neues Jahrb. Mineral. Monatsch*, H4, pp. 143–154.

Ruellan, A., 1968. Les horizons à individualisation et à accumulation du calcaire dans les sols. *Int. Cong. Soil. Sci.*, 9, Adelaide, IV, pp. 501–510.

Ruellan, A., 1971a. Les sols à profil calcaire différencié des plaines de la Basse Moulouya, Maroc Oriental. *Mém. Off. Rech. Sci. Tech. Outer-Mer, (ORSTOM)*, 54, p. 302.

Ruellan, A., 1971b. The history of soils: Some problems of definition and interpretation. In: *Paleopedology, Origin, Nature and Dating of Paleosols*, Yaalon, D. H., Ed., Hebrew Univ., Jerusalem, pp. 3–13.

Russ, W., and Andrew, C. W., 1924. The phosphate deposits of Absokuta Province. *Nigeria Geol. Surv. Bull.*, 7, pp. 9–38.

Sarrazin, G, Ildefonse, P., and Muller, J. P., 1982. Contrôle de la solubilité du fer et de l'aluminium en milieu ferrallitique. *Geochim. Cosmochim. Acta*, 46, pp. 1267–1279.

Savin, S., 1977. The history of the earth's surface temperature during the past 100 million years. *Annual Rev. Earth Planet. Sci.*, 5, pp. 319–355.

Sawhney, B. L., and Voigt, G. K., 1969. Chemical and biological weathering in vermiculite from Transvaal. *Soil Sci. Soc. Amer. Proc.*, 33, pp. 625–629.

Schofield, R. K., 1935. The pF of the water in soil. Trans. *Int. Cong. Soil Sc., 3, Proc.*, 2, pp. 37–48.

Schott, J., and Berner, R. A., 1983. X-ray photoelectron studies of the mechanism of iron silicate dissolution during weathering. *Geochim. Cosmochim. Acta*, 47, pp. 2333–2340.

Schott, J., and Berner, R. A., 1985. Dissolution mechanisms of pyroxenes and olivines during weathering. In: *'The Chemistry of Weathering'*, Drever, J. I., Ed., D. Reidel Publ. Co., Dordrecht, The Netherlands, pp. 35–53.

Schott, J., Berner R. A., and Sjöberg, E. L., 1981. Mechanism of pyroxene and amphibole weathering. I. Experimental studies of iron-free minerals. *Geochim. Cosmochim. Acta*, 45, pp. 2133–2135.

Schott, J., and Petit, J. C., 1987. New evidence for the mechanisms of dissolution of silicate minerals. In *Aquatic Surface Chemistry: Chemical Processes at the Particle-Water Interface*, Stumm, W. Ed., John Wiley & Sons, Inc., New York, pp. 293–315.

Schwertmann, U., and Murad, E., 1983. Effect of pH on the formation of goethite and hematite from ferrihydrite. *Clays Clay Miner.*, 31, pp. 277–284.

Scott, A. D., and Smith, S. J., 1967. Visible changes in macro mica particles that occur with potassium depletion. *Clays and Clay Miner.*, 15, pp. 357–373.

Seddoh, F., 1973. Altération des roches cristallines du Morvan. Etude minéralogique, géochimique et micromorphologique. Thesis, Univ. Dijon, unpublished, p. 341.

Segalen, P., 1973. L'aluminium dans les sols. *Cah. Office Rech. Sci. Tech. Outre-Mer (ORSTOM), sér. Initiations*, Paris, 22, p. 282.

Serratosa, J. H., and Bradley, W. F., 1958. Determination of the orientation of OH bond axes in layer silicates by infrared absorption. *J. Phys. Chemistry*, 62, pp. 1164–1167.

Sherman, G. D., and Kanehiro, Y., 1954. Origin and development of ferruginous concretions in Hawaiian latosols. *Soil Sci.*, 77, pp. 1–8.

Slansky, M., Lallemand, A., and Millot, G., 1964. La sédimentation et l'altération latéritique des formations phosphatées du gisement de Taiba. *Bull. Serv. Carte Géol. Alsace Lorraine.*, 17, pp. 311–324.

Sleeman, J. R., 1963. Cracks, peds and their surfaces in some soils of the Riverine Plain, NSW. *Austr. J. Soil Res.*, 1, pp. 91–102.

Smalley, I. J., 1974. Fragmentation of granitic quartz in water: Discussion. *Sedimentology*, 21, pp. 633–635.

Smith, J. V., 1974. *Feldspar Minerals*. Vol. I: *Crystal Structure and Physical Properties*. Vol. II: *Chemical and Textural Properties*. Springer Verlag, New York.

Sornein, J. F., 1980. Altération supergène de minéralisations ferrifères; les chapeaux de fer du gisement de sidérite de Batere (Pyrénées Orientales) et du gisement de pyrite de Sain-Bel (Rhône). Thesis, Ecole Nationale Sup. Mines, Paris, unpublished, p. 171.

Stevens, R. E., and Carron, M. K., 1948. Simple field test for distinguishing minerals by abrasion pH. *Am. Mineral.*, 33, pp. 31–49.

Stoops, G., 1967. Le profil d'altération au Bas Congo (Kinshasa). *Pédologie*, Ghent, 17, pp. 60–105.

Stoops, G., 1968. Micromorphology of some characteristic soils of the Lower Congo (Kinshasa). *Pédologie*, Ghent, 1, pp. 110–147.

Strakhov, N. M., 1969. Principles of lithogenesis. Oliver and Boyd, Publ., Edinburg, p. 609.

Stumm, W., Furrer, G., Wieland, E., and Zinder, B., 1985. The effects of complex-forming ligands on the dissolution of oxides and alumino-silicates. *In: The Chemistry of Weathering*, Drever, J. I., Ed., D. Reidel Publ. Co., Dordrecht, The Netherlands, p. 55–74.

Swindale, L. D., and Jackson, M. L., 1956. Genetic processes in some residual podzolised soils of New Zealand. *Int. Cong. Soil Sci., 6, Proc., Paris*, pp. 233–239.

Tamini, Y. N., Kanehird, Y., and Sherman, G. D., 1964. Reactions of ammonium phosphate with gibbsite and with montmorillonitic and kaolinitic soils. *Soil. Sci.*, 98, pp. 249–255.

Tamm, O., 1934. The oxalate method of soil analysis. *Medd. Skogsforsoksanst*, Stockholm, 27, pp. 1–20.

Tardy, Y., 1969. Géochimie des altérations. Etude des arènes et des eaux de quelques massifs cristallins d'Europe et d'Afrique. *Mém. Serv. Carte Géol. Alsace Lorraine*, Strasbourg, 31, p. 199.

Tardy, Y., 1986. *Le Cycle de l'Eau. Climats, Paléoclimats et Géochimie Globale.* Masson, Paris, p. 338.

Tardy, Y., Bocquier, G., Paquet, H., and Millot, G., 1973. Formation of clay from granite and its distribution in relation to climate and topography. *Geoderma*, 10, pp. 271–284.

Tardy, Y., and Garrels, R. M., 1976. Prediction of Gibbs free energies of formation. I. Relationships among Gibbs free energies of formation of hydroxides, oxides and aqueous ions. *Geochim. Cosmochim. Acta*, 40, pp. 1051–1056.

Tardy, Y., and Garrels, R. M., 1977. Prediction of Gibbs free energies of formation of compounds from the elements. II. Monovalent and divalent metal silicates. *Geochim. Cosmochim. Acta*, 41, pp. 87–92.

Tardy, Y., and Gartner, L., 1977. Relationships among Gibbs free energies of formation of sulfates, nitrates, carbonates, oxides and aqueous ions. *Contrib. Mineral. Petrol.*, 63, pp. 89–102.

Tardy, Y., Duplay, J., and Fritz, B., 1986. Stability fields of smectites and illites as a function of temperature and chemical composition. *Int. Meeting Geoch. Earth Surface and Process. Miner. Formation, 1, Proc.*, Granada, Spain, pp. 461–494.

Tardy, Y., Melfi, A. J., and Valeton, I., 1988. Climats et paléoclimats tropicaux périatlantiques. Rôle des facteurs climatiques et thermodynamiques: Température et activité de l'eau, sur la répartition et la composition minéralogique des bauxites et des cuirasses ferrugineuses au Brésil et en Afrique. *C. R. Acad. Sci. Paris*, 306, pp. 289–295.

Tardy, Y., and Monnin, C., 1983. Recherches sur les mécanismes du concrétionnement. Coefficients d'activité des ions, solubilité des sels et de la silice dans les pores de petite taille. *Bull. Minéral.*, 106, pp. 321–328.

Tardy, Y., and Nahon, D., 1985. Geochemistry of laterites. Stability of Al-goethite, Al-hematite and Fe^{3+}-kaolinite in bauxites and ferricretes. An approach to the mechanism of concretion formation. *Am. J. Sci.*, 285, 10, pp. 865–903.

Tardy, Y., N'Kounkou, R., and Probst, J. L., 1989. The global water cycle and continental erosion during Phanerozoic Time (570 my). *Am. J. Sci.*, 289, pp. 455–483.

Tardy, Y., and Novikoff, A., 1988. Activité de l'eau et déplacement des équilibres gibbsite-kaolinite dans les profils latéritiques. *C. R. Acad. Sci. Paris*, 306, pp. 39–44.

Tardy, Y., and Vieillard, P., 1977. Relationships among Gibbs free energies and enthalpies of formation of phosphates, oxides and aqueous ions. *Contrib. Mineral. Petrol.*, 63, pp. 75–88.

Taylor, A. W., and Gurney, E. L., 1965. Precipitation of phosphate by iron oxide and aluminum hydroxide from solution containing calcium and potassium. *Soil Sci. Soc. Amer. Proc.*, 29, pp. 18–22.

Tessier, D., 1975. Recherches expérimentales sur l'organisation des particules dans les argiles. Thesis, Ing. CNAM, Paris, unpublished, p. 231.

Tessier, D., 1978a. Etude de l'organisation des argiles calciques. Evolution au cours de la dessiccation. *Ann. Agron.*, 29, 4, pp. 319–355.

Tessier, D., 1978b. Technique d'études de l'orientation des particules utilisables sur des échantillons secs ou humides. *Ann. Agron.*, 29, 2, pp. 193–207.

Tessier, D., 1980. Sur la signification de la limite de retrait dans les argiles. *C. R. Acad. Sci., Paris*, 291, pp. 377–380.

Tessier, D., 1984. Etude expérimentale de l'organisation des matériaux argileux. Hydratation, gonflement et structuration au cours de la dessiccation et de la rehumectation. Thesis, Univ. Paris, Public. Documents INRA, Paris, p. 361.

Tessier, D., and Berrier, J., 1979. Utilisation de la microscopie électronique à balayage dans l'étude des sols. Observation de sols humides soumis à différents pF. *Sci. Sol*, 1, pp. 67–82.

Tessier D., and Pédro, G., 1976. Les modalités de l'organisation des particules dans les matériaux argileux. *Sci. Sol*, 2, pp. 85–99.

Tessier, D., and Quirk, J. P., 1979. Sur l'apport de la microscopie électronique dans la connaissance du gonflement des matériaux argileux. *C. R. Acad. Sci., Paris*, 288, pp. 1375–1378.

Tessier, F., 1950. Age des phosphates et des latéritoïdes phosphatés de l'ouest du plateau de Thiès (Sénégal). *C. R. Acad. Sci. Paris*, 230, pp. 981–983.

Tessier, F., 1952. Contribution à la stratigraphie et à la paléontologie de la partie ouest du Sénégal (Crétacé et Tertiaire). 1e et 2e parties: Historique et Stratigraphie. Thesis, Univ. Marseille (1950), and *Bull. Dir. Mines Géol., A.O.F.*, 1, 14, p. 267.

Tessier, F., 1954. Notice explicative sur la feuille Thiès Ouest au 1/200.000. *Dir. Féd. Mines Géol. Afr. Occid. Fr., Dakar*, p. 86.

Tessier, F., 1965. Les niveaux latéritiques du Sénégal. *Ann. Fac. Sci. Marseille*, 38, pp. 221–237.

Tessier, F., Flicoteaux, R., Lappartient, J. -R., Nahon, D., and Triat, J. -M., 1975. Reform of the concept of "Continental Terminal" in the coastal sedimentary basins of West Africa. *Trav. Lab. Sci. Terre, St-Jérôme, Marseille*, A, 8, p. 6.

Thiry, M., 1977. Structures en "coiffe" résultant de lessivages verticaux de formations conglomératiques. *Bull. B.R.G.M.*, IV, 2, pp. 105–120.

Thiry, M., 1978. Silicification des sédiments sablo-argileux de l'Yprésien du sud-est du bassin de Paris. Genèse et évolution des dalles quartzitiques et silcrètes. *Bull. BRGM*, I, 1, pp. 19–46.

Thiry, M., 1981. Sédimentation continentale et altérations associées: calcitisations, ferruginisations et silicifications. Les argiles plastiques du Sparnacien du Bassin de Paris. *Mém. Sci. Géol.*, Strasbourg, 64, p. 173.

Thiry, M., and Schmitt, J. M., 1983. Silicifications and paleosilcretes in the Tertiary detrital series of the Paris Basin. *Int. Colloqium CNRS on the Petrology of Weathering and Soils, Field Trip Guidebook*, Paris, p. 33.

Touray, J. C., 1980. *La Dissolution des Minéraux. Aspects Cinétiques*. Masson, Paris, p. 109.

Towner, G. D., 1981. The correction of *in situ* tensiometer readings for overburden pressures in swelling soils. *J. Soil Sci.*, 32, pp. 499–504.

Trescases, J. J., 1975. L'évolution supergène des roches ultrabasiques en zone tropicale. Formation des gisements nickelifères de Nouvelle Calédonie. *Mém. Off. Rech. Sci. Tech. Outre-Mer (ORSTOM)*, Paris, 78, p. 259.

Tsao, S. T., and Pask, J. A., 1982. Reaction of glasses with hydrofluoric acid solution. *J. Am. Ceram. Soc.*, 65, pp. 360–362.

Turenne, J. -F., 1977. Mode d'humification et différenciation podzolique dans deux toposéquences guyanaises. *Mém. Office Rech. Sci. Tech. Outre-Mer (ORSTOM)*, Paris, 84, p. 173.

U.S.D.A., 1951. *Soil Survey Manual*. U.S. Dept. Agr. Handbook, 18. U.S. Dept. Agriculture, Washington D.C.

Valeton, D., 1972. *Bauxites. Developments in Soil Science*, Elsevier Publish. Co. Amsterdam, p. 226.

Van Eldik, R., and Palmer, D. A., 1982. Effects of pressure on the kinetics of the dehydration of carbonic acid and the hydrolysis of CO_2 in aqueous solution. *J. Solution Chem.*, 11, pp. 339.

Van Olphen, H., 1962. Unit layer interaction in hydrous montmorillonite systems. *J. Colloid Sci.*, 17, pp. 660–667.

Van Olphen, H., 1963. *An Introduction to Clay Colloid Chemistry*, *Interscience*, Wiley, New-York.

Van Olphen, H., 1977. *An Introduction to Clay Colloid Chemistry*, 2nd ed., *Interscience*, Wiley, New-York.

Veblen, D. R., and Buseck, P. R., 1980. Microstructures and reaction mechanisms in biopyriboles. *Am. Mineral.*, 65, pp. 599–623.

Vedder, W., and Mac Donald, R. S., 1963. Vibrations of the OH ions in muscovite. *J. Chem. Phys.*, 38, pp. 1583–1590.

Velbel, M. A., 1983. A dissolution-reprecipitation mechanism for the pseudomorphous replacement of plagioclase feldspars by clay minerals during weathering.

In: *Pétrologie des Altérations et des Sols*, vol. I, Nahon, D., and Noack, Y., Eds. *Mém. Sci. Géol.*, Strasbourg, 71, pp. 139–147.

Velbel, M. A., 1984a. Mineral transformations during rock weathering and geochemical mass-balances in forested watersheds of the Southern Appalachians. Ph. D. Thesis, Yale University, unpublished, p. 175.

Velbel, M. A., 1984b. Weathering processes of rock-forming minerals. In: *Environmental Geochemistry*, Fleet, M. E., Ed., *Mineralogical Association of Canada Short Course Handbook*, Mineral. Assoc. Canada, London, Ontario, 10, pp. 67–111.

Velbel, M. A., 1985. Geochemical mass balances and weathering rates in forested watersheds of the southern Blue Ridge. *Am. J. Sci.*, 285, pp. 904–930.

Velbel, M. A., 1986. The mathematical basis for determining rates of geochemical and geomorphic processes in small forested watersheds by mass-balance: examples and implications. In: *Rates of Chemical Weathering of Rocks and Minerals*, Coleman, S. M. and Dethier, D. P., Eds., Academic Press, Inc., San Diego, ch. 18, p. 439–451.

Velbel, M. A., 1989. Weathering of hornblende to ferruginous products by a dissolution-precipitation mechanism: Petrography and stoichiometry. *Clays Clay Minerals*, 37, pp. 515–524.

Velbel, M. A., and Dowd, J. F., 1983. Distribution of weathering products in weathered bedrock, western Fairfield County, Connecticut. *Geol. Soc. Amer. Abstract*, 15, p. 200.

Verges, V., Madon, M., Bruand, A., and Bocquier, G., 1982. Morphologie et cristallogenèse de microcristaux supergènes de calcite en aiguilles. *Bull. Minéral.*, 105, pp. 351–356.

Vieillard, P., 1978. Géochimie des Phosphates. Etude thermodynamique; application à la genèse et à l'altération des apatites. *Mém. Sci. Géol.*, Strasbourg, 51, p. 181.

Vieillard, P., Tardy, Y., and Nahon, D., 1979. Stability fields of clays and aluminum phosphates: Parageneses in lateritic weathering of argillaceous phosphatic sediments. *Am. Mineral.*, 64, pp. 626–634.

Vizier, J. F., 1983. Etude des phénomènes d'hydromorphie dans les sols des régions tropicales à saisons contrastées. Dynamique du fer et différenciation des profils. *Trav. Documents Off. Rech. Sci. Techn. Outre-Mer (ORSTOM)*, 165, p. 294.

Wada, K., 1959. Reaction of phosphate with allophane and halloysite. *Soil Sci.*, 87, pp. 325–330.

Walker, G. F., 1949. The decomposition of biotite in the soil. *Min. Mag.*, 28, pp. 693–703.

Webber, M. D., and Clarke, J. S., 1969. Reactions of phosphate by aluminum and Wyoming bentonite. *Can. J. Soil. Sci.*, 49, 2, pp. 231–240.

Wey, R., 1953. Sur l'adsorption de l'anion phosphorique par la montmorillonite. *C.R. Acad. Sci. Paris*, 236, pp. 1298–1300.

Wey, R., 1955. Sur l'adsorption en milieu acide d'ions PO_4H_2 par la montmorillonite. *Bull. Groupe Fr. Argiles*, 6, pp. 31–34.

White, W. B., 1974. The Carbonate Minerals, In: *The Infrared Spectra of Minerals*, Farmer, V. C., Ed., Mineralogical Society, London, p. 227–284.

Whittaker, E. J. W., 1971. Madelung energies and site preferences in amphiboles. *Am. Mineral.*, 56, pp. 980–996.

Wilson, M. J., 1970. A study of weathering in a soil derived from a biotite-hornblende rock. I. Weathering of biotite. *Clay Min.*, 8, pp. 291–303.

Wilson, M. J., 1975. Chemical weathering of some primary rock-forming minerals. *Soil Sci.*, 119, pp. 349–355.

Wolkewitz, H. von, 1958. Cracking of drying clays and its significance. *Z. Pflernahr. Dung. Bodenkunde*, 82, p. 17.

Wollast, R., 1967. Kinetics of the alteration of K-feldspar in buffered solutions at low temperature. *Geochim. Cosmochim. Acta*, 31, pp. 635–648.

Yamasaki, T., and Yoshizawa, T., 1961. Concretion of ferrous carbonate (siderite) in paddy soils in Japan. I. Occurrence and constituents of ferrous carbonate and the mechanism of its formation. *Bull. Hokuriku Agric. Exp. Sta.* 2, pp. 1–14.

Zanin, Yu. N., 1968. Zones of laterite weathering of secondary phosphorites of Altay-Sayan region. *Trans. from Geologiya i Geofizika*, (1967), 9, p. 56–65, *Inter. Geol. Rev.*, 10, pp. 1119–1127.

Ziegler, A. M., Scotese, C. R., and Barrett, S. F., 1982. Mesozoic and Cenozoic paleogeographic maps. In: *Tidal Friction and the Earth's Rotation II*, Brosche, P., and Sündermann, J., Eds., Springer Verlag, New York, Heidelberg, pp. 240–252.

INDEX

Accumulation, 90, 92, 142, 147–150, 154, 162, 167–175, 187–191, 199, 205, 206, 229, 238, 240, 242, 248, 252, 257
absolute, 63, 68, 70, 102, 238
accumulation transfers, 92
argillaceous, 137, 141, 142
calcite, 231, 241, 242, 245
centripetal, 150, 151
clay, 248
continuous, 139
features of, 139
front of, 257
gibbsite, 69
glaebular, 149, 150, 152, 154, 156, 163, 167, 168, 187, 199
hematite, 152
illuvial accumulation, 141, 248
kaolinite, 57, 215, 221
organic matter, 219, 220
oxyhydroxides (or oxides), 68, 215, 216, 228, 253
phase of, 257
process, 236, 257
quartz, 220
relative, 68, 70, 92, 95, 136
secondary products, 63
selective, 90
siliceous, 142, 145
smectite, 257
subsequent, 68
types of, 236, 253
zones, 229
Aggradation, 200
banded structures of, 200
Alterability:
differences of, 70
rate of, 2
sequence of minerals, 9
Alterite(s), 133, 136, 160, 161, 162, 170, 172, 178, 207, 259, 262, 263, 264, 267, 268, 269, 270
ferralitic, 102
fossil, 263
isovolumetric, 98
kaolinitic, 133
migmatite, 102
smectitic, 133
structures of, 98, 190, 259
transformation of, 97
Amphibole, 45, 67
dissolution of, 15
dissolution of tremolite, 16, 18, 21
monoclinic amphibole, 15
structure of, 15, 17
tremolite, 15
weathering of, 42
Arrangement, 9, 117, 118, 128, 135, 136, 137, 141, 147, 148, 170, 172, 175, 187, 193
crystallite, 119, 120, 125
disorganized, 125
grains, 117
intracrystalline, 114

Arrangement (*Continued*)
layers, 123
microdomains, 113
particles, 111, 114, 124, 126, 127, 128
plasmic, 122
primary, 125
smectite, 76
tactoids, 113
of volumes, 127

Biotite, 54, 55, 57, 67, 82, 89, 137, 181, 241
bronzed, 56
ferruginization of, 56, 106
hydration of, 53
lamellae, 105
layers of, 56, 57, 115, 123
modifications of, 75
rate of biotite weathering, 3
transformation of, 44
trioctahedral micas, 27

Chad, 6, 136, 139, 141, 142, 147
Clay minerals, 54, 64, 73, 84, 85, 104, 106, 111, 118, 122, 123, 152, 161, 186, 187
Ca-saturated, 116
crystallization of, 84
distribution of, 238
kaolinite type, 116
micaceous, 111
Na-saturated, 116, 124
nature of, 104, 124, 131
organization of, 106, 116
paragenesis of, 72
property of, 108
pseudomorphoses by, 70
shrinking of, 252
smectite type, 116
structure of, 132
swelling of, 123
weathering of, 104, 105, 106
Cleavage, 53, 54
micas, 27
opening of, 27
Climate, 268
change of climatic factors, 248, 254, 256, 259
climatic cause, 257
climatic changes, 222, 228, 260, 264
climatic conditions, 221, 223, 227, 245, 249, 256, 260
climatic parameters, 262
climatic transitions, 260
climatic zones, 2, 51, 257
humid, 229, 252, 268
oceanic, 245
temperate, 169, 170, 245
tropical, 231
CO_2, 8, 38, 40, 265, 266, 268, 269
aqueous, 38, 40
content of the atmosphere, 265
control of, budgets, 268
dissolved, 7, 38, 40
molality:
of CO_2, 38
of pCO_2, 38
partial pressure of, 28
production of, 1, 38, 265
time variation of, 360
Coatings, 136, 155, 170–172, 178, 183, 185–189. *See also* Cutan
argillaceous, 137, 139, 202
argilloferruginous, 102, 142, 185, 201
clay, 236. *See also* Coatings, argillaceous
gibbsitic, 185
kaolinite, 193
millisite, 196
plasmic concentration, 170
precipitated silica, 26
protective, 11, 21
talc, 43
wavellite, 192
Concentration:
Al_2O_3, 162
calcitic (or carbonate), 154, 165, 168
calcium, 241
centripetal, 149
CO_2, 38. *See also* CO_2
concentration profiles, 251
critical, 61
dissymmetry of, 48
equilibrium, 29, 38, 46
evaporation, 81
ferrous ions, 161
ferruginous, 140, 142, 154, 180, 189, 228
glaebular, 149, 152, 154, 155, 156, 158, 160, 165, 189
gradients of, 13, 48
hydrogen ions, 21, 32, 33
increasing, 116
initial, 61, 149, 150, 158
log, 204
organic matter, 137
oxygen, 25
plasmic, 148, 149, 150, 151, 154, 155, 156, 158, 160, 165, 170, 172, 175, 176, 187, 190, 196, 198, 205
process of, 70
relative, 149

respective, 48
silica, 29, 61
solutions, 168, 180, 186, 190, 199
spatial, discontinuities, 251, 253
Constituent:
argillaceous, 108, 125, 131, 132
constituent particle, 111
ferruginous, 204
major, 225
mineralogical (or mineral), 149, 150, 156, 160, 175, 180
nature of, 207
non-argillaceous, 126
opaque, 122
plasmic, 118, 122, 126, 149, 163, 224
Contacts:
degree of opening of, 48
intergranular (or grain), 48, 137, 187, 188
mineral-solution, 267, 269
nature of, 86, 87
between plasmas, 101
reactional (or reaction), 115, 126
reciprocal, 145
between tactoids, 113
Continental areas, 262, 264
distribution of, 265
Cortex, 142, 154, 156, 157, 158, 160, 199, 200, 204, 205, 253
banded, 156–159, 253
ferruginous, 142, 158
induration of, 156
pisolites, 203
weathering, 102
Cover, 209, 223, 240, 259–264
ancient, 264
calcitic, 261
clay, 236
ferralitic, 262
geological history of, 362
pedological, 139, 245, 259, 260, 262, 263, 264, 269
smectitic, 260, 261
surficial, 217
Crust, 142, 143, 163, 165, 166, 173, 198, 201, 204, 229, 244
bauxitic, 182
calcareous, 162
ferruginous (or iron), 142, 143, 162, 187, 199, 200, 204, 228, 229, 242, 244, 245, 253, 262
manganesiferous, 162, 199, 204, 205
siliceous, 162
Crystallites, 108–111, 114–120, 125, 126
anisotropy of, 118, 120
argillaceous, 99, 121, 123
arrangement of, 120, 125, 126
association of, 118, 119, 132
charges of, 111
dimensions, 116, 117, 118, 119, 124
goethite, 172
juxtaposition of, 142, 149, 153
kaolinite, 111, 114, 116, 123, 124, 130, 162, 225
nature of, 131
packages of, 115
particles consisting of, 109, 111
structure of, 131, 133
thickness, 109
Crystallorelict, 97
Cutan, 136, 170–184, 189, 193, 194. *See also* Coatings
argillaceous, 143, 176
argilloferruginous, 139, 181
calcite, 174, 175
constraint, 155, 178
detrital cutans, 171
evolution of, 181
fissuration of, 178
gibbsite, 183
goethite, 174, 175, 191
hematitic, 183
illuviation, 139
kaolinite, 181, 191
neoformed, 172, 181

Degradation, 44, 56, 62, 82, 87, 199, 200, 201, 202, 205, 208
banded structures of, 204
phyllosilicates by, 44
transformation by, 44
Differentiation, 14, 21, 106, 207, 208, 219, 223, 228, 238, 245, 252
complete, 215
lateral, 215, 217, 219, 231, 234, 241, 256, 257
latosols, 215
pedologic, 207, 208, 241, 248, 251, 253
polyphased, 248
type of, 240
vertical, 216, 219, 220, 221, 228, 231, 234, 241, 253, 254, 256, 258
weathering minerals, 43
Diffusion, 11, 175, 180, 250, 253
avenues, 54
cations, 24
centripetal, 156, 180
chemical, 48, 84
elements, 14

Diffusion (*Continued*)
reactants, 14
water, 21
Dissolution, 11, 13, 14, 24, 30, 33, 37, 41, 52, 53, 59, 60, 61, 63, 64
albite, 15. *See also* Dissolution, feldspars
aluminous hematite, 200
amphibole, 15
bronzite, 24
cavities, 167. *See also* Dissolution, voids
clay minerals, 161
complete, 3, 11
congruent, 20, 33, 40, 42, 43, 45, 46
continuous, 101
curves, 20
diffusion, 11
diopside, 16, 18, 20, 21
dissolution-precipitation, 166, 167
enstatite, 16, 18, 21. *See also* Dissolution, pyroxene
experimental, 4
fayalite, 24
features, 269
feldspars, 9, 11, 12, 13, 14, 36, 37, 39
forsterite, 27, 34
front, 36, 249
granite, 82
incongruent, 40, 43, 46, 52, 53, 63, 90
isotropic, 21
kaolinite, 161, 181, 187, 228, 252
kinetic inhibitors of, 45
kinetics of, 9, 11, 21
magnesian and calcic silicate, 15, 36. *See also* Dissolution, pyroxene
micas, 27, 28, 30
modes of, 8
olivine type, 24, 27. *See also* Dissolution, fayalite; Dissolution, forsterite
oxyhydroxides, 252
phlogopite, 29, 30. *See also* Dissolution, micas
preferential, 11, 20
progressive, 11
pyroxene, 16, 18, 21
quartz, 59, 60, 62, 173
rate, 3, 6, 14, 33, 37, 38, 40, 41
reactions, 8, 14, 16, 18, 21, 31, 33, 36, 37, 38, 40, 41, 46
simple, 57, 60, 61, 102
stages of, 18
stoichiometric, 9, 10, 14, 18, 33, 34, 57
striae, 78
structure, 11
tremolite, 16, 18, 21
types of, 43
voids, 68
wollastonite, 41
Distribution, 136, 137, 147, 148, 150, 207, 260
clay minerals, 238
hematite and goethite, 189
horizons, 208, 254
morphology of, 86
plasmic concentration, 190
sequences, 262
Domain, 72, 83, 84, 101, 110, 113–121, 162, 189, 193, 228, 234, 245
arrangement, 113
associated into, 116, 121
crandallite, 193
dimensions of, 123
downslope, 236, 238, 245, 248
Fe-, 84
ferralitic, 114, 214
humid, 228
illite, 132
initial concentration, 158
isotropic and anisotropic, 121, 122
kaolinitic, 240
limits of, 115
micaceous clays, 114, 115
Ni-, 84
nodules, 158
nontronitelike, 83
oriented, 101
pedoturbated, 105
pimelitelike, 83
prefered orientation of, 117. *See also* Domain, oriented
segregation, 107
smectitic, 257
upslope, 234, 238, 246
weathering, 258
Drainage, 115, 161, 215, 221, 222, 246, 248, 257
conditions, 209, 215, 220, 221
lateral, 220, 221, 227, 248
local, 229
modification of, 226
tilting of, 248
vertical, 214, 220, 221, 226, 248
Drying, 114, 115, 116, 123, 125, 129, 131, 133, 135, 155, 170, 209, 226, 252
contraction, 123, 129, 131
effects of, 111, 122, 123, 129
maximum, 115, 116, 123
phases of, 102, 116, 126, 129, 132
previous state of, 111, 124

wetting and drying phases, 170, 176, 257
Duricrusts, 162. *See also* Crust

Exchange sites:
accessibility of, 36, 37
stability of, 34

Fayalite, 24, 25, 26
dissolution of, 24
structure of, 24
surface of, 26
Feedback, 14, 223, 267
reactive-infiltration feedback, 250
Feldspar, 9–15, 34, 37, 39, 42, 71, 99, 101, 105, 181
albite, 9, 11, 13
alkaline, 13, 34
amelia, 13
anorthite, 9
areas of, 71
dissolution of, 9, 11, 12, 14, 15, 41. *See also* Feldspar, weathering
dissolution structures, 11
experimental dissolution, (*see* Feldspar, dissolution of)
fragment of, 153
grains of, 11, 146, 153, 155
hybla, 13, 14
microcline, 9. *See also* Feldspar, potassic
orthoclase, 9, 11
plagioclase, 41
potassic, 9, 35, 82, 99
relicts, 102
samples, 11
sanidine, 9. *See also* Feldspar, potassic
structure of, 9, 42
weathering, 2, 12, 13, 14, 89
Forsterite, 7, 27, 34, 75, 76, 78
Fragmentation, 101, 132, 176
geochemical, 200, 205
mechanical, 178
Front, 249, 256
accumulation, 215, 257
alteroplasmation, 249
dissolution, 36, 249
illuviation, 257
oxidation, 265
podzolization, 215
reaction, 250
sinking of, 267
weathering, 250, 260, 267, 268

Glaebular concentration, 149–160. *See also* Glaebulization
Glaebules, 149, 150, 152, 158, 160, 162, 165, 166, 167, 168, 189, 191, 192, 194, 226, 228
continuous glaebular structures, 162
goethitic, 193
hematitic, 189
millisite, 191
soil, 165
stages of glaebular microstructure, 158, 159
wavellite, 192
Glaebulization, 149, 160, 163, 175, 200
calcitic, 165
centripetal, 156
epigenetic, 206
ferruginous, 161, 200
hematitic, 198
processes of, 160
Global evolution, 191
Granite, 4, 72, 136
dissolution of, 82
granite–gneiss, 188
Middle Precambrian granite, 182
muscovite–biotite granite, 181
porphyroidal granite, 240
Proterozoic granite, 61
structure of, 181
theorical, 82
weathering of, 5
Gruss, 4, 152, 173, 231, 236, 240, 241
granitic, 151
migmatitic, 173
sandy, 231
weathering, 152

Horizons, 249–256
accumulation, 142
A horizons, 106
A′2 horizon, 236, 240
aluminum phosphate, 121
argillaceous, 106, 199, 240, 252
association of, 207
bauxitic, 181, 182
B horizon, 101, 102, 234, 240, 241
B′2 horizons, 236
BC horizons, 99, 101, 106, 128, 241
calcrete horizon, 173
C horizons, 98, 99, 128
degradation of, 199
ferralitic, 98, 214, 240
ferruginous, 142, 174, 228
ferruginous nodules, *see* Horizons, ferruginous; Horizons, nodular
gruss, 236
humic (Bh), 217
illuviated, 248

Horizons (*Continued*)
indurated, 162, 163, 167
kaolinitic, 256
leached, 221, 229, 238, 240, 246
microaggregated, 219, 220, 221, 223, 225, 226
mottled clay, 162, 229
nodular, 142, 143, 220
original, 249, 251, 252, 254, 257
pedologic, 98, 249, 256
pedoturbated, 99, 102, 105, 106
podzolic, 219
sandy, 200, 220
sandy-argillaceous, 221
saprolite, 209, 229
self-development of, 253
soil, 122, 127, 245
succession of, 240
surficial, 136, 217, 224
tongue-shaped, 236, 238, 240
unconsolidated, 210
vertisolic, 98, 101, 104
weathering, 11, 181, 187, 189, 191
Hydrolysis, 7, 8, 24, 31, 45, 161, 187. *See also* Dissolution
albite, 8
crystalline framework, 31
equilibria, 8
forsterite, 7
Oxidation-, 161
processes of, 7, 24, 34
reactions, 7, 8, 11, 15, 21, 31, 34, 36
silicon, 42

Isostasy, 268
Isovolume, 94, 105, 190
concept, 105
Ivory Coast, 6, 43, 71, 75, 105, 173, 180, 181, 182, 199, 204

Kaolinite, 111–116, 118, 123, 130, 132, 142, 143, 151, 152, 161, 162, 178–181, 185, 186, 202, 208–215, 221–226
accumulation of, 215, 221
association of, 212
booklets of, 152. *See also* Kaolinite, vermicular
coatings of, 193
crystallites of, 111, 114, 116, 123, 162, 225
cutans, 181. *See also* Kaolinite, coatings of
dissolution of, 161, 181, 187, 228
ferralitic soils, 145
ferritization of, 143
ferritized kaolinite, 142
ferruginous (or ferriferous), 152
formation of, 4, 8
generation of, 229
illuviation of, 172
immobilized, 213
microaggregates of, 208
particles, 214
plasma, 152, 210
portion of, 220
profile, 241
properties of, 145
secondary concentration of, 208, 209, 222
separation of, 228
structure, 111
vermicular, 71, 120, 152, 189
Kinetic, 9, 51, 56
CO_2, 38
dissolution, 11, 28, 31
factors of weathering, 2
inhibitors of dissolution, 45
linear, 11, 21
non-linear, of dissolution, 21
parabolic kinetics, 11, 21
reasons, 168, 251

Landscape, 6, 57, 180, 206, 209, 228, 248, 254, 260
flat, 260
geochemical evolution of, 167
geomorphological features of, 163
Savanna, 228
tropical, 162
Layer, 24, 26, 27, 41, 44, 48, 56, 62, 83, 84, 85, 108, 109, 110, 115, 116, 120, 123
adherent, 27
aqueous, 40
arrangement of, 115, 123
biotite, 57
brucitic, 44
charge of, 109, 115
cortical, 200
deformation in layer structure, 108
diffused, 115
dioctahedral, 84
double, of water, 108, 115
elementary, 108
expandable, 123, 132
heterogeneous, 8
hydrated, 26, 27
illite-smectite mixed, 238
interlayer, 54
ions, 46
iron-rich, 26
lateral extent of, 108

leached, 15
level of, 132
mica, *see* Layer, biotite; Layer, micaceous
micaceous, 57
nontronitelike, 84
octahedral, 56
package of, 116
pimelitelike, 84
protective, 10, 14
protonated, 42
residual, 12, 13, 14
silicate, 44
stacking (or stack) of, 56, 109, 114
superficial, 33, 46
surface, 15, 21, 26
thickness of, 27, 116
trioctahedral, 84
type of, 29, 108
water, 45, 48
Lithorelict, 97

Mantle:
indurated, 242
original, 221
pedologic, 6, 207, 208, 209, 228, 231, 240, 245, 251, 255
podzolic, 219
smectitic, 240
surficial, 208, 209, 220, 228, 231, 240
weathering, 1, 6, 263, 267
Mass balance, 3
Micas, 27–31, 41, 53, 54, 56, 57, 61, 62, 63, 89
annite, 27
biotite, 27, 54, 56
black, 56. *See also* Micas, biotite
cleavage of, 27
dioctahedral, 27, 28, 30
dissolution of, 27, 28, 30
ferruginized micas, 56
grains of, 153
layers of, 27, 56
margarite, 27
muscovite, 27
nature of, 27
phlogopite, 27
structure of, 27
trioctahedral, 27, 28, 30
weathering of, 27, 42
Microaggregate, 114, 124, 132, 178, 208, 209, 210, 211, 212, 215, 216, 223, 225, 226
argillaceous particles of, 225
assemblage of, 225
collapse of, 224, 226
destruction of, 220, 221, 225, 228
disorganization of, 227
generation of, 212
horizons, 216
lowering of, 220
organization into, 214
relicts of, 225
ultradessication of, 226
Microdeposit, 171
alternating (or alteration of), 171
Microdomains, 110, 111, 113, 115, 121
arrangements of, 113
assemblages in, 111
dimensions of, 116
juxtaposition, 121
structure of, 121
Microstructures, 139, 141, 142, 145, 148, 205, 206
accumulation, 206
destruction of, 141
differentiation of, 136
disturbance of, 149
effect of, 117
glaebular, 158
hierarchy of, 147
inherited, 205
leaching, 206
major type of, 117
nodular, 182
petrographic, 148
second order, 116, 117, 125, 128
survival of, 139
transformation, 206
Modifications:
biochemical, 221
biotite, 57
crystallochemical, 181, 252
drainage, 226
geochemical, 89, 97
geochemical system, 254
hydric regim, 256
initial, 54
mineralogical, 152
optically invisible, 53
polarization colors, 27, 54
structural, 215
surface slope, 89
visible, of optical properties, 62

Neocutan, 170, 174, 175, 176, 177
position of, 175, 176
Neoformation, 43, 52, 57, 63, 85, 89, 180
argilliplasma, 89

Neoformation (*Continued*)
crystalliplasma, 89
cutans, 181
mineralogical, 190
phyllosilicate, 85
talc, 43
Network, 46, 65, 75, 105, 110, 112, 113, 115, 116, 121, 135, 152, 156, 193, 201, 202, 227
anastomosed, 162, 201
bidimensional, 108
cavity network, 193
deformation of, 40
hydrographic, 220, 221, 223, 227, 260
infinite, 9
lizardite, 75
outlines, 121
parent crystalline, 48
rectilinear, 116
reticulated, 234
smectites, 132
tactoids, 113, 115, 116, 118, 121, 132
three-dimensional, 113
tight, 193
New Caledonia, 6
Nodules, 140, 141, 142, 153, 155, 156, 158, 160, 200
calcareous, 160, 167, 173
calcite (or calcitic), 151, 152, 154, 155, 231, 241, 242
evolution of, 156
ferruginous, 140, 141, 142, 143, 151, 157, 160, 224, 228, 252
friable, 242
gibbsitic, 215
hematite (or hematitic), 150, 157, 200
indurated, 253
iron (oxyhydroxides), 209, 217, 252
lithiophorite, 152
mineralogical constituents of, 156
size of, 156, 160

Olivine, 8, 11, 24, 40, 67, 75, 78. *See also* Fayalite; Forsterite
congruent dissolution of, 27
dissolution of, 24, 27
hydrolysis of, 7
Mg-, 7
structure of, 24
Oxidation, 2, 56
front of, 265
iron (or Fe^{2+}) oxidation, 26
organic matter, 2
oxidation-hydrolysis, 161
oxidizing conditions, 11, 24, 25
process of, 267
Oxyhydroxides, 49, 54, 57, 64, 65, 68, 70, 78, 79, 89, 91, 143, 149, 162, 175, 181, 185, 200, 208, 252
aluminous goethite, 157, 163, 200, 203
aluminous hematite, 151, 152, 157, 161–163
cementation by, 140
crystallization of, 214
destruction of, 220
effects of opaque, 125, 126
iron, 104, 121, 122, 139, 140, 141, 143, 148, 162, 175, 178, 185, 189, 204, 208, 209, 211, 212, 214, 216, 219, 220, 221, 223, 224, 226, 228, 229, 252, 260
manganese, 154, 162, 175
metallic, 97
nodules, 217
reorganization between, 104
secondary, 220
separation of, 126
septa of, 64–69, 77, 79, 88, 89, 97, 100

Parent mineral:
alteroplasmation of, 240
chemical memory of, 87, 92
composition of, 42
control by, 44
debris of, 105
degradation and transformation of, 45
different, 48
dissolution of, 43, 46, 53, 57, 65, 68
dissolution rates of, 4
epigenic replacement of, 242
framework of, 42
mixture of, 82
nature and structure of, 249
oxygen of, 43
peripheral pseudomorphosis of, 70, 91
precipitation from, 238
preservation, 89
pseudomorphosis of, 97, 209
reactivity, 249, 268
relicts, 63, 64, 70, 73, 99, 208
restructuration, 104
structure of, 9
surface of, 48, 52
transformation of, 53, 63, 89
weathering of, 52, 53, 99, 101, 105, 229, 231, 234, 238, 249, 252
of surface of, 7

Parent rock, 3, 51, 75, 191, 198, 204, 208, 220, 229, 234, 241, 242, 246, 254, 269
chemical memory of, 253
composition of, 81
evolution, 98
granitic, 128
migmatitic, 102, 231, 234
modification of, 97
of volume of, 92
nature of, 208, 242
original structure, 97, 133
pyroxenite, 79
sedimentary, 125, 216
shale, 242
structure of, 251, 252
structure preservation of, 133
texture of, 249
weathering of, 238
Particles, 109, 110, 111, 117, 120, 125, 126
accumulation of, 102
argillaceous, 82, 85, 86, 109, 111, 114, 123, 124, 126, 130, 131, 132, 137, 139, 140, 141, 143, 147, 160, 171, 176, 180, 225, 226
arrangement of, 111, 114, 124, 125
association of, 106, 108, 111, 116, 132
capacity of reorganization of, 125
detrital, 170
ferruginous (or goethite), 142, 244, 245
groups of, 132
illite, 111
kaolinite, 120, 214
orientation of, 125
phyllosilicate, 141
plasticity of, 125
reaction of, 114
rigid, 124, 125, 126, 130, 132
size of, 114, 116
smectite, 111
structure of, 123, 133
surfaces of, 108, 226
in suspension, 257
ultrafine, 11, 21
Pebbles, 145, 146, 147, 148, 169
Ped, 126, 127
elementary, 127
primary ped, 126
structure of, 106, 126, 127
volume, 127
Pedoturbation, 97, 101, 102, 104, 105, 106, 125
of altero-illuvial structure, 102
characteristics of, 104, 125
macroscopic expression of, 133
major features of, 106
of non-argillaceous horizons, 106
processes of, 101
role of argilloplasmas in, 98
structures plasmic of, 106
Phosphates, 186, 187, 196
alumino-calcic, 185, 187, 189, 190, 191, 192, 193, 194
aluminum, 171, 185, 186, 190, 192, 196
Bone Valley, 263
calcium, 171, 189, 191
clay–minerals–phosphate, 186
Phyllosilicates, 45, 75, 79, 85, 87, 89, 92, 141, 208
argillaceous, 44
magnesian, 36
neoformation of, 85
phyllosilicate secondary products, 57
structure of, 57, 87
transformation of, 44, 85
trioctahedral, 30
weathering, 52, 76, 92
Plasma:
argillaceous, 99, 101, 104, 106, 114, 117, 121, 122, 123, 124, 125, 129, 153, 155, 187, 224, 225, 234, 238, 256. *See also* Plasma, argilliplasma
argilliplasma, 70, 72, 73, 74, 81, 82, 84, 85, 86, 91, 92, 97, 98, 99, 101
argillo-ferruginous, 102, 142, 143, 178, 180
asepic, 102
calcitic, 153, 169
complex, 70
crystalliplasma, 64, 65, 69, 70, 88, 91, 92, 97, 100, 103
deferritized, 142
disassociation of, 225
ferruginous, 143, 200, 203
glaebular, 160
hematitic, 157
illitic, 113, 114, 124, 130
illuviation, 102, 141, 142
kaolinitic, 104, 113, 114, 124, 130, 143, 151, 152, 156, 161, 180, 191, 200, 210, 252
montmorillonitic, *see* Plasma, smectitic
orientation of, 101, 126
pedoturbation, 125
phosphatic, 193
redistribution of, 252
restructuration of, 209
secondary, 89
segregation of, 224
skelsepic, 101

Plasma (*Continued*)
smectitic, 81, 113, 124, 125, 130, 132, 152, 191
soil, 118
structure of argillaceous, 127
swelling capacity of, 124
weathering, 48, 63, 85, 86, 97, 98, 99, 102, 105, 106, 122, 125, 149, 170, 175, 189, 229
Plasmation, weathering, 85, 93, 98, 101, 102, 104
Preservation, 63, 66, 79, 81, 181, 191, 192, 248, 251, 252, 253
lithologic, 87
original volumes, 92
parent mineral, 89
structure, 51, 97, 133
texture, 92
Pyroxene, 16, 26, 34, 44, 45, 65, 75, 78
bronzite, 15
clinopyroxene, 15
diopside, 15, 20, 21
dissolution of, 15
bronzite, 24
diopside, 20, 21
enstatite, 21
enstatite, 15, 16, 18, 21
orthopyroxene, 15, 24
structure of, 15
bronzite, 24
surface of, 33
weathering of, 42, 75, 79

Quartz, 57, 59, 60, 62, 71, 102, 105, 146, 147, 153, 166, 167, 192, 215, 223, 224, 249
coating of, 172
corroded, 61, 191, 192, 203
debris, 105
dissolution of, 57, 59, 61, 173
fresh, 59, 61
parent, 60, 63
skeleton, 132, 143, 155
silica of, 57
surface aspects of, 59, 61
weathered, 59
Quasicutan, 170, 175, 176, 177

Reaction:
chemical or geochemical, 7, 32, 81, 208, 252
dissociation reaction, 42
dissolution, 8, 9, 10, 14, 16, 18, 21, 31, 36, 38, 46
durations of, 29
effect of, 7
evolutive reaction front, 249, 250
exchange, 32
hydrolysis, 7, 8, 11, 15, 21, 31, 34, 36
interface, 1
kinetics of, 249
reactions-transports, 251, 253, 254
reversible, 81
solid–solution interface, 42
steps of complex, 12
subsequent, 41
surface-controlled, 24, 41, 61
transfer, 81
weathering, 1, 3, 7, 14, 34, 35, 38, 45, 46, 265, 267, 269
Reorganization, 92, 125, 155, 160, 185, 199, 206, 226
argillaceous particles, 123
ferruginous plasma, 200
grains, 257
sequences, 223
simple, 104
structural, 98
Residual minerals, 4, 254

Saprolites, 4, 5, 6, 97, 99, 208, 209, 220, 221, 222, 223, 228, 229, 251
coarse, 209, 231, 241, 251
porosity, 221
Seasons, 2, 98, 169, 228
dry, 228, 229
rainy, 242, 246
Secondary minerals, 2, 51, 63, 64, 252
neoformation of, 63
precipitation of, 82, 269
sequences of, 81
Septa, 64–70, 79
ferruginous, 79
gibbsite, 67
goethitic, *see* Septa, ferruginous
oxides, 66
oxyhydroxides, 64, 79, 89, 97
peripheral, 70
Sequence:
alterability sequence of minerals, 9
climatic transition, 260
derived, 221, 223, 229, 257
distribution of, 262
evolution of, 208
garango, 155
garango I, 238
garango II, 234, 238
intermediate biopyriboles, 45
kosselili, 136
lateral, 245

normal, 231
of organization, 248
original, 219, 220, 221, 223, 229, 231, 234, 238, 245, 248, 254
pedological, 141, 207, 208, 209, 245, 254, 261, 262
reorganization of, 223
self-evolution of, 209
soils, 215
types of, 219, 231
vertical, 136, 245, 256, 257, 260
weathering, 215
of weathering of minerals, 4
Silica, 3, 14, 25, 26, 29, 41, 42, 67, 185, 192, 220, 229, 252, 254, 267
accumulation of, 147
amorphous silica, 27, 29, 57, 173
amounts of, 3
concentration of, 61
contents of, 4
dissolution rate of, 3
elimination of, 28
evacuation rate of, 4
inhibiting role of, 185
leaching of, 185
network of, 9
precipitated, 26
quartz, 57
Silicate minerals, 7, 8
dissolution processes of, 8
dissolution rate of, 41
hydrolysis reactions of, 8
structural framework of, 15, 32
weathering of, 269
S-matrix, 126, 132, 133, 175, 178, 190
accumulation in, 149, 150, 167, 199
altero-illuvial, 104
alteroplasmation, 234
argillaceous, 155, 156, 160, 187
calcrete, 173, 174
crandallite, 192, 193, 194
deferretized, 224
enclosing, 150, 155, 158, 174, 180
kaolinitic, 152, 158, 189
original, 167
pedoplasmation, 234
pedoturbated, 101, 102, 104, 135
porosity of, 161
rearrangement of, 178
relationships glaebule, 149
sandy-argillaceous, 163
structures of, 149, 158
surrounding, 149, 165, 172. *See also* S-matrix, enclosing
wavellite, 194
weathering, 99, 104, 135, 152, 154, 187, 189, 192
Solutes, 236, 250
organic, 40
transport of, 48
Solutions:
acidic, 21
alkaline, 245
alkalinity of, 7
aqueous, 15, 84, 85, 185, 249
chemistry of, 61
chocking of, 257
circulation of, 41, 52, 166, 225
composition of, 52, 172
concentration of, 61, 168, 186, 190
degree of renewal of, 41
desequilibrium with, 82
diffusing solution, 161
distribution of, 64, 65
elements in, 63, 82
elimination by, 12
equilibria with, 95
flux of, 269
infiltrating, 251
initial, 82
interstitial, 114–116
leaching, 43, 190
meteoric, 43, 259
migration of, 234
nature of, 63, 180
neutral, 42, 81
percolating, 4, 135, 170, 171, 180, 181, 196, 199, 209, 220, 249, 251, 259, 269
percolation of, 205
solid–solution, 64, 84, 85, 162
transfers of, 208
undersaturated, 3
weathering, 46, 48, 52, 60, 62, 68, 268
Structure:
accumulation, 191, 199, 205, 206
aggradation, 200
alterite, 190
altero-illuvial, 104
amphiboles, 15
argillaceous, 114, 117, 123, 126, 127, 130, 131
asepic, 99, 104
banded, 193, 194
beehive, 11
bronzites, 24
calcretes, 167
cap, 136, 145

Structure (*Continued*)
clay, 83
collapse of, 226
crystalline, 32, 43–45, 185, 251
crystallo chemical, 2
crystallographic, 84, 86, 89
differentiation of, 185, 205
evolutionary series of, 205
feldspars, 9, 42
fibroradiated, 174
first-order, 118, 121, 125, 126, 130–133
geopetal, 171
glaebular, 160–162, 168
inherited, 205, 206
initial, 9, 136, 141, 206
insepic, 118, 125
internal, 149
isalteric, 99
isotic, 125
lamellar, 137
layer, 108
leached, 132
leaching, 139, 141, 205, 206
levels of, 128
lithologic, 66, 87
macroscopic, 106, 126, 127, 129, 132, 259
masepic, 118, 125, 128
massive, 198, 201, 206
mega-structures, 106
micas, 27, 51, 54, 56
microaggregated, 104, 128, 133, 183, 214, 220, 221, 225, 246
microscopic, 106, 117, 122, 127, 209, 259
mineral, 66
mosepic, 118
nodular, 154
olivine, 24
original, 97, 98, 105, 106, 152, 181, 192, 194, 196, 205, 209, 223
original parent-rock, 133
parent minerals, 249
parent rock, 63, 208, 251, 252
pedologic, 252
pedologic mantle, 248
petrographic, 70, 98, 102, 136
phyllosilicate (or clay minerals), 57, 87, 132
pisolitic, 199, 200, 204
plasmic, 101, 106, 117–125, 127–130, 133, 149, 154
polyhedral, 127
cubic, 127, 141
lamellar, 127
platy, 127
polyhedric, 127, 209, 212, 215, 234, 241, 246
prismatic, 106, 127
porous, 193
preexisting, 199
preservation of, 97, 98, 105, 133, 191
preserved, 231, 234, 252
prismatic, 236, 241, 245, 246, 252
(pseudo) brecciated, 191–196, 198, 199, 206
(pseudo) conglomeratic, 198, 199, 206
pyroxenes, 15
receptive, 57, 68, 70, 139, 148, 257
reorganization, 199
residual product, 136
schistose, 242
secondary, 89
second-order, 117, 118, 121, 122, 124, 128–133
sepic, 117, 118, 122, 124, 125, 128
sequence of, 190
silicate, 15, 26
skeleton grains, 136
skelsepic, 105, 118, 125, 128
s-matrix, 149
soils, 190, 207
spheral, 127
lenticular, 128
polysphedral, 128
polyspheral, 128
spheroidal, 128
striated, 137
structure preserving weathering, 66, 90
sub-structure, 122
succession of, 147, 196
superposed, 136
tactoid, 115
third-order, 117, 128–133
three-dimensional, 150
transformation, 205, 206
types of, 127, 128
undulic, 125
vosepic, 118, 125, 128
zonation, 190
zoned, 141
Substitution, 9, 20, 32, 108, 109, 161, 163
cations (or ions), 26, 40, 42, 108
isomorphic, 108, 163
original, 109
system, 108
Superposition, 141, 158
horizons, 127
plasmic, 150
principle of, 171

System, 61, 92, 215, 250, 251, 256
argillaceous, 111
biogeochemical, 259
biogeodynamic, 249, 250
closed, 82
derived, 223
evolutive, 256
geochemical, 251, 253, 254, 256–258
hydrographic, 260
living, 258
micropores and macropores, 170
open, 82
original, 248, 256
parameters, 250
pedologic, 222, 223, 227, 242, 245, 248, 256, 258, 262
reaction-transport, 251, 253, 254
self-development of, 251, 259
transfer, 70
transformation, 240, 248
weathering, 70

Tactoid, 110–121
anisotropy of, 121
arrangement of, 113
assemblages of, 111
network of, 115, 116, 118, 121, 132
organization of, 116
structure of, 115
tactoids or quasicrystals, 111, 113, 115, 121, 132
thickening of, 116
Talc, 29, 30, 43, 44, 46, 108, 109
coatings, 43
formation of, 43, 89
generation of, 43
growth of, 44
hydrothermal, 43
isotopic values of, 43
meteoric, 43
mixture of, 44
moyango, 43
neoformation of, 43
precipitation of, 43
weathering of orthopyroxene into, 44, 46
Temperature, 7, 9, 11, 16, 38, 60, 84, 186, 265
factor, 267
of formation, 9
high temperature polymorph (feldspar), 9
low temperature polymorph (feldspar), 9
role of, on dissolution rates, 38
surface of continent, 267
Time, 3, 12, 67, 78, 85, 101, 111, 131, 142, 151, 167, 168, 181, 190, 207–209, 220, 221, 227–231, 236, 242, 251, 254, 259–269
accumulation, 206
factors in, 205
function of, 12, 29
residence, 7, 41
weathering, 27
Transfers, 85, 90–93, 98, 135, 136, 147, 175, 190, 194, 196, 199, 205, 206, 228, 229, 252, 254
alternation of, 199
anisotropic transfers, 166
chemical accumulative transfers, 92
chemical transfers, 92
direction of, 162
elements, 7, 92, 135, 158, 160
extracrystalline transfers, 92
intercrystalline transfers, 92
intraplasma, 165
long-distance, 165
mass, 81
material, 48, 136, 149, 229
particulate, 136
reactions, 81
short, 85, 92
solution, 208
subtractive, 194, 199
system of, 70
Transformation, 135, 181, 206
accumulating, 257
additive, 181, 186
advanced, 253
biochemical, 222
biotite, 44
chlorite, 44
cutanic, 181
enantiotropic, 60
environment, 139
ferromagnesian minerals, 67
front, 256
hypogene, 56
independence of, 74
intracrystalline, 62
lateral, 228, 254–257
lateritic plateaus, 223
latosols, 215, 223
leaching, 256
limited, 252, 253
mechanism of, 52
microstructural, 97, 98, 104, 141, 205, 206
new organizations of, 269

Transformation (*Continued*)
organization, 257
pedological system, 223, 229, 253, 256
phyllosilicate minerals, 44, 75, 85, 87, 89
plasma, 72
process of, 43, 51, 52, 63, 175, 225
product, 53, 54, 73, 79, 170
simple, 56, 57, 63
stage of, 63
structural units, 209
systems of, 240, 248
transformation by degradation, 44
type of, 54, 176, 177
vertical, 256

Ultrastructures, 126
first-order, 106, 108, 117, 125

Water, 7, 30, 32, 38, 106–109, 114, 115, 122, 129–133, 196, 259, 268
activity of, 162, 189, 204, 269
anisotropy of water circulation, 140
behavior of ions in, 47, 48
composition of, 81
content, 115, 116
diffusion, 21
distilled, 28
double layer of, 115
elimination, 129
flushing of, 6, 268
index, 129
initial, 81
interaction minerals and, 51
interlayer, 43, 45
intertactoid, 115
intratactoid, 115
layer, 48
meteoric, 1
molecular, 21, 32, 33
natural, 3
percolating, 2, 269
phase, 158
polarizing power on, 47
potential of, 131, 132
protons of water molecules, 46
pure, 46
rain, 2
saturated, 129, 180
sea, 43
undersaturated, 204
vapor, 45
variation of, 115, 116
watershed, 3
Weathering:
absence of, 61
agent, 1
albite, 12. *See also* Weathering, feldspars
analysis of, 131
anorthite, 3. *See also* Weathering, feldspars
argilliplasmas, 63, 64, 70, 80–82, 85, 86, 89, 99, 101
basic rock, 262
beidellitic, 74. *See also* Weathering, smectitic
biotite, 3. *See also* Weathering, micas; Weathering, phyllosilicates
carbonate rock, 268
centripetal progression of, 86. *See also* Weathering, progress (or progression) of
chemical, 2–6, 9, 53, 57, 79, 101, 105, 220, 249, 256, 268, 269
chlorite, 204
clays, 43
conditions of, 70
congruent, 52
cortex, 102
crystalliplasma, 64, 70
differential, 2, 63, 189, 192
domains, 258
dunites, 81
enstatite, 89
environment of, 11, 31, 54, 57, 70
evolution, 61
evolutive, 269
experiment, 13
factors of, 2, 267
feldspars, 2, 12, 71, 89, 181
ferralitic domain, 214
finger-like weathering plasmation, 102
front of, 220, 250, 251, 260, 267, 268
gabbro, 44
garnet, 3, 204
geochemical, *see* Weathering, chemical
geochemical composition of, 97
geochemical processes of, 1
geochemical system of, 2
global, 267
gradient, 256
granite, 5
gruss, 152
hornblende, 24, 89
incongruent, 52
initial phase of, 97
initial sequence, 215
initial stage of, 51, 97
inosilicates, 44
isovolume (or isovolumetric), 4, 190, 191
k-feldspar, 35. *See also* Weathering, feldspars
kinetic factors of, 2

lateritic, 151, 152, 171, 189, 263
lizardite, 78
mantles, 1, 6, 263, 267
mechanism, 52
micas, 42, 56
microenvironment, 53, 89
minerals, 1, 4, 31, 40, 42, 51, 52, 63, 64, 75, 79–82, 252, 254, 259, 260
normal weathering conditions, 267
orthoenstatite, 44. *See also* Weathering, orthopyroxene; Weathering, pyroxenes
orthopyroxene, 44, 75, 80. *See also* Weathering, pyroxenes
oxidizing, 267
parent mineral, 7, 92, 101, 229, 231, 234, 238, 252. *See also* Weathering, minerals
parent relicts, 209
parent rock, 238
past weathering products, 262
pathway, 52
phases, 8, 52, 75, 87, 126
phyllosilicates, 51, 76, 87, 92. *See also* Weathering, micas
plagioclase, 3, 71–73. *See also* Weathering, feldspars
plasmas, 48, 97–106, 122, 125, 149, 170, 175, 229
plasmic structures, 106
porosity, 105
process of, 4, 31, 45, 63, 90, 95, 191, 264, 268, 270
product, 3
products of, 42–45, 48, 51–53, 61–64, 70, 74, 76, 86–92, 97, 136, 208, 251, 260, 262, 269
progress (or progression) of, 42, 86
pyritic shales, 67
pyroxenes, 32, 43, 44
pyroxenites, 82
rates of, 2, 6, 33, 57, 262, 263, 267–270
  of mineral, 3
  of rocks, 4
reaction of minerals, 3
reactions of, 1, 7, 13, 14, 27, 34, 35, 38, 40, 45, 46, 267
  of mineral, 265
reactivity of rock to, 6
recent, 262, 263
resistance to, 27
restructuration of, 199
rock, 81, 87
sensitivity to, 27
sequence of, 215
silicate mineral, 41, 269
simultaneous, 75, 79, 81
sites, 48, 52, 63
s-matrix, 99, 104, 135, 152, 154, 187–192, 198
smectitic, 260
  profiles, 73–78
solutions of, 38, 43, 46, 48, 60, 62, 68, 268
stages of, 79, 86
structure-preserving weathering, 63, 93
system of, 70, 82
tectosilicate, 89
thermodynamic simulation of, 3
time of, 27
tool, 242
transformation, 63, 67
type of, 53, 75, 192
ultrabasic rocks, 6, 78
zones of, 45, 75
Wetting, 113, 114, 123, 125, 126, 130–133
  alternation of, 170, 209
  amplitude of, 102
  conditions of rewetting, 129
  cycle, 214
  effects of rewetting, 130
  phases of, 125, 132, 226, 252, 257
  processes of, 176
  reactivity to, 212, 213
  wetting–drying phases, 102, 106, 111, 113, 116, 122, 129, 131, 170